Computer Calculation
of Phase Diagrams

WITH SPECIAL REFERENCE TO REFRACTORY METALS

REFRACTORY MATERIALS

A SERIES OF MONOGRAPHS

John L. Margrave, *Editor*
DEPARTMENT OF CHEMISTRY
RICE UNIVERSITY, HOUSTON, TEXAS

COMPUTER CALCULATION OF PHASE DIAGRAMS

With Special Reference to Refractory Metals

LARRY KAUFMAN and HAROLD BERNSTEIN

ManLabs, Inc.,
Cambridge, Massachusetts

A C A D E M I C   P R E S S *New York and London*   *1970*

ACADEMIC PRESS, INC.
111 Fifth Avenue, New York, New York 10003

United Kingdom Edition published by
ACADEMIC PRESS, INC. (LONDON) LTD.
Berkeley Square House, London W1X 6BA

LIBRARY OF CONGRESS CATALOG CARD NUMBER: 70-84155

PRINTED IN THE UNITED STATES OF AMERICA

To
S. K.

Preface

This monograph describes a formalism for computing phase equilibria and stability in metallic systems. The development and implementation of this framework has evolved through analysis and synthesis stages over the past 15 years. The keystone of this approach is quantitative definition of the lattice stability of metals, which specifies the relative free energies of stable and unstable crystallographic forms of metals. Recognition of the vital importance of this factor in controlling phase equilibria and appreciation of its broad applicability arose during the senior author's thesis studies under Professor Morris Cohen of the Department of Metallurgy at M.I.T. These studies illustrated the importance of the lattice stability concept in characterizing martensitic transformations. In addition, the association with Professor Cohen provided first-hand insight into Lord Kelvin's assertion that to understand a phenomenon, one must be able to describe it in quantitative terms.

Recognition of the problem was an important first step which led to considerable insight and frustration. Careful appraisal of the literature disclosed that thermodynamicists cleverly sidestepped the problem by deft manipulation of standard state definitions. Moreover, first-principle quantum mechanical approaches proved to be impotent in providing quantitative data. The only available route left open was the painfully slow experimental path. Late in 1958, the Metallurgy Branch of the Office of Naval Research, directed by Julius Harwood, provided support for a study of the problem at ManLabs, Inc. under Contract NONR 2600(00). Continued sponsorship and guidance have been provided by ONR under the direction of Drs. Edward I. Salkovitz and W. G. Rauch. This support provided a means for carrying out thermo-dynamic measurements and transformation studies under high pressure which permitted analysis of the lattice stability of a few transition metals.

In the early 1960's support for additional studies of phase stability at high temperatures and high pressures was provided. These studies were per-formed under Contracts AF-33-(657)-9826 and AF-33-(615)-6837, F. Vahldiek, S. Worcester, K. Kojola, and I. Perlmutter providing technical liaison. Addi-tional funding by the Air Force Materials Laboratory for a study of refractory borides in the mid-1960's under AF-33-(657)-8635 guided by J. Krochmal, J. Latva, and W. Ramke provided further opportunities to conduct experi-mental studies of high-temperature phenomena relevant to the subject of this monograph. Individual publications resulting from these studies provided clear evidence of the general utility of the "lattice stability" concept as a vehicle for correlating diverse phenomena and providing a useful means for predicting phase equilibria and transformation at high pressure and high temperatures in metallic systems. Nevertheless, the specific cases where such illustrations could be provided was limited.

A means for extending the scope of these considerations to the synthesis stage was afforded through application of computer techniques by Harold Bernstein. These methods were developed under AF-33-(615)-2352 sponsored by the Air Force Materials Laboratory under the guidance of H. Marcus. The results of this study, reported in Technical Documentary Reports 67-108 and 67-397, entitled "Development and Application of Computational Methods for Predicting the Temperature-Compositional Stability of Re-fractory Compounds," and published in 1967, provide the matrix for this monograph.

The authors have had substantial assistance and support from their col-leagues at ManLabs. Drs. S. A. Kulin, E. V. Clougherty, S. V. Radcliffe, G. Stepakoff, and Professor Morris Cohen have contributed to many aspects of this study. In addition, P. A. Kulin, J. Elling, and W. Lindonen have pro-vided encouragement and assistance in performing this research. Discussions and correspondence with Dr. R. J. Weiss, Army Mechanics and Materials Research Center, Watertown, Massachusetts, Professor Mats Hillert, Royal Technical University, Stockholm, Professor Peter Rudman, Technion, Haifa, Professor Leo Brewer, University of California, Berkeley, and the late Professor William Hume-Rothery, Oxford, have influenced this work. Although the views presented here are those of the authors, and are in some instances at variance with the views of these outstanding research workers, many benefits have been derived from these interactions. (Computer soft-ware for all programs is available from ManLabs, Inc.)

Cambridge, Massachusetts   LARRY KAUFMAN
December, 1969   HAROLD BERNSTEIN

Contents

ix

Computer Calculation of Phase Diagrams

WITH SPECIAL REFERENCE TO REFRACTORY METALS

CHAPTER I

Introduction

In an elegant series of papers written sixty years ago, J. J. Van Laar presented an explicit mathematical synthesis of phase equilibria in binary systems (*1*).* These papers described monotectic (-oid), eutectic (-oid), and peritectic (-oid) formation, retrograde solubility, and miscibility gap, as well as spinodal formation in terms of the regular solution model for binary phases. The liquid phase was described by an ideal entropy of mixing and a temperature-independent heat of mixing which was parabolically dependent on composition. Multiphase equilibria were illustrated by specifying the thermodynamic properties of each solid phase in terms of five quantities. These quantities included the enthalpy of mixing, represented by a temperature-independent function which was parabolically dependent on composition, as well as the heat of fusion and melting points of each component. The entropy of mixing was assumed to be ideal. By performing a parameterization study of these five quantities for each of the competing phases in binary systems, Van Laar illustrated how all of the characteristic features of phase diagrams could be computed. Comparison of the generalized predictions (*1*) with typical phase diagrams observed for metallic systems (*2, 3*) illustrates the success of Van Laar's simple development. The rediscovery of retrograde solubility 20 years later (*4, 5*) on the Zn-rich portion of the Zn–Cd diagram was considered to be so abnormal than an allotropic transformation in Zn was postulated. It took some 40 years to produce a sufficient number of examples and an awareness of Van Laar's work to recognize the validity of his result (*6*).

The above-mentioned thermodynamic description of solution phases provided a simple calculation of miscibility gap formation for the case where

* Italic numbers denote references.

1

the enthalpy of mixing is positive for a solution phase which is more stable than competing phases. Miscibility gap formation and spinodal decomposition phenomena have been subjects of exhaustive research in recent years. Indeed more complex solution models allowing for departures from ideal entropy and parabolic temperature-independent enthalpies of mixing have been employed. These refinements (i.e., quasi-chemical theory, subregular solution theory, etc.) have been introduced to deal with the observed departures from regular solution behavior encountered in experimental studies of the thermodynamic properties of solution phases (7). Departures from regular mixing due to excess entropy (8), size effects (9), strain energy (10), valence differences (11), vibrational and electronic specific heat contributions (12, 13), solubility parameter differences (14), and magnetic contributions (15) have been considered at length in the literature.

In contrast to the extensive effort directed toward elucidating the mixing properties which were represented by one of the five quantities in Van Laar's simple model, little attention has been paid to the other terms relating to the melting point and heat of fusion of the pure components (16). Measurements have been made of the melting points and heats of fusion of the stable forms of many metals. However, prediction of phase equilibria requires a knowledge of these quantities for the unstable forms as well (1, 16)! Thus, if calculations are to be made of equilibrium between the liquid and bcc phases in a binary system, the melting point and heat of fusion of the bcc form of both components must be known. If one of the components is stable up to its melting point in some other configuration (i.e., hcp), then the required data cannot be measured. This information could be obtained if methods were available for computing the relative stability of different crystallographic forms of the pure components from first principles. Unfortunately, however, such calculations cannot be performed at present (17). Nevertheless, some progress has been made along empirical lines by employing thermodynamic data on solid solutions. Lumsden (18) has used this method to compute the free energy difference between the fcc form of thallium, which is not stable at 1 atm, and the bcc and hcp forms that are observed. This information was subsequently employed successfully (19, 20) to predict the occurrence of the fcc form at high pressure. In considering the extremely small differences in enthalpy between the bcc, hcp, and fcc forms of thallium (10–100 cal/g-atom), Lumsden (18) observed that " Much has been written in explanation of the relative stability of different crystalline forms in intermediate phases of alloy systems. Surprisingly, little attention has been paid to the simplest aspect of this problem, the stable lattices of metals themselves. For thallium, the free energy of interconversion of the three typical metallic structures, face-centered cubic, close-packed hexagonal, and body-centered cubic, can be calculated with an accuracy that would seem extraordinarily high if expressed in electron volts per atom. Any theory which could give a quantitative explanation of the heat and entropy of transformation between these forms of thallium would merit credence when applied to the more difficult problem of explaining the free energy of alloys of complex structure."

As indicated above, an adequate theory for quantitatively predicting the relative stability of metallic structures is not available. However, extension of the thermodynamic approach employed by Lumsden (18), coupled with information generated by studies of phase transformations at high pressure (16, 20, 21), has permitted estimates to be made of the lattice stability (enthalpy and entropy differences) of various crystallographic forms of many metals. In turn, this information can be employed to compute the enthalpy of fusion and melting point for the unstable phases to derive the requirements of Van Laar's simple model.

The lattice-stability information (16) has been combined with an estimation procedure for evaluating the regular-solution heat of mixing for a series of 72 binary systems between refractory transition metals. In each system, a complete description (four explicit parameters) is given for the liquid, bcc, hcp, and fcc phases. Computer programs for solving the equilibrium equations have been developed and applied in considering the competition between the above-mentioned solution phases and two classes of compound phases in each system. This approach permits definition of the complete phase diagram and all of the thermodynamic properties. In this sense, it represents the first complete application of Van Laar's model to a large number of metallic systems. The refractory transition metals were chosen for consideration because many of the phase diagrams are available for comparison with the computations, magnetic factors are absent, and there are currently no thermodynamic data available for these systems. In this regard the calculations provide predictions of thermodynamic properties which can be tested in future experimental studies. Finally, the formulation is extended to provide an explicit description of ternary systems composed of the refractory transition metals and complemented by computer programs for dealing with ternary equilibrium.

Thus, the development presented here provides an explicit and operational framework for dealing with multiphase equilibrium in systems containing as many as three components. Estimation procedures and simple models have been employed to obtain data required for solution of the general problem that are currently unavailable. Notwithstanding the fact that these estimates will be refined and revised in the future, the current development provides the first complete treatment of multiphase competition. Many authors have depicted the dependence of phase equilibria at a fixed temperature on the relative free energy–composition curves for stable and unstable phases over the complete range of binary and ternary compositions [e.g., cf.

Fig. 1 of Kubaschewski and Chart (22)]. Moreover, it has long been recognized that " a solubility is not a property of the solution, but depends also on the thermodynamic properties of the coexisting phase " (23). Nevertheless, many of the current theories of phase stability are based on the occurrence of a phase at unique electron/atom ratios. With the exception of the recent work of Brewer (24, 25), most of these approaches ignore the aspect of competition. They are based instead on the inference that electronic contributions which can exhibit rapid increases in electron energy due to Fermi surface–Brillouin zone interactions *near phase boundaries are controlling*. This concept has dominated the field, in spite of the fact that rational examination of many physical and electronic properties near phase boundaries suggests no discontinuous change in the shape of the Fermi surface (26). The present work, dealing with phase competition as the central theme, offers an alternative viewpoint of the subject.

CHAPTER II

The Lattice Stability of Metals

The most direct representation of the competition between phases as a basis for phase stability is obtained by considering the behavior of pure metals. Consequently, it is appropriate to consider initially the effects of temperature and pressure on phase transformations in a pure metal i. If the stable crystal modification of i at 1 atm between absolute zero and the melting temperature $\overline{T}_i{}^\alpha$ is the α form,* then the free energy of αi at atmospheric pressure is $F_i{}^\alpha[T]$. At some pressure P the free energy of αi is †

$$F_i{}^\alpha[T, P] = F_i{}^\alpha[T] + \int_{P_0}^{P} V_i{}^\alpha \, dP \quad \text{cal/g-atom} \tag{1}$$

where P_0 is atmospheric pressure. If the α modification is the most dense or close packed form, then the melting curve of pure i under pressure, i.e., $\overline{T}_i{}^\alpha[P]$ may be represented by writing an equation for the free energy of liquid i:

$$F_i{}^L[T, P] = F_i{}^L[T] + \int_{P_0}^{P} V_i{}^L \, dP \quad \text{cal/mole} \tag{2}$$

* The subject of the symbolism of metallic phases has a rather long and confused history (2, 3). Although it is customary to use Greek letters to denote the crystal structure of a phase, there is no standard scheme in which a given letter is used for a particular phase (e.g., β for bcc, α for fcc, etc.). In most cases the stable room temperature, 1 atm form, is called α, irrespective of the structure, and other phases are given various designations as they arise. This haphazard situation naturally raises havoc with any attempt to discuss the crystal structures of different metals in a systematic way. At the risk of further confusing the issue, the following symbolism will be used here. The bcc crystal structure will be denoted by β, the hcp crystal structure will be denoted by ϵ, and the fcc crystal structure will be denoted by α. This notation will be employed wherever possible and any departures will be clearly specified.

† Square brackets will be used to denote the argument of a function. Thus $F_i{}^\alpha[T]$ is the free energy of αi, which is a function of temperature.

and subtracting Eq. (1) from it to obtain

$$\varDelta F_i^{\alpha \to L}[T, P] = \varDelta F_i^{\alpha \to L}[T] + \int_{P_0}^{P} [\varDelta V_i^{\alpha \to L}] \, dP \quad \text{cal/mole} \qquad (3)$$

where $\varDelta V_i^{\alpha \to L}$ will in general be temperature and pressure dependent. The melting curve is then defined by the condition that

$$\varDelta F_i^{\alpha \to L}[\overline{T}_i^{\alpha}, P] = 0 \qquad (4)$$

In general the free energy and volumetric differences between the α and liquid phases are not known over a wide range of temperature and pressure so that $\overline{T}_i^{\alpha}[P]$ cannot be predicted precisely. However, Eq. (3) can be differentiated to obtain the Clapeyron equation

$$\left(\frac{dT}{dP}\right)_{\overline{T}_i^{\alpha}}^{\alpha \to L} = \left(\frac{\varDelta V}{\varDelta S}\right)_{\overline{T}_i^{\alpha}}^{\alpha \to L} = \overline{T}_i^{\alpha}\left(\frac{\varDelta V}{\varDelta H}\right)_{\overline{T}_i^{\alpha}}^{\alpha \to L} \qquad (5)$$

Equation (5) can be used in cases where the entropy and volume change associated with the melting of α have been measured to compute the initial slope of the melting curve.

The above-mentioned case in which the α modification is the stable form for $0°\text{K} \leqslant T \leqslant \overline{T}_i^{\alpha}$ is the simplest situation. In order to illustrate a more complex case, consider the phase relations between the fcc (α), bcc (β), and liquid (L) phases of metal i. If the entropy, enthalpy, and volume differences between these phases, as defined by Eqs. (6)–(8), are approximated by constants at temperatures that are well removed from the third-law range, the corresponding free energy differences can be formulated as

$$S_i^L - S_i^\alpha \equiv \varDelta S_i^{\alpha \to L}, \quad S_i^L - S_i^\beta \equiv \varDelta S_i^{\beta \to L}; \quad S_i^\beta - S_i^\alpha \equiv \varDelta S_i^{\alpha \to \beta} \qquad (6)$$
$$H_i^L - H_i^\alpha \equiv \varDelta H_i^{\alpha \to L}, \quad H_i^L - H_i^\beta \equiv \varDelta H_i^{\beta \to L}; \quad H_i^\beta - H_i^\alpha \equiv \varDelta H_i^{\alpha \to \beta} \qquad (7)$$
$$V_i^L - V_i^\alpha \equiv \varDelta V_i^{\alpha \to L}, \quad V_i^L - V_i^\beta \equiv \varDelta V_i^{\beta \to L}; \quad V_i^\beta - V_i^\alpha \equiv \varDelta V_i^{\alpha \to \beta} \qquad (8)$$

Hence, the free energy differences are

$$F_i^L - F_i^\alpha \equiv \varDelta F_i^{\alpha \to L} = \varDelta H_i^{\alpha \to L} - T\varDelta S_i^{\alpha \to L} + 23.9P\,\varDelta V_i^{\alpha \to L} \qquad (9)$$
$$F_i^L - F_i^\beta \equiv \varDelta F_i^{\beta \to L} = \varDelta H_i^{\beta \to L} - T\varDelta S_i^{\beta \to L} + 23.9P\,\varDelta V_i^{\beta \to L} \qquad (10)$$
$$F_i^\beta - F_i^\alpha \equiv \varDelta F_i^{\alpha \to \beta} = \varDelta H_i^{\alpha \to \beta} - T\varDelta S_i^{\alpha \to \beta} + 23.9P\,\varDelta V_i^{\alpha \to \beta} \qquad (11)$$

where the free energy differences $\varDelta F_i$ and the enthalpy differences $\varDelta H_i$ are in calories per mole, the entropy differences $\varDelta S_i$ are in calories per mole per degree Kelvin, and the volume differences $\varDelta V_i$ are in cubic centimeters per mole. In Eqs. (9)–(11), the pressure P is in kilobars and the temperature T is in degrees Kelvin. Figures 1–4 illustrate alternate situations for metal i as a function of pressure.

Case 1. Metal i exhibits polymorphism at atmospheric pressure with the α

phase stable for $0°K \leqslant T \leqslant T_0$ and the β form stable between $T_0 \leqslant T \leqslant \bar{T}_i^\beta$. This case is shown schematically in Fig. 1. The free energies of the α, β, and liquid forms of i are shown as a function of temperature and pressure. The slopes of these curves vanish at temperatures approaching $0°K$ in accordance with third-law requirements.

The most stable phase at any temperature and pressure is the one that exhibits the lowest free energy as seen in Fig. 1. The free energy difference

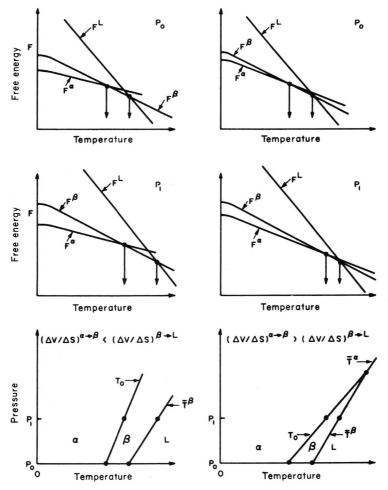

FIG. 1. Schematic representation of high-pressure transitions in a metal exhibiting 1-atm polymorphism where $\Delta V^{\alpha \rightarrow L} > \Delta V^{\beta \rightarrow L} > 0$, and the volume changes are constant. The upper panels show free energy relations that result in the phase diagrams below (20).

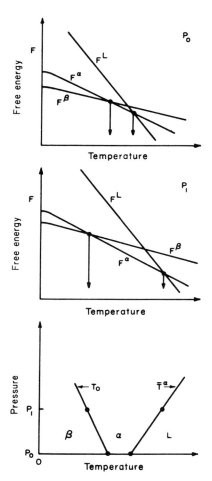

FIG. 2. Schematic representation of high-pressure transitions in a metal exhibiting 1-atm polymorphism to a denser high-temperature phase where $\Delta V^{\alpha \to L} > \Delta V^{\beta \to L} > 0$, and the volume changes are constant. The upper panels show free energy curves leading to the phase diagram below (20).

between any two phases is merely the difference between the respective curves. These differences go to zero at transition points. The free energy differences can be estimated from measurements of the latent heat of transformation, $\Delta H_i^{\alpha \to \beta}[T_0]$, the heat of fusion, $\Delta H_i^{\beta \to L}[\overline{T}_i^{\beta}]$, and the volume changes attending the $\alpha \to \beta$ and $\beta \to L$ reactions. These data would be sufficient to apply the Clapeyron equation for the purpose of calculating $(dT/dP)^{\alpha \to \beta}$, $(d\overline{T}/dP)^{\beta \to L}$, and $(d\overline{T}/dP)^{\alpha \to L}$ and to predict, with some confidence, the

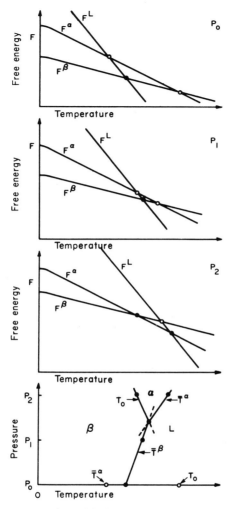

FIG. 3. Schematic representation of high-pressure transitions in a metal exhibiting no transitions at 1 atm, where $\Delta S^{\alpha \to \beta} < 0$ and $\Delta V^{\alpha \to L} > \Delta V^{\beta \to L} > 0$ (volume changes are constant). The free energy curves show the origin of the high-pressure phase in the phase diagram below (20).

qualitative features of the T–P diagram. A more accurate diagram could be calculated if specific heat, volumetric, and compressibility data were available for each phase. As shown in Fig. 1, a triple point will occur only if

$$(\Delta V / \Delta S)^{\alpha \to \beta} > (\Delta V / \Delta S)^{\beta \to L}.$$

Case 2. Metal i exhibits polymorphism at atmospheric pressure with the

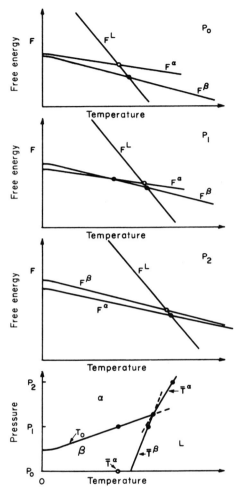

FIG. 4. Schematic representation of high-pressure transitions in a metal exhibiting no transitions at 1 atm, where $\Delta S^{\alpha \to \beta} > 0$ and $\Delta V^{\alpha \to L} > \Delta V^{\beta \to L} > 0$ (volume changes are constant). The free energy curves in the upper panels show the development of the phase diagram below (20).

β phase stable for $0°K \leqslant T \leqslant T_0$ and α phase stable from $T_0 \leqslant T \leqslant \overline{T}_i{}^\alpha$. This situation is similar to Case 1 in that atmospheric pressure calorimetric and volumetric data can be used to predict quantitatively the features of the $T–P$ diagram. In this case, as shown in Fig. 2, triple-point phenomena are not indicated.

   Case 3. Metal i is stable as βi for $0°K \leqslant T \leqslant \overline{T}_i{}^\beta$ at atmospheric pressure i.e., no polymorphism at 1 atm. Figures 3 and 4 show the alternative possibilities for this case in the situations where $\Delta S^{\alpha \rightarrow \beta}$ at atmospheric pressure is positive and negative. In this case triple-point phenomena are observed. However, apart from calculations of the $\overline{T}_i^{\beta}[P]$ curve, other characteristics of the P–T diagram cannot be directly estimated from 1-atm data concerning pure i.

The cases described above are idealized descriptions of what might occur when a real metal is exposed to hydrostatic pressure. Restricting our consideration to situations where the liquid is least dense, the volume differences are constant, and the 1-atm free energy curves nearly linear is not overly realistic. However, the behavior indicated by these simple examples should be representative of some of the more complex behavior of real systems. We will now turn to a consideration of real metallic systems.

1. Phase Equilibria in Metals at High Pressure

The occurrence of phase transformations in bismuth at high pressures has been the subject of much study. Although the phases which form when the rhombohedral (A7) structure is compressed have not been determined, the features of the phase diagram have been established. Figure 5 shows the diagram presented by Klement and Jayaraman (27) in their recent review of high-pressure equilibria. This diagram can be employed in combination with 1-atm thermodynamic and volumetric data to describe the lattice stability of the various polymorphs of bismuth along the lines indicated by Eqs. (9)–(11), where the enthalpy, entropy, and volume differences between polymorphs are treated as being constants. Based on such a first approximation, the following equations, obtained from observed volume changes and solid/liquid equilibria, can be derived for the lattice stability of the five polymorphs of bismuth (21).

$$\Delta F_{\text{Bi}}^{\text{I}\rightarrow\text{L}} = 2610 - 4.8T - 23.9P(0.98) \quad \text{cal/g-atom} \tag{12}$$

$$\Delta F_{\text{Bi}}^{\text{I}\rightarrow\text{II}} = 945 - 1.2T - 23.9P(0.98) \tag{13}$$

$$\Delta F_{\text{Bi}}^{\text{II}\rightarrow\text{III}} = 505 - 0.3T - 23.9P(0.64) \tag{14}$$

$$\Delta F_{\text{Bi}}^{\text{III}\rightarrow\text{IV}} = 545 - 1.2T \tag{15}$$

$$\Delta F_{\text{Bi}}^{\text{IV}\rightarrow\text{V}} = 365 - 0.2T - 23.9P(0.25) \tag{16}*$$

These simple equations (which presuppose volume and entropy changes that are independent of pressure and temperature) are sufficient to generate the phase diagram shown in Fig. 6. This computed diagram is in accordance with the experimental diagram shown in Fig. 5 to within 5%. As a consequence, Eqs. (12)–(16) are probably reasonable first approximations of the

* Equation (16) has been changed from the previous representation (21) to reflect alteration of the pressure scale.

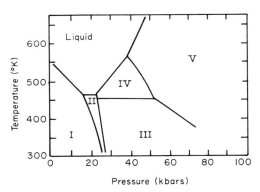

FIG. 5. Observed diagram for bismuth (27).

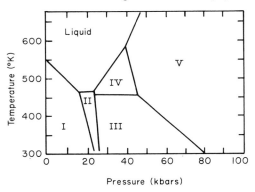

FIG. 6. Computed phase diagram for bismuth based on Eqs. (12)–(16).

lattice stability of the various polymorphs of bismuth. Klement et al. (28) have suggested that Bi^{II} corresponds to an ω structure, with III, IV, and V corresponding to hcp, bcc, and fcc, respectively. While the entropies of fusion indicated by Eqs. (12)–(16) are consistent with these suggestions, verification of these awaits further x-ray studies.

Additional examples are provided by considering the behavior of thallium, titanium, and zirconium under pressure as shown in Figs. 7–10. The observed phase diagrams (27) are shown in Figs. 7 and 8, while the calculated phase diagrams (21) stem from simple relations of the form described by Eqs. (9)–(11). For the case of thallium (Fig. 9), the relative stability of the L (Liquid), β (bcc), ϵ (hcp), and α (fcc) phases is defined by Eqs. (17)–(20).

$$\Delta F_{Tl}^{\beta \to L} = 1025 - 1.78T + 23.9P(0.39) \quad \text{cal/g-atom} \tag{17}$$

$$\Delta F_{Tl}^{\epsilon \to \beta} = \quad 90 - 0.18T - 23.9P(0.019) \tag{18}$$

$$\Delta F_{Tl}^{\alpha \to \epsilon} = -76 - 0.04T + 23.9P(0.101) \tag{19}$$

$$\Delta F_{Tl}^{\alpha \to \beta} = \quad 14 - 0.22T + 23.9P(0.082) \tag{20}$$

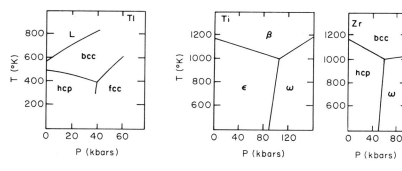

FIG. 7. Observed diagram for thallium (27).

FIG. 8. Observed phase diagrams for titanium and zirconium (27).

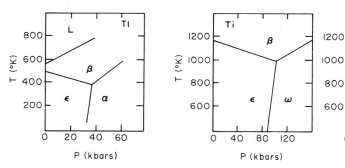

FIG. 9. Calculated diagram for thallium, based on Eqs. (17)–(20).

FIG. 10. Calculated phase diagrams for titanium and zirconium, based on Eqs. (21)–(26).

For the titanium and zirconium cases (Fig. 10) the high-pressure phase has the ω structure, which is a distortion of the bcc structure into a hexagonal lattice. This structure forms at 1 atm in Ti–V, Ti–Fe, and Ti–Mn alloys as the result of cooling β below the range of 720–620°K, depending on alloy content. Equations (21)–(23) describe the relative stability of the β, ϵ (hcp), and ω forms of titanium.

$$\Delta F_{\text{Ti}}^{\epsilon \to \beta} \approx 1050 - 0.91T - 23.9P(0.060) \quad \text{cal/g-atom} \tag{21}$$

$$\Delta F_{\text{Ti}}^{\omega \to \epsilon} \approx -360 - 0.08T + 23.9P(0.190) \tag{22}$$

$$\Delta F_{\text{Ti}}^{\omega \to \beta} \approx 690 - 0.99T + 23.9P(0.130) \tag{23}$$

As shown in Fig. 10, these equations yield a triple point at 1000°K and 97 kbar, an $\epsilon \to \omega$ transition pressure of 85 kbar at 300°K, and a $\beta \to \omega$ metastable transition temperature of 695°K at 1 atm. The computed dT/dP values are -1.6°K/kbar for $\epsilon \to \beta$, $+57$°K/kbar for $\epsilon \to \omega$, and $+3.1$°K/kbar for $\omega \to \beta$. In the case of zirconium, a similar set of results is obtained (21) as shown by Eqs. (24)–(26).

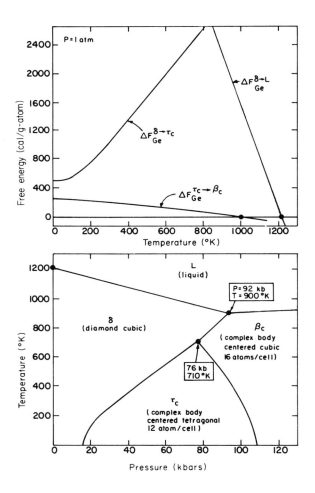

FIG. 11. Computed free energy differences and phase relations for germanium, based on Eqs. (34)–(38).

$$\Delta F_{Zr}^{\epsilon \to \beta} \approx \quad 1040 - 0.91T - 23.9P(0.096) \quad \text{cal/g-atom} \tag{24}$$

$$\Delta F_{Zr}^{\omega \to \epsilon} \approx \; -360 - 0.08T + 23.9P(0.320) \tag{25}$$

$$\Delta F_{Zr}^{\omega \to \beta} \approx \quad 680 - 0.99T + 23.9P(0.224) \tag{26}$$

As indicated in Fig. 10, these equations predict a triple point at 1000°K and 57 kbar, an $\epsilon \to \omega$ transition pressure of 50 kbar at 300°K, and dT/dP values of -2.5°K/kbar for $\epsilon \to \beta$, $+96$°K/kbar for $\epsilon \to \omega$ and $+5.4$°K/kbar for $\omega \to \beta$. Table I lists free energy differences for $\alpha \to \beta$, $\epsilon \to \beta$ and $\alpha \to \epsilon$ transitions in several other metals as well as the diamond ⇌ graphite transition in carbon.

Needless to say, the simple description afforded by Eqs. (9)–(11) are of limited applicability if a complete description is required covering cases

TABLE I

FREE ENERGY DIFFERENCES BETWEEN bcc (β), hcp (ϵ), AND fcc (α) FORMS OF SEVERAL METALS AS A FUNCTION OF TEMPERATURE AND PRESSURE [a,b]

$$\Delta F_{\text{Ca}}^{\alpha \rightarrow \beta} = 58 - 0.08T + 23.9P(0.011) \quad \text{cal/g-atom} \tag{27}$$
$$\Delta F_{\text{Sr}}^{\alpha \rightarrow \beta} = 200 - 0.23T - 23.9P(0.10) \tag{28}$$
$$\Delta F_{\text{Be}}^{\epsilon \rightarrow \beta} = 1100 - 0.72T - 23.9P(0.18) \tag{29}$$
$$\Delta F_{\text{Yb}}^{\alpha \rightarrow \beta} = 760 - 0.71T - 23.9P(0.8) \tag{30}$$
$$\Delta F_{\text{Ba}}^{\beta \rightarrow \epsilon} = 1040 + 0.80T - 23.9P(0.74) \tag{31}$$
$$\Delta F_{\text{Pb}}^{\alpha \rightarrow \epsilon} = 600 - 0.10T - 23.9P(0.18) \,^d \tag{32}$$
$$\Delta F_{\text{C}}^{\delta \rightarrow \gamma}[T, P] = -300 - 1.14T + 23.9P(1.76) \quad \text{cal/g -atom} \,^e \tag{33}$$

[a] Applicable at temperatures above 300°K.

[b] Kaufman (21) and Kaufman and Clougherty (29).

[e] δ = diamond, γ = graphite.

[d] Based on results of T. Takahashi, H. K. Mao, and W. Bassett, *Science* **165**, 1352 (1967).

where the volume and entropy changes are not constant. Nevertheless, the simple description can be employed as a first approximation and improved by a more detailed description. This procedure can be described by considering the relative stability of the diamond cubic (δ), complex body-centered tetragonal (τ_c), and complex body-centered cubic (β_c) forms of germanium. Bates *et al.* (30, 31) have determined the phase diagram and the fields of stability of these forms. The phase relations and free energy differences between the various forms are shown in Fig. 11 and Table II (32).

TABLE II

LATTICE STABILITY OF VARIOUS STRUCTURES OF GERMANIUM

Designation	Structure
L	Liquid
δ	Diamond cubic
τc	Complex body-centered tetragonal (12 atoms/unit cell)
β_c	Complex body-centered cubic (16 atoms/unit cell)
τ	Body-centered tetragonal (white tin structure)

Free energy differences

$$\Delta F_{\text{Ge}}^{\delta \rightarrow L} = 8830 - 7.30T - 23.9P(1.00) \quad \text{cal/g-atom} \tag{34}$$
$$\Delta F_{\text{Ge}}^{\delta \rightarrow \tau c} = 500 + F[\Theta^{\tau c}/T] - F[\Theta^{\delta}/T] - 23.9P(1.30) \quad \text{cal/g-atom} \tag{35}$$
where $\Theta^{\delta} = 340°K$ and $\Theta_{\tau c} = 600°K$
$$\Delta F_{\text{Ge}}^{\delta \rightarrow \tau c} = 30 + 3.30T - 23.9P(1.30) \quad \text{for} \quad T \geq 400°K \tag{36}$$
$$\Delta F_{\text{Ge}}^{\tau c \rightarrow \beta c} = 250 + F[\Theta^{\beta c}/T] - F[\Theta^{\tau c}/T] - 23.9P(-0.08 + 1.65 \times 10^{-3}P) \quad \text{cal/g-atom} \tag{37}$$
where $\Theta^{\beta c} = 570°K$ and $\Theta^{\tau c} = 600°K$
$$\Delta F_{\text{Ge}}^{\tau c \rightarrow \beta c} \simeq 300 - 0.30T - 23.9P(-0.08 + 1.65 \times 10^{-3}P) \quad \text{for} \quad T \geq 200°K \tag{38}$$
where P is in kilobars and T is in degrees Kelvin.

Reference to Table II shows that Eqs. (34) and (36), which describe the free energy differences between the δ and L forms and the δ and τ_c forms as a function of temperature and pressure, are applicable at high temperatures. In order to describe the free energy difference between the δ and τ_c forms at low temperatures (where third-law restrictions require that ΔS approach zero), Debye functions are employed. Finally, Eqs. (37) and (38) illustrate the description of the volume difference between the β_c and τ_c forms as a pressure-dependent function.

This description permits specification of the free energy differences between the various forms of germanium and calculation of the phase diagram as shown in Fig. 11.

The value of $\Theta^\delta = 340°K$ was chosen (32) on the basis of $S^\delta_{Ge} = 7.43$ cal/g-atom-°K at 298°K and 1 atm. The symbol $F[\Theta^\delta/T]$ in Table II is the Debye free energy function (33) defined by $\Theta^\delta = 340°K$ (Table III). Figure 11 shows the free energy differences between the various forms of germanium at 1 atm and the computed phase diagram, which closely resembles the diagram suggested by Bates et al. (30, 31). The volume of the τ_c form was found to be 1.30 cm³/g-atom less than that of the δ form and 0.08 cm³/g-atom less than the β_c form at 1 atm. Since the β_c form becomes more stable than the τ_c form as the pressure is increased, the difference in volume between τ_c and β_c form must be dependent upon pressure as shown in Table II. Finally, it should be noted that the volume of the white tin form, τ, is 2.28 cm³/g-atom less than the δ form near 120 kbar. This would indicate that the τ form may become the most stable form at very high pressures (32).

The most complex description of the relative stability of crystal forms as a function of temperature and pressure involves enthalpy, entropy, and volume differences which are dependent upon temperature and pressure. Such a description has been developed for the purpose of characterizing the relative stability of bcc (α), fcc (γ), and hcp (ϵ) iron (34–36). The complex description required in this case is due to the magnetic phenomena characteristic of iron. A similar analysis has been performed for manganese by Tauer and Weiss (37).

The description of the heat capacity of ϵ iron is considered in terms of lattice, electronic, and magnetic contributions:

$$C_p^\epsilon[T] = C_v[\Theta^\epsilon/T](1 + 10^{-4}T) + \gamma^\epsilon T + C_p^{\epsilon\mu}[T] \tag{39}$$

where Θ^ϵ is the Debye temperature of ϵ iron, γ^ϵ is the electronic specific-heat coefficient, and $C_p^{\epsilon\mu}$ is the magnetic contribution to the heat capacity. This relation can then be integrated to yield an expression for the free energy of ϵ iron:

$$F^\epsilon[T] = H_0^\epsilon + F^\epsilon[\Theta^\epsilon/T] + F^{\epsilon c}[\Theta^\epsilon, T] + F^{\epsilon\mu}[T] - \tfrac{1}{2}\gamma^\epsilon T^2 \tag{40}$$

where H_0^ϵ is the enthalpy at 0°K, based on $H_0^\alpha = 0$.

TABLE III

DEBYE FREE ENERGY FUNCTION

$$-F/RT = R^{-1}(S - UT^{-1})\,{}^{a}$$

Θ/T	T									
	0.00	0.01	0.02	0.03	0.04	0.05	0.06	0.07	0.08	0.09
0.00	—	—	—	11.5534	10.7015	10.0433	9.5075	9.0562	8.6667	8.3245
0.10	8.0195	7.7447	7.4947	7.2656	7.0544	6.8584	6.6758	6.5050	6.3445	6.1932
0.20	6.0503	5.9149	5.7863	5.6638	5.5470	5.4354	5.3287	5.2263	5.1280	5.0336
0.30	4.9427	4.8551	4.7706	4.6891	4.6103	4.5340	4.4602	4.3887	4.3194	4.2522
0.40	4.1869	4.1234	4.0618	4.0018	3.9434	3.8866	3.8312	3.7773	3.7247	3.6733
0.50	3.6232	3.5743	3.5265	3.4799	3.4342	3.3896	3.3460	3.3033	3.2615	3.2206
0.60	3.1805	3.1413	3.1028	3.0652	3.0282	2.9920	2.9565	2.9216	2.8874	2.8538
0.70	2.8209	2.7885	2.7568	2.7255	2.6949	2.6647	2.6351	2.6060	2.5774	2.5493
0.80	2.5216	2.4944	2.4676	2.4413	2.4153	2.3898	2.3647	2.3400	2.3157	2.2917
0.90	2.2681	2.2449	2.2220	2.1994	2.1772	2.1553	2.1337	2.1125	2.0915	2.0708
1.00	2.0505	—	—	—	—	—	—	—	—	—

	1.10	1.20	1.30	1.40	1.50	1.60	1.70	1.80	1.90	2.00
	1.8573	1.6896	1.5465	1.4160	1.3022	1.1990	1.1059	1.0200	0.9406	0.8750

^{a} $R = 1.9873$ cal/g-atom-°K.

Similar equations are defined for the α (bcc) and γ (fcc) forms of iron accounting for the enthalpy H_0 at $0°K$, the vibrational free energy $F[\Theta/T]$, the $(C_p - C_v)$ correction term $F^c[\Theta, T]$, and the magnetic free energy $F^\mu[T]$. The approximation employed to compute $F^c[\Theta, T]$ is given by Eq. (41).

$$F^c[\Theta, T] = 10^{-4}(3R\Theta^2\{(T^2/2\Theta^2) - 0.025 \ln[(20T^2/\Theta^2) + 1]\} - TU[\Theta/T]) \quad (41)$$

where $U[\Theta/T]$ is the Debye energy and $R = 1.987$ cal/g-atom-$°K$ is the gas constant (33).

As indicated in Table IV, the magnetic specific heat and free energy of ϵ iron are negligible. However, this is not the case for the α and γ forms as shown in Figs. 12 and 13. The magnetic free energy of α iron is obtained through separation of $C_p^{\alpha\mu}[T]$ as shown in Fig. 13, followed by graphical integration. Table V illustrates the results. The magnetic free energy and

TABLE IV

THERMODYNAMIC PROPERTIES OF α, γ, AND ϵ IRON [a]

Phase	$V[T = 0°K]$ (cm³/mole) [b]	H_0 (cal/mole)	Θ (°K)	γ [cal/mole-(°K)²]	F^μ (cal/mole)
α	7.061	0	432	12×10^{-4}	Table V
γ	7.216 [c] 6.695 [c]	1303	432	12×10^{-4}	Eq. [46]
ϵ	6.731	1150	385	14×10^{-4}	0

[a] Kaufman et al. (34), Blackburn et al. (35), Stepakoff and Kaufman (36).
[b] $V[T] = V[T = 0] (1 + 2.043 \times 10^{-5}T + 1.520 \times 10^{-8}T^2)$ cm³/mole.
[c] The two values at $V[T = 0°K]$ given for fcc correspond to a high-moment–high-volume and a low-moment–low-volume state (34).

specific heat of the γ phase are described in terms of a two-spin-state system unique to the fcc form of iron, which is characterized in terms of a "Schottky anomaly." This model considers two energy levels, E_0 and E_1, characterized by degeneracies g_0 and g_1. If y is the fraction of atoms present in the upper *level*, then

$$y/(1 - y) = (g_1/g_0) \exp(-\Delta E/RT) \quad (42)$$

where $\Delta E = E_1 - E_0$ or

$$y = [1 + (g_0/g_1) \exp(\Delta E/RT)]^{-1} \quad (43)$$

The energy E of the system (or the enthalpy H at zero pressure) is given by

$$E = E_0^{\gamma\mu} + y \Delta E = E_0^{\gamma\mu} + H_{Fe}^{\gamma\mu}[T] \quad (44)$$

TABLE V

FREE ENERGY AND ENTHALPY DIFFERENCES BETWEEN bcc (α) AND fcc (γ) IRON [a]

$$\Delta H^{\alpha \to \gamma} = \Delta H^{\alpha \to \gamma}[0°K] + H^{\gamma \mu}[T] - H^{\alpha \mu}[T] \quad \text{cal/mole}$$
$$\Delta F^{\alpha \to \gamma} = \Delta H^{\alpha \to \gamma}[0°K] + F^{\gamma \mu}[T] - F^{\alpha \mu}[T] \quad \text{cal/mole}$$
$$\Delta H^{\alpha \to \gamma}[0°K] = 1303 \quad \text{cal/mole}$$

T (°K)	$H^{\alpha\mu}$	$F^{\alpha\mu}$	$\Delta H^{\alpha \to \gamma}$	$\Delta F^{\alpha \to \gamma}$	T (°K)	$H^{\alpha\mu}$	$F^{\alpha\mu}$	$\Delta H^{\alpha \to \gamma}$	$\Delta F^{\alpha \to \gamma}$
0	0	0	1303	1303	740	262	−100	1454	375
100	1	0	1325	1297	60	290	−110	1430	342
200	3	−1	1451	1223	80	320	−121	1401	322
300	7	−5	1551	1085	800	352	−134	1374	283
20	8	−6	1567	1043	20	386	−148	1343	258
40	10	−7	1578	1021	40	423	−162	1308	233
60	13	−8	1587	990	60	463	−179	1271	205
80	16	−9	1595	956	80	507	−197	1229	183
400	20	−11	1603	921	900	556	−216	1182	165
20	25	−13	1607	888	20	610	−236	1131	142
40	31	−14	1609	855	40	672	−257	1061	120
60	38	−17	1611	819	60	744	−277	999	105
80	46	−19	1610	784	80	827	−300	918	89
500	55	−23	1609	749	1000	926	−325	811	76
20	66	−25	1604	706	20	1050	−350	699	63
40	77	−30	1600	681	40	1225	−375	526	50
60	88	−36	1592	649	60	1312	−406	440	39
80	102	−41	1584	619	80	1368	−439	387	29
600	117	−47	1574	586	1100	1415	−472	341	20
20	133	−53	1562	553	20	1456	−504	301	13
40	151	−59	1550	517	40	1491	−538	268	8
60	170	−66	1533	491	60	1521	−573	238	4
80	190	−74	1516	462	80	1547	−609	213	1
700	212	−82	1500	424					
20	236	−91	1478	400	1183	1550	−615	210	0

[a] This set of values corrects the error that was noted (38) in the earlier tabulation (34).

On the basis of Eq. (44)

$$C_p^{\gamma\mu} = \frac{R(\Delta E/RT)^2(g_0/g_1) \exp(\Delta E/RT)}{[(g_0/g_1) \exp(\Delta E/RT) + 1]^2} \tag{45}$$

and

$$F_{Fe}^{\gamma\mu} = E_0^{\gamma\mu} + F_{Fe}^{\gamma\mu}[T] = E_0^{\gamma\mu} + RT \ln(1 - y[T])$$

or

$$F_{Fe}^{\gamma\mu} = E_0^{\gamma\mu} - RT \ln[1 + (g_1/g_0) \exp(-\Delta E/RT)] \tag{46}$$

Since the energy $E_0^{\gamma\mu}$ can be absorbed along with the difference in lattice energy, $\Delta E^{\alpha \to \gamma}$, into the zero point enthalpy of γ iron, i.e., $H_{Fe}^{\gamma}[0°K] = E_0^{\gamma\mu} + \Delta E^{\alpha \to \gamma}$, the description of the magnetic free energy and enthalpy in explicit terms involves the evaluation of the difference between energy

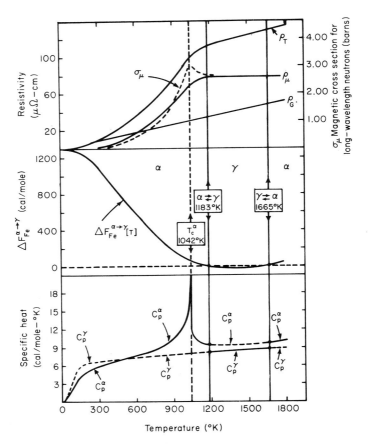

FIG. 12. Specific heat, electrical resistivity, magnetic cross section, and difference in free energy for bcc (α) and fcc (γ) iron at 1 atm, illustrating origin of complex lattice stability of iron.

levels ΔE, the degeneracy ratio (g_0/g_1), and $H_{\text{Fe}}^\gamma[0^\circ\text{K}]$. As indicated in Table IV, $H_{\text{Fe}}^\gamma[0^\circ\text{K}] = 1303$ cal/mole.* The energy level difference ΔE and the degeneracy ratio (g_0/g_1) were found to equal 820 cal/mole and 0.559, respectively (34).

Reference to Table IV indicates that the volume of α iron is given by Eq. (47) as

$$V_{\text{Fe}}^\alpha[T] = 7.061(1 + 2.043 \times 10^{-5}T + 1.520 \times 10^{-8}T^2) \quad \text{cm}^3/\text{mole} \qquad (47)$$

The representation of $V_{\text{Fe}}^\gamma[T]$ is somewhat complicated by the necessity

* The most unambiguous unit is calories per gram atom. In the case of a pure metal, calories per gram atom and calories per mole are identical. This is the only case in which the latter unit will be used.

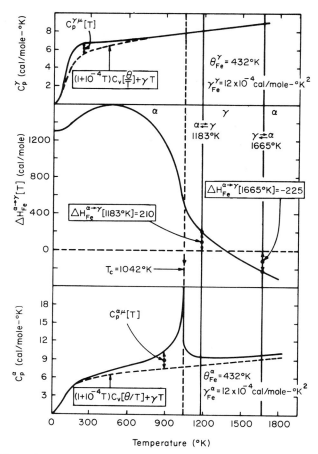

Fig. 13. Magnetic, electronic and vibrational specific heats of α and γ iron and the enthalpy difference as a function of temperature, showing separation of magnetic specific heat components.

to describe two spin states having different volumes. In principal, however,

$$V_{\text{Fe}}^{\gamma}[T] = (1 - y)V_{\text{Fe}}^{\gamma_0}[T] + y\, V_{\text{Fe}}^{\gamma_1}[T] \tag{48}$$

where $V_{\text{Fe}}^{\gamma}[T]$ is the observed volume of γ iron, $V_{\text{Fe}}^{\gamma_0}[T]$ and $V_{\text{Fe}}^{\gamma_1}[T]$ are the molar volumes of the low-spin–low-volume state and high-spin–high-volume states, respectively, and y is the population fraction of the high-volume state. Reference to Table IV indicates that these volumes are given by Eqs. (49a) and (49b) as

$$V_{\text{Fe}}^{\gamma_0}[T] = 6.695(1 + 2.043 \times 10^{-5}T + 1.520 \times 10^{-8}T^2) \tag{49a}$$

$$V_{\text{Fe}}^{\gamma_1}[T] = 7.216(1 + 2.043 \times 10^{-5}T + 1.520 \times 10^{-8}T^2) \tag{49b}$$

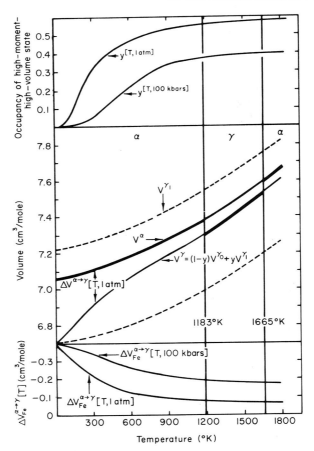

FIG. 14. Relative population of high- and low-moment spin states in γ iron and volumetric relations as a function of temperature. Relative population and volumetric relations as a function of temperature and pressure are based on Eqs. (48)–(53).

Figure 14 illustrates the temperature dependence of the population y and the volume of γ iron at 1 atm and 100 kbar.

This description permits specification of the free energy differences between the liquid, bcc, hcp, and fcc forms of iron as a function of temperature at 1 atm shown in Fig. 15. Representation of the free energy differences at high pressure is complicated by the variation in the population of the high-volume–low-volume states in γ iron with temperature and pressure. However, the present description permits this variation to be specified. Thus, the difference in volume between these states is

$$\Delta V_{\text{Fe}}^{\gamma_0 \rightarrow \gamma_1}[T] = V_{\text{Fe}}^{\gamma_1}[T] - V_{\text{Fe}}^{\gamma_0}[T] \tag{50}$$

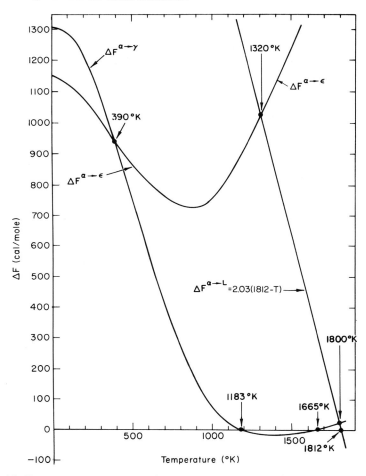

FIG. 15. Free energy–temperature relations for the bcc (α), fcc (γ), hcp (ϵ), and liquid (L), forms of iron at 1 atm, illustrating transition temperatures and melting temperatures.

The thermodynamic equations for the specific heat, enthalpy, and free energy at a pressure P are obtained directly from Eqs. (42), (43), and (50) by setting

$$\Delta E[P] = \Delta E + 23.9P\ \Delta V_{\text{Fe}}^{\gamma_0 \to \gamma_1}[T] \tag{51}$$

and

$$E_0^{\gamma\mu}[P] = E_0^{\gamma\mu} + 23.9P\ V_{\text{Fe}}^{\gamma_0}[T] \tag{52}$$

Hence

$$y[T, P] = \{1 + (g_0/g_1)\exp[(\Delta E + 23.9P\ \Delta V_{\text{Fe}}^{\gamma_0 \to \gamma_1}[T])/RT]\}^{-1} \tag{53}$$

and

$$F_{\text{Fe}}^{\gamma\mu}[T, P] = E_0^{\gamma\mu} + 23.9P\ V_{\text{Fe}}^{\gamma_0}[T] + RT\ln(1 - y[T, P]) \tag{54}$$

so that

$$\Delta F^{\alpha \rightarrow \gamma}[T, P] \approx 1303 - F^{\alpha\mu}_{Fe}[T] + RT \ln(1 - y[T, P])$$
$$+ 23.9P(V^{\gamma_0}_{Fe}[T] - V^{\alpha}_{Fe}[T]) \tag{55}$$

It should be noted that Eq. (55) meets the requirement that

$$\left(\frac{\partial \Delta F^{\alpha \rightarrow \gamma}_{Fe}}{\partial P}\right)_T = (V^{\gamma_0}_{Fe}[T] + y[P, T] \Delta V^{\gamma_0 \rightarrow \gamma_1}_{Fe}[T]) - V^{\alpha}_{Fe} \equiv \Delta V^{\alpha \rightarrow \gamma}[T, P] \tag{56}$$

Similarly, the free energy difference between the α and ϵ forms is defined as

$$\Delta F^{\alpha \rightarrow \epsilon}_{Fe}[T, P] = \Delta F^{\alpha \rightarrow \epsilon}_{Fe}[T] + 23.9P \Delta V^{\alpha \rightarrow \epsilon}_{Fe}[T] \tag{57}$$

Finally,

$$\Delta F^{\gamma \rightarrow \epsilon}_{Fe}[T, P] = \Delta F^{\gamma \rightarrow \alpha}_{Fe}[T, P] + \Delta F^{\alpha \rightarrow \epsilon}_{Fe}[T, P] \tag{58}$$

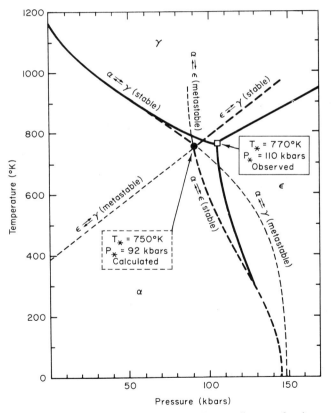

FIG. 16. Comparison of computed and observed T–P diagram for iron. Computed diagram is based on Eqs. (55)–(58) (*21*).

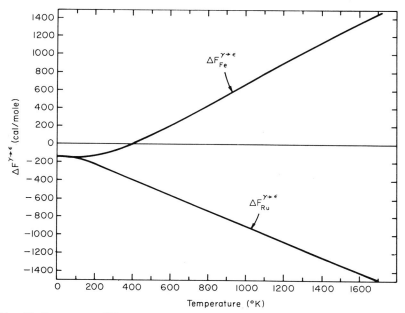

FIG. 17. Free energy difference between hcp (ϵ) and fcc (γ) forms of iron and ruthenium. These curves show how the magnetic contributions to the fcc form of iron stabilize it at high temperatures. These contributions are absent in ruthenium, leading to stability of the hcp phase.

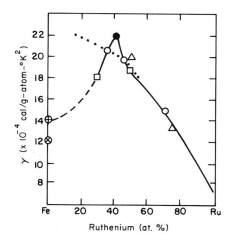

FIG. 18. Electronic specific heat coefficients of hcp Fe–Ru alloys. Measurements for alloys with 30–100% Ru indicate that the estimated values for pure iron are correct. [Data points: ●, (42) C; ○, (40) (C 1967); □, (41) (ABMS); △, (39) (HBM); ⊗ (35) (BKC); ⊕, (36) (SK).]

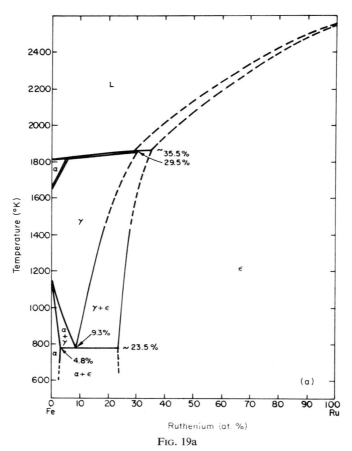

FIG. 19a

FIG. 19. Comparison of computed and observed phase equilibria in the iron–ruthenium system. (a) Observed iron–ruthenium phase diagram showing bcc (α), fcc (γ), and hcp (ϵ) fields. (b) Computed iron–ruthenium phase diagram based on regular-solution models [see discussion of Eqs. (103) and (104)]. (c) Comparison of observed and computed triple point pressures and temperatures in iron–ruthenium alloys (43). These temperatures and pressures are the binary analogs of the triple point for iron shown in Fig. 16. Solid lines are computed (21).

The phase boundaries for the $\alpha \rightleftharpoons \gamma$, $\alpha \rightleftharpoons \epsilon$, and $\gamma \rightleftharpoons \epsilon$ reactions are defined by setting Eqs. (55), (57), and (58) equal to zero and solving for T and P. The results are compared with experiment in Fig. 16, while Fig. 17 shows the difference in free energy between the γ and ϵ forms of iron at 1 atm.

Recent measurement of the electronic specific-heat coefficients of hcp Fe–Ru alloys shown in Fig. 18 have provided additional verification of the current description of iron presented in Table IV. Values of 12×10^{-4}

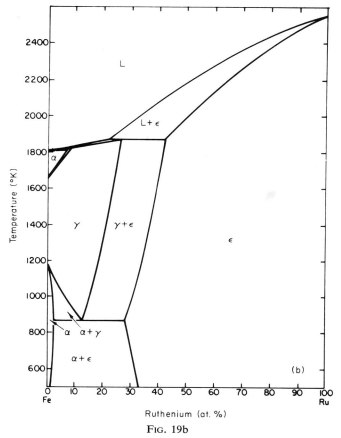

FIG. 19b

cal/mole-°K² (35) and 14×10^{-4} cal/mole-°K² (36) were deduced from phase
equilibria and specific heat data obtained at temperatures above 60°K. These
results are denoted by (BKC 1967) (35) and (SK 1967) (36) in Fig. 18.
Measurements by Heiniger, Bucher, and Muller (39) and Claus (40) of alloys
containing 36, 42, 50, and 72 at.% Ru designated (HBM 1966) and (C 1967)
in Fig. 18 suggested that the electronic specific heat coefficient for hcp iron
was near 23×10^{-4} cal/g-atom-°K² as indicated by the extension of the dotted
curve to pure iron shown in Fig. 18. Recently however, Andres et al. (41)
(ABMS) and Claus (42) (C) have measured the electronic specific heat
coefficient of alloys containing 30, 40, and 50 at.% ruthenium as shown in
Fig. 18. The maximum in the γ vs. Ru curve supports the value of γ for
pure hcp iron shown in Table IV.

The thermodynamic description of iron shown in Table IV indicates that
the vibrational entropy of hcp iron exceeds that of bcc iron (36). This finding

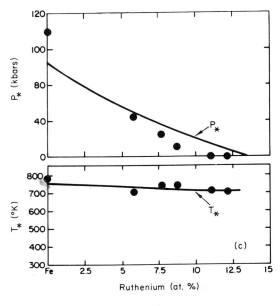

FIG. 19c

is in contrast to the commonly held view that the vibrational entropy of a bcc lattice must exceed that of a close-packed form.

Subsequent to the generation of specific information on the relative stability of bcc, hcp, and fcc iron, an analysis was performed of the Fe–Ru system (43). This analysis considered the relative stability of the bcc, hcp, and fcc forms of ruthenium (which occupies the same column of the periodic table) to be identical to that of iron except for the magnetic contributions. The foregoing description for iron (Eqs. (31)–(51) and Table IV) was employed to specify the relative stability of β, ϵ, and α ruthenium, the only difference being that the magnetic contributions to the bcc and fcc phases were set equal to zero. Under these conditions, the hcp phase is stable up to its melting point as indicated by Fig. 17. Figure 19 shows a comparison of computed and observed phase equilibria in the Fe–Ru system based upon the above-mentioned description of ruthenium (43).

The foregoing discussion of phase stability in bismuth, thallium, titanium, zirconium, and iron illustrates the behavior idealized in Figs. 1–4. Only one of these idealized characteristics has not been considered thus far. This behavior is the " pinchoff " of a stable phase at high pressure shown in the right of Fig. 1. Such a pinchoff occurs when a phase that is stable at 1 atm becomes unstable at high pressure. Uranium and manganese are real metals that exhibit such phenomena. In the case of the former metal, the absence of complicating magnetic phenomena permit a simplified representation of the relative stability of the orthorhombic (α), tetragonal (β), and bcc (γ) phases on the basis of thermodynamic and volumetric data (44) obtained under atmospheric conditions as shown in Eqs. (59) and (60).

$$\Delta F_U^{\alpha \to \beta} = 714 - 0.76T + 23.9P(0.142) \quad \text{cal/g-atom} \tag{59}$$

$$\Delta F_U^{\beta \to \gamma} = 1166 - 1.12T + 23.9P(0.095) \tag{60}$$

However, in the case of manganese, complex magnetic contributions require a more detailed description along the lines employed for iron. As indicated earlier, Tauer and Weiss (37) have performed such an analysis of the relative stability of the complex cubic (α), primitive cubic (β), fcc (γ), and bcc (δ) forms. Their results are shown in Fig. 20.

At high temperatures, the relative stability of these solid phases (and liquid manganese) can be approximated by Eqs. (61)–(64),

$$\Delta F_{Mn}^{\alpha \to \beta} = 540 - 0.54T + 23.9P(0.31) \quad \text{cal/g-atom} \tag{61}$$

$$\Delta F_{Mn}^{\beta \to \gamma} = 530 - 0.39T + 23.9P(0.07) \tag{62}$$

$$\Delta F_{Mn}^{\gamma \to \delta} = 420 - 0.30T + 23.9P(0.09) \tag{63}$$

$$\Delta F_{Mn}^{\delta \to L} = 3500 - 2.30T + 23.9P(0.36) \tag{64}$$

which result from 1-atm thermodynamic (7, 37) and volumetric data (45). Equations (59)–(64) have been employed to compute the phase equilibria in uranium and manganese, which are compared with observed behavior (27) in Figs. 21 and 22. In both cases, good agreement is obtained and the high-pressure pinchoff of stable phases is observed.

2. First-Principle Calculations of the Lattice Stability of Metallic Phases

The foregoing description of the lattice stability of a number of metals illustrates the current status of quantitative information determined along empirical or thermodynamic lines. This route is very slow and is limited to cases where a metal exhibits polymorphism under atmospheric conditions or at high pressure. Ideally, it would be desirable to obtain quantitative information on lattice stability from first-principle calculations. Although a number of such calculations have been presented (46–48), reliability (46) and current means of application (49) of these methods are questionable.

Lomer (17) has suggested that the formal path for fundamental calculation of phase stability should consist of six stages as follows: "1, Assume structure; 2, Evaluate self-consistent fields for the component-free atoms; 3, Superpose these potentials to get band structure, and use the resulting wave

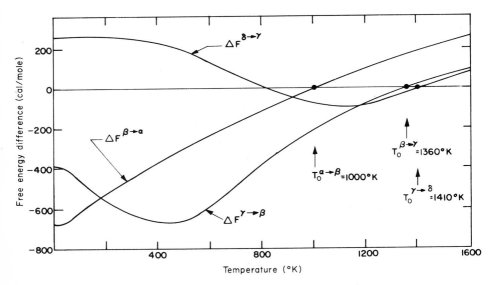

FIG. 20. Free energy differences between the α (complex cubic), β (primitive cubic), γ (fcc), and δ (bcc) forms of manganese as a function of temperature. Transition points are indicated. [After Tauer and Weiss (37).]

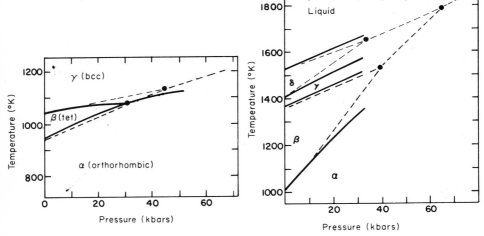

FIG. 21. Comparison of calculated and observed phase diagram for uranium. Calculations are based on Eqs. (59) and (60). Solid curves are observed, broken ones are calculated.

FIG. 22. Comparison of calculated and observed phase diagram for manganese. Calculations are based on Eqs. (61)–(64). Solid curves are observed; broken ones are calculated; α is complex cubic, β is primitive cubic, γ is fcc, δ is bcc.

function to get self-consistent solution for solid; 4, Evaluate correlation energy correction; 5, Evaluate total energy; and 6, Repeat for all suspected rival structures." Although this procedure is clearly the most desirable, it is almost never followed.

Instead, some particular term in the energy expression is investigated on the *presumption* that it is varying rapidly as a function of some structural parameter. Such a discussion is usually based on the importance of the outer electrons or the details of the Fermi surface (e.g., Hume-Rothery rules, magnetic structures, etc.). The distortion of structures or their local stability can usually be managed in this manner. However, comparisons between distinct structures on this basis are less trustworthy (*17*). Moreover, Stringer (*26*) has shown that for the "classical" fcc solid solutions in Cu–Zn and Cu–Ge there is *no discontinuous change* in the Fermi surface at any value of the electron/atom ratio.

In the opinion of the authors, the source of the current difficulty is quite clear. Lomer's recipe is undoubtedly correct. However, the most important component (Step No. 6), which requires comparison of competing structures, is never performed. Instead, the "phase competition criterion" is erroneously replaced by the *presumption* that phase stability can be predicted by locating the point at which a single property exhibits a rapid variation with composition.

The method of pseudopotentials has been applied with variable success (*46–49*) to the problem of computing the stability of various structures of pure metals. This method could provide relevant information on the enthalpy difference between the principal crystal forms at 0°K. Harrison (*46, 50, 51*) has computed the relative energies of the β, ϵ, and α forms of Na, Mg, Al, and Zn. Table VI summarizes the results. For Na, Mg, and Al, the correct structure was calculated, since Na becomes hcp at low temperature. However, the calculations indicate that fcc Zn is more stable than the hcp

TABLE VI

SUMMARY OF ENTHALPY DIFFERENCES BETWEEN THE bcc (β), hcp (ϵ), AND fcc (α) STRUCTURES OF VARIOUS METALS CALCULATED BY THE PSEUDOPOTENTIAL METHOD

Metal	$\Delta H^{\beta \to \epsilon}$ (cal/g-atom)	$\Delta H^{\alpha \to \epsilon}$ (cal/g-atom)	Ref.
Zn	−1700	+2325	*50*
Al	−3200	+1170	*51*
Mg	−1345	−715	*51*
Na	−119	−13	*51*
Na	117	−36	*48*
Li	−66	−32	*48*
K	−46	−13	*48*

form, contrary to observation. Pick has performed similar calculations for Na, Li, and K as shown in Table VI (48). His values are in relatively good agreement with Harrison's results for Na. In addition, his results for Li are in agreement with observations of the ϵ and α forms at low temperatures. However, Pick's results do indicate stability of the hcp form of K, in contrast to present observations. Harrison's current appraisal of the pseudopotential calculations is that " the prediction of structure with the theory is not reliable, although this may change with improved pseudopotentials " (46).

Although the questionable reliability (46) of fundamental phase-stability calculations (which include the competition criterion) complicates the problem, progress can be made along experimental lines. In particular, judicious use of thermodynamic data and observations of phase equilibria over a wide range of conditions permit description of phase stability on a competitive basis as indicated in the foregoing discussion of phase equilibria at high pressure. Accordingly this approach will be developed in subsequent chapters.

CHAPTER III

Phase Equilibria in Binary Metallic Systems

As indicated in Chapter II, quantitative definition of the relative stability of various crystal forms of a pure metal, as well as the stability of the liquid phase relative to the solid forms, is required in order to calculate the pressure–temperature diagram. Similarly, a binary phase diagram is based on competition between phases. Thus a phase diagram represents the temperature–composition relations corresponding to the minimum free energy of all possible competing phases. In order to display the interactions between competing phases in quantitative fashion, it is instructive to consider Fig. 23 which compares computed and observed phase diagrams for the Zr–Nb, Zr–Ta, Hf–Nb, and Hf–Ta systems. (References in the legend denote the sources of the experimental phase diagrams, starting in the upper left corner and proceeding clockwise.) Figures 24 and 25 show the relevant free energy–composition curves for L (Liquid), β (bcc), ϵ (hcp), and α (fcc) phases in the Hf–Ta system at various temperatures. Detailed description of the calculations of these curves will be presented later. However, it is important to note several salient features of these curves as a precursor to the discussion of specific factors. At 1200°K (Fig. 24) the β and ϵ phases are more stable than the α and L phases, since their free energies (F^β and F^ϵ) lie below the free energies of the fcc and liquid phases (F^α and F^L). The free energies of the β and ϵ phases are equal at $x_0^{\beta\epsilon} = 0.23$ at 1200°K. However, the most stable situation is formation of a two-phase field bounded by the "common tangent rule" (equivalent to equilibration of the partial molar free energies) at $x_\epsilon = 0.03$ and $x_\beta = 0.94$. Figure 24 illustrates the important role played by $\Delta F_{Hf}^{\beta \to \epsilon}$ and $\Delta F_{Ta}^{\beta \to \epsilon}$ in controlling the phase stability (16). Moreover, Fig. 25 shows that the other lattice stability differences (involving the α and L phases) determine the relative free energies and hence the stability of these

33

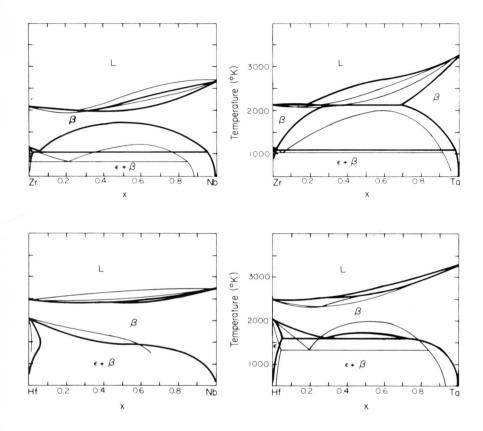

Fig. 23. Comparison of regular-solution computed and experimental phase diagrams (3), (3), (3), (52).

phases. The partial molar free energies of hafnium and tantalum in the β phase at $x = 0.9$ and $1800°K$ (i.e., $\bar{F}^{\beta}_{Hf} [x = 0.9]$ and $\bar{F}^{\beta}_{Ta} [x = 0.9]$) are illustrated in Fig. 25. Another example of competition is shown in Figs. 26 and 27. The former compares the computed and observed phase diagrams in the Mo–Rh, Mo–Ir, W–Rh, and W–Ir systems [i.e., the (Mo/W) vs. (Rh/Ir) set], while the latter illustrates the free energy–composition relations in the W–Rh system at $2400°K$. At this temperature, interactions between the β, ϵ, α, and L occur. Tungsten is most stable in the β form, while Rh is most stable in the fcc form. However, an ϵ phase (hcp) forms at intermediate compositions. Figure 27 illustrates the requirement for defining the lattice stability of all of the phases of W and Rh in order to compute the phase diagram (16).

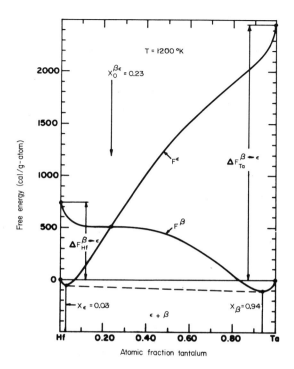

FIG. 24. Calculated free energy–composition relations in the hafnium–tantalum system at 1200°K. Common tangent location of phase boundaries and $x_0^{\beta\epsilon}$ are illustrated.

These examples illustrate how the lattice stability parameters contribute to the relative *position* of the free energy–composition curves. The relative *shape* of these curves is controlled by the free energy of mixing. Consideration of Figs. 23–27 in the light of the discussion presented in Chapter II defines the full magnitude of the problem at hand. Since lattice stability data, which are required for all phases (stable and unstable), cannot be obtained currently from first-principle methods, other routes must be sought. In addition, procedures must be evolved for assessing the free energy of mixing of stable *and unstable* phases under complete ranges of temperature and composition. Although experimental evaluation of the free energy of mixing is relevant to the problem at hand, it *cannot* provide the complete description required. This is due to the fact that a description of the *stable* and *unstable* phases are needed at all temperatures and compositions for each of the systems involved (55). Subsequent portions of this chapter will be directed toward developing such a description.

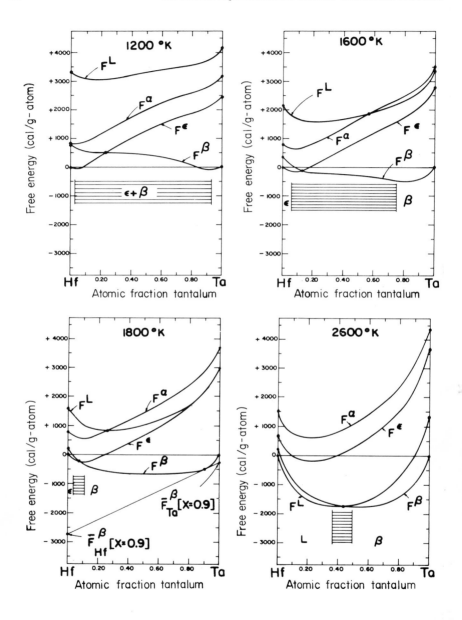

FIG. 25. Computed free energy–composition relations in the hafnium–tantalum system. Phase boundaries are shown. The panel at lower left illustrates partial molar free energies (or chemical potentials).

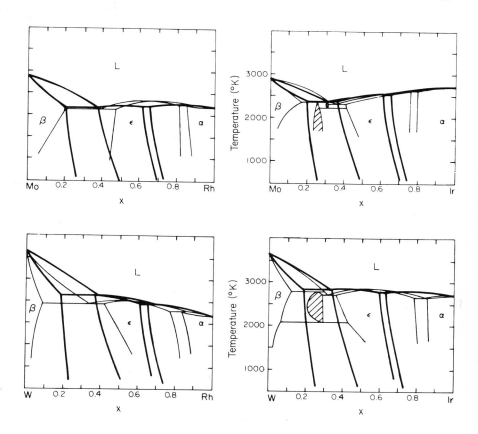

FIG. 26. Comparison of regular solution computed and experimental phase diagrams (3), (53), (54), (54).

1. THERMODYNAMIC DEFINITION OF BINARY EQUILIBRIUM

The free energy of a liquid phase L in the i–j system (see Figs. 25 and 27) is given by

$$F^L = (1 - x)F_i^L + xF_j^L + RT[x \ln x + (1 - x) \ln(1 - x)] + F_E^L \quad \text{cal/g-atom} \quad (65)$$

where x is the atomic fraction of metal j, F_i^L and F_j^L are the free energies of pure liquid i and pure liquid j, and F_E^L is the excess free energy of mixing, which is zero when $x = 0$ and 1. In the regular-solution approximation, F_E^L is given by

$$F_E^L = Lx(1 - x) \quad \text{cal/g-atom} \quad (66)$$

where the interaction parameter L is a constant. When $L = 0$, the solution is

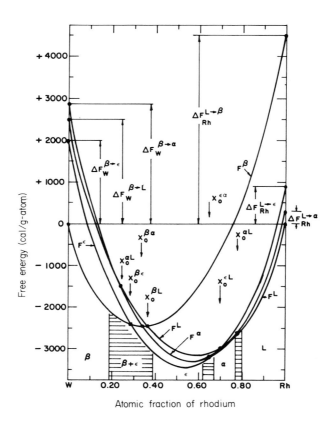

FIG. 27. Calculated free energy–composition relations in the tungsten–rhodium system at 2400°K. Phase boundaries, x_0, and lattice stability values are shown.

ideal. The partial molar free energies (Fig. 25) of components i and j in the i–j binary are defined as

$$\bar{F}_i^{\mathrm{L}} = F^{\mathrm{L}} - x(\partial F^{\mathrm{L}}/\partial x) \tag{67}$$

and

$$\bar{F}_j^{\mathrm{L}} = F^{\mathrm{L}} + (1 - x)(\partial F^{\mathrm{L}}/\partial x) \tag{68}$$

For the regular-solution case, Eqs. (67) and (68) reduce to (69) and (70):

$$\bar{F}_i^{\mathrm{L}} = F_i^{\mathrm{L}} + RT \ln(1 - x) + x^2 L \tag{69}$$

and

$$\bar{F}_j^{\mathrm{L}} = F_j^{\mathrm{L}} + RT \ln x + (1 - x)^2 L \tag{70}$$

When L is positive, a miscibility gap forms below a temperature T_c, and the liquid decomposes into two liquids having compositions x_1 and x_2 as defined by Eqs. (72) and (73):

$$\bar{F}_i^L\big|_{x_1} = \bar{F}_i^L\big|_{x_2} \quad \text{and} \quad \bar{F}_j^L\big|_{x_1} = \bar{F}_j^L\big|_{x_2} \tag{71}$$

For the regular-solution case,

$$RT \ln(1 - x_1) + x_1^2 L = RT \ln(1 - x_2) + x_2^2 L \tag{72}$$

and

$$RT \ln x_1 + (1 - x_1)^2 L = RT \ln x_2 + (1 - x_2)^2 L \tag{73}$$

Equations (72) and (73) yield the symmetrical miscibility gap, since

$$x_2 = 1 - x_1 \tag{74}$$

and

$$(L/RT) = (1 - 2x_1)^{-1} \ln(1 - x_1)/x_1 \tag{75}$$

The critical temperature T_c coincides with $x = 0.5$ and $(\partial \bar{F}_i/\partial x) = (\partial \bar{F}_j/\partial x) = (\partial^2 \bar{F}_i/\partial x^2) = (\partial^2 \bar{F}_j/\partial x^2) = 0$, under these conditions:

$$L = 2RT_c \quad \text{cal/g-atom} \tag{76}$$

Thus, formation of a miscibility gap requires positive values of L. The free energy of a bcc phase β is defined in a manner similar to that of the liquid, as shown in Figs. 24, 25, and 27. For a regular solution,

$$F^\beta = (1 - x)F_i^\beta + xF_j^\beta + RT(x \ln x + (1 - x) \ln(1 - x))$$
$$+ Bx(1 - x) \quad \text{cal/g-atom} \tag{77}$$

where x is the atom fraction of j, F_i^β, and F_j^β are the free energies of the bcc forms of pure i and j, and B is the regular-solution interaction parameter for the bcc phase in the i–j system. The partial molar free energies of i and j in the bcc phase are

$$\bar{F}_i^\beta = F_i^\beta + RT \ln(1 - x) + Bx^2 \tag{78}$$

and

$$\bar{F}_j^\beta = F_j^\beta + RT \ln x + B(1 - x)^2 \tag{79}$$

Two-phase equilibria between the β and liquid phases at a fixed temperature require that

$$\bar{F}_i^\beta\big|_{x_\beta} = \bar{F}_i^L\big|_{x_L} \tag{80}$$

and

$$\bar{F}_j^\beta\big|_{x_\beta} = \bar{F}_j^L\big|_{x_L} \tag{81}$$

where x_β is the composition of the $\beta/\beta + L$ boundary and x_L is the composition of the $\beta + L/L$ boundary. Combination of Eqs. (69), (70), and (78)–(81) yields

$$\Delta F_i^{\beta \rightarrow L} + RT \ln(1 - x_L)/(1 - x_\beta) = x_\beta^2 B - x_L^2 L \tag{82}$$

and

$$\Delta F_j^{\beta \to L} + RT \ln x_L/x_\beta = (1 - x_\beta)^2 B - (1 - x_L)^2 L \tag{83}$$

where

$$\Delta F_i^{\beta \to L} = F_i^L - F_i^\beta \tag{84}$$

and

$$\Delta F_j^{\beta \to L} = F_j^L - F_j^\beta \tag{85}$$

are the differences in free energy between the bcc and liquid forms of the pure metals i and j. Finally, the $x_0[T]$ or $T_0[x]$ curve for the β/L equilibria is defined by

$$\Delta F^{\beta \to L}[T_0, x] = (1 - x) \Delta F_i^{\beta \to L}[T_0]$$
$$+ x \Delta F_j^{\beta \to L}[T_0] + (L - B)x(1 - x) = 0 \tag{86}$$

or

$$\Delta F^{\beta \to L}[T, x_0] = (1 - x_0) \Delta F_i^{\beta \to L}[T]$$
$$+ x_0 \Delta F_j^{\beta \to L}[T] + (L - B)x_0(1 - x_0) = 0 \tag{87}$$

The $x_0[T]$ or $T_0[x]$ curve defines the locus of points along which $F^\beta = F^L$ (Figs. 24 and 27). This curve must lie within the two-phase field. Thus, for a given temperature, x_0 must lie between x_β and x_L. When $x = 0$, T_0 (for the β/L case) coincides with the melting point of the bcc form of element i, $\overline{T}_i^\beta$. When $x = 1$, T_0 (for the β/L case) coincides with the melting point of the bcc form of element j, $\overline{T}_j^\beta$. There is no distinction between $T_0[x]$ and $x_0[T]$ except in the manner in which they are computed. The former is obtained by solving Eq. (86) at various compositions, while the latter is obtained by solving Eq. (87) at various temperatures.

Inclusion of the hcp (ϵ) phase and the fcc (α) phase leads to the generation of five additional sets of equations similar to Eqs. (82) and (83) covering the ϵ/L, α/L, β/ϵ, ϵ/α, and β/α equilibria. For the regular-solution model,

$$F_E^\epsilon = Ex(1 - x) \quad \text{cal/g-atom} \tag{88}$$

and

$$F_E^\alpha = Ax(1 - x) \quad \text{cal/g-atom} \tag{89}$$

In addition, five more $T_0[x]$ or $x_0[T]$ curves result which are defined in a manner similar to Eqs. (86) and (87). For the pure metals, the free energy differences are related by identities similar to Eq. (90).

$$\Delta F_i^{\epsilon \to L} \equiv \Delta F_i^{\epsilon \to \beta} + \Delta F_i^{\beta \to L} \tag{90}$$

Reference to Eqs. (82) and (83) shows that for the special case of ideal solutions (i.e., $L = B = E = A = 0$) the phase boundaries depend only on the free energy differences between the bcc, hcp, fcc, and liquid forms of the elemental partners.

In order to consider the competition between L, β, ϵ, and α phases with a compound phase Ψ, the latter is considered to be a fully ordered line compound existing at composition x_* as a first approximation. On this basis, the free energy of the Ψ phase is defined as

$$F^\Psi = (1 - x_*)F_i^\theta + x_* F_j^\theta + x_*(1 - x_*)(L - C) \quad \text{cal/g-atom} \tag{91}$$

where L is the interaction parameter of the liquid phase in the i–j system, $\theta = \beta$, ϵ, or α is the *base phase* selected for the compound, and C is a constant. In order to compute the phase boundary composition of the liquid phase $x_{L\Psi}$ in a system i–j in equilibrium with the Ψ phase, the partial molar free energies are equilibrated, yielding

$$\left.\bar{F}_i^L\right|_{x_{L\Psi}} = \left.\bar{F}_i^\Psi\right|_{x_{\Psi L}} = \left.\bar{F}_i^\Psi\right|_{x = x_*} \tag{92}$$

and

$$\left.\bar{F}_j^L\right|_{x_{L\Psi}} = \left.\bar{F}_j^\Psi\right|_{x_{\Psi L}} = \left.\bar{F}_j^\Psi\right|_{x = x_*} \tag{93}$$

where $\bar{F}_i^L$ and $\bar{F}_j^L$ are defined by Eqs. (68) and (69). Multiplication of Eq. (29) by $(1 - x_*)$ and Eq. (93) by x_* and summing the results yields

$$F^\Psi = (1 - x_*)F_i^L + x_* F_j^L + RT[(1 - x_*)\ln(1 - x_{L\Psi}) + x_* \ln x_{L\Psi}]$$
$$+ L[x_* - 2x_* x_{L\Psi} + x_{L\Psi}^2] \tag{94}$$

Equations (91) and (94) define $x_{L\Psi}$ as a function of temperature. Similar equations can be written for $x_{\beta\Psi}$, $x_{\epsilon\Psi}$, and $x_{\alpha\Psi}$, representing the compositions of the bcc, hcp, and fcc phases in equilibrium with the compound. These equations are obtained by substituting β (B), ϵ (E), and α (A) for L (L) in Eq. (94).

When a compound or alloy is exposed to very high temperatures below its melting point, vaporization of the components can occur. If the vaporization of individual components occurs at differing rates, the composition of the alloy or compound phase can change. Under certain conditions this can lead to melting or changes in phase constitution. The thermodynamic description presented here permits computation of the vapor pressure of component elements in all of the binary systems as a function of temperature and composition. On the basis of Eqs. (65)–(70), the vapor pressure and activity of i and j in the L, β, ϵ, and α phases of an i–j binary system can be expressed by Eqs. (95)–(102).

$$RT \ln p_i^L/p_i^\circ = \Delta F_i^{\circ \to L} + RT \ln(1 - x) + Lx^2 = RT \ln a_i^L \tag{95}$$

$$RT \ln p_j^L/p_j^\circ = \Delta F_j^{\circ \to L} + RT \ln x + L(1 - x)^2 = RT \ln a_j^L \tag{96}$$

$$RT \ln p_i^\beta/p_i^\circ = \Delta F_i^{\circ \to \beta} + RT \ln(1 - x) + Bx^2 = RT \ln a_i^\beta \tag{97}$$

$$RT \ln p_i^\epsilon/p_i^\circ = \Delta F_i^{\circ \to \epsilon} + RT \ln(1 - x) + Ex^2 = RT \ln a_i^\epsilon \tag{98}$$

$$RT \ln p_i^\alpha/p_i^\circ = \Delta F_i^{\circ \to \alpha} + RT \ln(1 - x) + Ax^2 = RT \ln a_i^\alpha \tag{99}$$

$$RT \ln p_j{}^\beta / p_j{}^\circ = \Delta F_j^{\circ \to \beta} + RT \ln x + B(1 - x)^2 = RT \ln a_j{}^\beta \qquad (100)$$

$$RT \ln p_j{}^\epsilon / p_j{}^\circ = \Delta F_j^{\circ \to \epsilon} + RT \ln x + E(1 - x)^2 = RT \ln a_j{}^\epsilon \qquad (101)$$

$$RT \ln p_j{}^\alpha / p_j{}^\circ = \Delta F_j^{\circ \to \alpha} + RT \ln x + A(1 - x)^2 = RT \ln a_j{}^\alpha \qquad (102)$$

where x is the atomic fraction of j, and $p_i{}^\beta$, $p_i{}^\epsilon$, $p_i{}^\alpha$, $p_j{}^\beta$, $p_j{}^\epsilon$, and $p_j{}^\alpha$ are the vapor pressures of i and j over the β, ϵ, and α phase at a composition x and a temperature T. The pressure $p_i{}^\circ$ is the vapor pressure (in atmospheres) of the stable form (β, ϵ, α, or L) of metal i, while $p_j{}^\circ$ has the same meaning for metal j. The pressures over the liquid are $p_i{}^L$ and $p_j{}^L$.

In each case, a_i and a_j are the activities of elements i and j in the phase indicated relative to the stable form of i or j at the temperature of interest. As an example of Eqs. (95)–(102) it is instructive to consider recent measurements of the activity of iron which have been performed for a series of Fe–Ru alloys at 1600°K (36). Reference to Fig. 19 shows that the γ (fcc) form of iron is the stable form at this temperature, but the Fe–Ru alloys can be fcc (γ) or hcp (ϵ), depending upon the composition. Figure 28 shows the experimental results. Application of Eqs. (95)–(102) suggests that for the case at

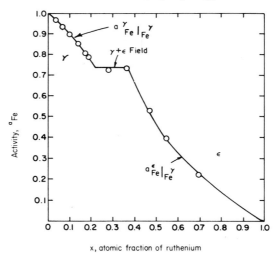

FIG. 28. Activity of iron in the Fe–Ru system at 1600°K. Measurements were performed within the fcc (γ) and hcp (ϵ) fields indicated in Fig. 19.

hand, the activity of iron in the γ phase relative to γ iron $a_{\mathrm{Fe}}^\gamma|_{\mathrm{Fe}}{}^\gamma$, is approximated by Eq. (103) on the basis of regular solutions as

$$RT \ln a_{\mathrm{Fe}}^\gamma|_{\mathrm{Fe}}{}^\gamma = RT \ln(1 - x) + Gx^2 \qquad (103)$$

where x is the atomic fraction of ruthenium and G is the interaction parameter for the γ phase. Similarly for the ϵ (hcp) phase, the activity of iron in the ϵ phase relative to γ iron is approximated by Eq. (104) for regular solutions as

$$RT \ln a_{Fe}^{\epsilon}|_{Fe}{}^{\gamma} = \Delta F_{Fe}^{\gamma \to \epsilon} + RT \ln(1 - x) + Ex^2 \qquad (104)$$

where $\Delta F_{Fe}^{\gamma \to \epsilon}$ is the free energy difference between the ϵ and γ forms of iron and E is the interaction parameter of the ϵ (hcp) phase. Figure 29 shows a plot of $RT \ln a/(1 - x)$ for iron in the hcp field of the Fe–Ru systems vs. x^2. This plot indicates a value of 1200 cal/mole for the free energy difference

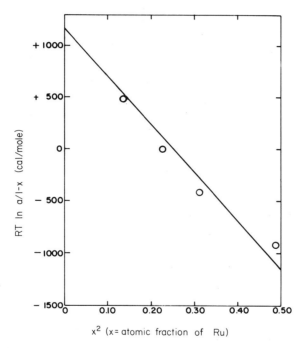

FIG. 29. $RT \ln a/(1 - x)$ vs. x^2 for ϵ-phase Fe–Ru alloys at 1600 °K. The solid line represents

$$RT \ln \frac{a_{Fe}^{\epsilon}}{(1 - x)}\bigg|_{Fe}{}^{\gamma} = \Delta F_{Fe}^{\gamma \to \epsilon} + F_{E}^{\epsilon} - x \frac{\partial F_{E}^{\epsilon}}{\partial x} = 1200 - 4600x^2 \quad \text{cal/mole}$$

where $F_{E} = -4600x(1 - x)$ cal/g-atom and $\Delta F_{Fe}^{\gamma \to \epsilon} = 1200$ cal/mole. Location of intercept at $x = 0$ provides an independent estimate of $\Delta F_{Fe}^{\gamma \to \epsilon}$ for comparison with Figs. 15 and 17.

between γ and ϵ iron and $E = -4600$ cal/g-atom. The former value compares favorably with the value $\Delta F_{Fe}^{\gamma \to \epsilon} = 1340$ cal/mole obtained independently from Fig. 15 on the basis of specific heat and transformation data. The interaction parameters for the liquid, γ (fcc), and α (bcc) phases in the

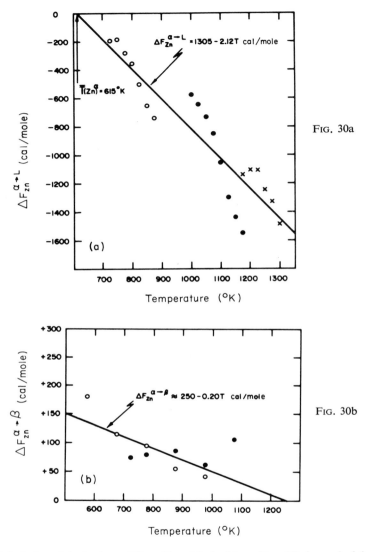

FIG. 30a

FIG. 30b

FIG. 30. Calculation of the lattice stability of hcp (η), fcc (α), and bcc (β) zinc and of the fcc (α) and bcc (β) forms of copper and silver (56). The results stem from individual analyses of the Al–Zn, Cu–Zn, and Ag–Zn systems summarized in Table VII. (a) The difference in free energy between fcc and liquid zinc as a function of temperature, calculated from the α/L equilibria of Al–Zn ($\bigcirc$), Cu–Zn ($\times$), and Ag–Zn ($\bullet$). The solid line is based on activity data for Al–Zn. (b) The difference in free energy between bcc and fcc zinc, calculated from the α/β equilibrium in Cu–Zn ($\bullet$) and Ag–Zn ($\bigcirc$). (c) The difference in free energy between the hcp, fcc, and bcc modifications of zinc as a function of temperature. (d) The difference in free energy between the bcc and fcc modifications of copper and silver as a function of temperature.

44

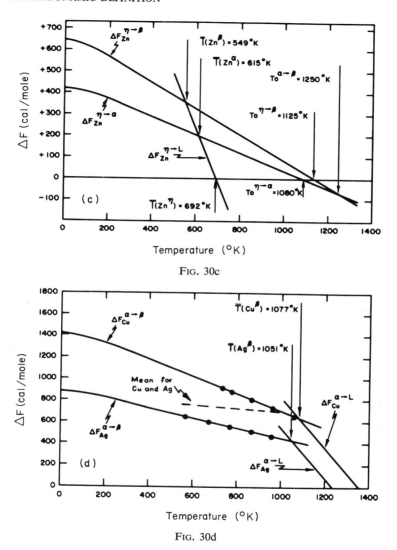

FIG. 30c

FIG. 30d

Fe–Ru system were found to be -1800 cal/g-atom, -2600 cal/g-atom, and $1200 - 2.5T$ cal/g-atom respectively. These values, when combined with the lattice stability description of iron and ruthenium (35, 36, 43) shown in Figs. 15 and 17, permit application of Eqs. (80)–(90) for the purpose of computing the entire Fe–Ru phase diagram as shown in Fig. 19.

The same method has been applied to an analysis of the Zn–Al, Zn–Cu, and Zn–Ag systems (56) to derive estimates of the lattice stability of the hcp

(η), fcc (α), and bcc (β) forms of zinc and the fcc (α) and bcc (β) forms of copper and silver. This analysis was performed by applying the solution thermodynamics of Eqs. (80)–(85) and (95)–(102) to fcc–liquid, fcc–bcc, and bcc–liquid equilibria in these systems. In addition, relevant activity data for these systems was employed in the analysis. The results are shown in Fig. 30 and Table VII.

TABLE VII

Summary of Computed Lattice Stability Parameters for Zinc, Copper, and Silver[a,b]

Metal		Enthalpy difference at 0°K	Debye temperature (°K)
Zn: hcp $= \eta$, fcc $= \alpha$, bcc $= \beta$			
$\Delta F^{\eta \to L} = 1765 - 2.55T$	(105)	$\Delta H^{\eta \to \alpha} = 460$	$\Theta^\eta = 235$
$\Delta F^{\eta \to \alpha} = 440 - 0.40T$	(106)	$\Delta H^{\alpha \to \beta} = 250$	$\Theta^\alpha = 219$
$\Delta F^{\alpha \to \beta} = 250 - 0.20T$	(107)		$\Theta^\beta = 212$
Cu: fcc $= \alpha$, bcc $= \beta$			
$\Delta F^{\alpha \to L} = 3120 - 2.30T$	(108)	$\Delta H^{\alpha \to \beta} = 1500$	$\Theta^\alpha = 315$
$\Delta F^{\alpha \to \beta} = 1500 - 0.80T$	(109)		$\Theta^\beta = 276$
Ag: fcc $= \alpha$, bcc $= \beta$			
$\Delta F^{\alpha \to L} = 2855 - 2.31T$	(110)	$\Delta H^{\alpha \to \beta} = 905$	$\Theta^\alpha = 220$
$\Delta F^{\alpha \to \beta} = 900 - 0.45T$	(111)		$\Theta^\beta = 204$
Mean Value for Cu and Ag			
$\Delta F^{\alpha \to \beta} = 850 - 0.20T$	(112)		

[a] In calories per mole.   [b] Kaufman (56).

Zinc is stable at all temperatures up to its melting point (692°K) in the hcp (η) form at 1 atm as indicated by Eqs. (105)–(107) in Table VII. Figure 30a shows values for the free energy difference between the α and L forms obtained from activity data and from analyses of α/L equilibria. Similar results for the free energy differences between the fcc (α) and bcc (β) forms of zinc which were derived from α/β equilibria are shown in Fig. 30b, while Fig. 30c summarizes the results for the lattice stability of zinc. Approximate representation of these differences are given by Eqs. (105)–(107) in Table VII. In addition, Table VII shows enthalpy differences at 0°K and Debye temperatures for the various forms that can be employed to represent the lattice stability of zinc at low temperatures along the lines presented earlier in Eqs. (35) and (37). Application of the results for zinc to α/β and β/L equilibria in the Zn–Cu and Zn–Ag systems on the basis of Eqs. (80)–(85) yield the differences in free energy between the α and β forms of Cu and Ag shown in Fig. 30d. These values can be described analytically on the basis of Eqs. (108)–(111) in Table VII.

2. ZERO-ORDER CALCULATION OF PHASE EQUILIBRIA IN REFRACTORY METAL SYSTEMS

Reference to Eqs. (82)–(90) shows that for the special case of ideal solutions (i.e., $L = B = E = A = 0$) the phase boundaries depend only on the free energy differences between the bcc, hcp, fcc, and liquid forms of the elemental components. In order to illustrate how such a zero-order calculation is performed, estimates of the lattice stability parameters for the transition metals have been derived. The results are shown graphically in Figs. 31 and 2 with numerical values given in Table VIII. In each case, the entropy of fusion of the stable form of these metals is taken to be 2.0 cal/g-atom-°K. Details of the estimation procedure are presented below.

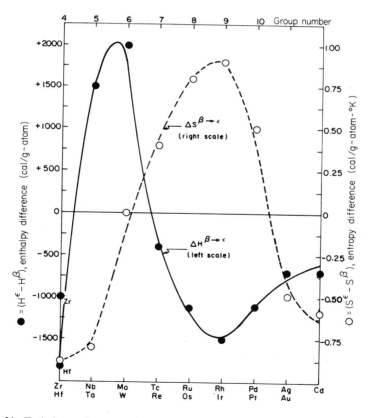

FIG. 31. Enthalpy and entropy differences between the hcp (ϵ) and bcc (β) forms of the transition metals. The lattice stability parameters for these metals, which are shown as a function of group number, are given in Table VIII.

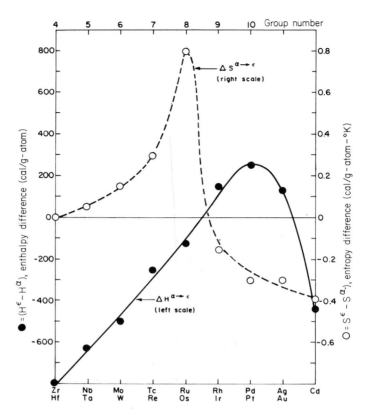

FIG. 32. Enthalpy and entropy differences between the hcp (ϵ) and fcc (α) forms of the transition metals. The lattice stability parameters for these metals, which are shown as a function of group number, are given in Table VIII.

The zero-order calculation ($L = B = E = A = 0$) is illustrated in Table IX and in Fig. 33 using the Mo–Ru and W–Ir systems as examples. In the Mo–Ru case, reference to Table VIII indicates that the melting temperatures of the bcc, hcp, and fcc forms of Mo are 2900, 1900, and 1530°K, respectively. These temperatures are laid off on the temperature axis at 0 at.% Ru. The melting points of hcp, fcc, and bcc Ru are then plotted at 2550, 1780, and 1420°K on the temperature axis at 100% Ru. The next step is to compute the T_0–x (or x_0–T) curve for the β/L, ϵ/L, and α/L cases with the aid of Eqs. (86)–(90). This procedure yields the three curves which show that the T_0–x locus for the α/L case lies below those for the β/L and ϵ/L over the entire composition range, indicating that the fcc phase does not appear in the Mo–Ru ideal-solution phase diagram. The T_0–x curves for the β/L and ϵ/L cases are then

TABLE VIII

Summary of Free Energy Differences for the Pure Metals

Metal	Free energy difference (cal/g-atom)	Temperature (°K)	Metal	Free energy difference (cal/g-atom)	Temperature (°K)
Zr	$\Delta F^{\beta \to L} = 4250 - 2.0T$	$\bar{T}^\beta = 2125$	Nb	$\Delta F^{\beta \to L} = 5480 - 2.0T$	$\bar{T}^\beta = 2740$
	$\Delta F^{\epsilon \to L} = 5280 - 2.9T$	$\bar{T}^\epsilon = 1820$		$\Delta F^{\epsilon \to L} = 3980 - 2.8T$	$\bar{T}^\epsilon = 1420$
	$\Delta F^{\alpha \to L} = 4480 - 2.9T$	$\bar{T}^\alpha = 1544$		$\Delta F^{\alpha \to L} = 3330 - 2.85T$	$\bar{T}^\alpha = 1170$
	$\Delta F^{\beta \to \epsilon} = -1030 + 0.90T$	$T_0^{\beta\epsilon} = 1144$		$\Delta F^{\beta \to \epsilon} = +1500 + 0.80T$	
	$\Delta F^{\alpha \to \epsilon} = -800$			$\Delta F^{\alpha \to \epsilon} = -650 - 0.05T$	
	$\Delta F^{\alpha \to \beta} = +230 - 0.90T$	$T_0^{\alpha\beta} = 255$		$\Delta F^{\alpha \to \beta} = -2150 - 0.85T$	
Hf	$\Delta F^{\beta \to L} = 4990 - 2.0T$	$\bar{T}^\beta = 2495$	Ta	$\Delta F^{\beta \to L} = 6540 - 2.0T$	$\bar{T}^\beta = 3270$
	$\Delta F^{\epsilon \to L} = 6820 - 2.9T$	$\bar{T}^\epsilon = 2351$		$\Delta F^{\epsilon \to L} = 5040 - 2.8T$	$\bar{T}^\epsilon = 1800$
	$\Delta F^{\alpha \to L} = 6020 - 2.9T$	$\bar{T}^\alpha = 2076$		$\Delta F^{\alpha \to L} = 4390 - 2.85T$	$\bar{T}^\alpha = 1540$
	$\Delta F^{\beta \to \epsilon} = -1830 + 0.90T$	$T_0^{\beta\epsilon} = 2033$		$\Delta F^{\beta \to \epsilon} = +1500 + 0.80T$	
	$\Delta F^{\alpha \to \epsilon} = -800$			$\Delta F^{\alpha \to \epsilon} = -650 - 0.05T$	
	$\Delta F^{\alpha \to \beta} = +1030 - 0.90T$	$T_0^{\alpha\beta} = 1144$		$\Delta F^{\alpha \to \beta} = -2150 - 0.85T$	
Mo	$\Delta F^{\beta \to L} = 5800 - 2.0T$	$\bar{T}^\beta = 2900$	Ru	$\Delta F^{\beta \to L} = 3980 - 2.8T$	$\bar{T}^\beta = 1420$
	$\Delta F^{\epsilon \to L} = 3800 - 2.0T$	$\bar{T}^\epsilon = 1900$		$\Delta F^{\epsilon \to L} = 5100 - 2.0T$	$\bar{T}^\epsilon = 2550$
	$\Delta F^{\alpha \to L} = 3300 - 2.15T$	$\bar{T}^\alpha = 1530$		$\Delta F^{\alpha \to L} = 4980 - 2.8T$	$\bar{T}^\alpha = 1780$
	$\Delta F^{\beta \to \epsilon} = 2000$			$\Delta F^{\beta \to \epsilon} = -1120 - 0.8T$	
	$\Delta F^{\alpha \to \epsilon} = -500 - 0.15T$			$\Delta F^{\alpha \to \epsilon} = -120 - 0.8T$	
	$\Delta F^{\alpha \to \beta} = -2500 - 0.15T$			$\Delta F^{\alpha \to \beta} = +1000$	
W	$\Delta F^{\beta \to L} = 7300 - 2.0T$	$\bar{T}^\beta = 3650$	Os	$\Delta F^{\beta \to L} = 5480 - 2.8T$	$\bar{T}^\beta = 1960$
	$\Delta F^{\epsilon \to L} = 5300 - 2.0T$	$\bar{T}^\epsilon = 2650$		$\Delta F^{\epsilon \to L} = 6600 - 2.0T$	$\bar{T}^\epsilon = 3300$
	$\Delta F^{\alpha \to L} = 4800 - 2.15T$	$\bar{T}^\alpha = 2230$		$\Delta F^{\alpha \to L} = 6480 - 2.8T$	$\bar{T}^\alpha = 2310$

TABLE VIII (*continued*)

Metal	Free energy difference (cal/g-atom)	Temperature (°K)	Metal	Free energy difference (cal/g-atom)	Temperature (°K)
W	$\Delta F^{\beta \to \epsilon} = 2000$ $\Delta F^{\alpha \to \epsilon} = -500 - 0.15T$ $\Delta F^{\alpha \to \beta} = -2500 - 0.15T$		Os	$\Delta F^{\beta \to \epsilon} = -1120 - 0.8T$ $\Delta F^{\alpha \to \epsilon} = -120 - 0.8T$ $\Delta F^{\alpha \to \beta} = +1000$	
Re	$\Delta F^{\beta \to L} = 6500 - 2.4T$ $\Delta F^{\epsilon \to L} = 6900 - 2.0T$ $\Delta F^{\alpha \to L} = 6650 - 2.3T$ $\Delta F^{\beta \to \epsilon} = -400 - 0.4T$ $\Delta F^{\alpha \to \epsilon} = -250 - 0.30T$ $\Delta F^{\alpha \to \beta} = +150 + 0.10T$	$\bar{T}^{\beta} = 2710$ $\bar{T}^{\epsilon} = 3450$ $\bar{T}^{\alpha} = 2890$	Pd	$\Delta F^{\beta \to L} = 2290 - 2.8T$ $\Delta F^{\epsilon \to L} = 3390 - 2.3T$ $\Delta F^{\alpha \to L} = 3640 - 2.0T$ $\Delta F^{\beta \to \epsilon} = -1100 - 0.50T$ $\Delta F^{\alpha \to \epsilon} = +250 + 0.30T$ $\Delta F^{\alpha \to \beta} = 1350 + 0.8T$	$\bar{T}^{\beta} = 820$ $\bar{T}^{\epsilon} = 1470$ $\bar{T}^{\alpha} = 1820$
Rh	$\Delta F^{\beta \to L} = 2830 - 3.05T$ $\Delta F^{\epsilon \to L} = 4330 - 2.15T$ $\Delta F^{\alpha \to L} = 4480 - 2.0T$ $\Delta F^{\beta \to \epsilon} = -1500 - 0.90T$ $\Delta F^{\alpha \to \epsilon} = +150 + 0.15T$ $\Delta F^{\alpha \to \beta} = 1650 + 1.05T$	$\bar{T}^{\beta} = 930$ $\bar{T}^{\epsilon} = 2010$ $\bar{T}^{\alpha} = 2240$	Pt	$\Delta F^{\beta \to L} = 2730 - 2.8T$ $\Delta F^{\epsilon \to L} = 3830 - 2.3T$ $\Delta F^{\alpha \to L} = 4080 - 2.0T$ $\Delta F^{\beta \to \epsilon} = -1100 - 0.50T$ $\Delta F^{\alpha \to \epsilon} = +250 + 0.30T$ $\Delta F^{\alpha \to \beta} = 1350 + 0.80T$	$\bar{T}^{\beta} = 980$ $\bar{T}^{\epsilon} = 1670$ $\bar{T}^{\alpha} = 2040$
Ir	$\Delta F^{\beta \to L} = 3850 - 3.05T$ $\Delta F^{\epsilon \to L} = 5350 - 2.15T$ $\Delta F^{\alpha \to L} = 5500 - 2.0T$ $\Delta F^{\beta \to \epsilon} = -1500 - 0.90T$ $\Delta F^{\alpha \to \epsilon} = +150 + 0.15T$ $\Delta F^{\alpha \to \beta} = 1650 + 1.05T$	$\bar{T}^{\beta} = 1260$ $\bar{T}^{\epsilon} = 2490$ $\bar{T}^{\alpha} = 2750$			

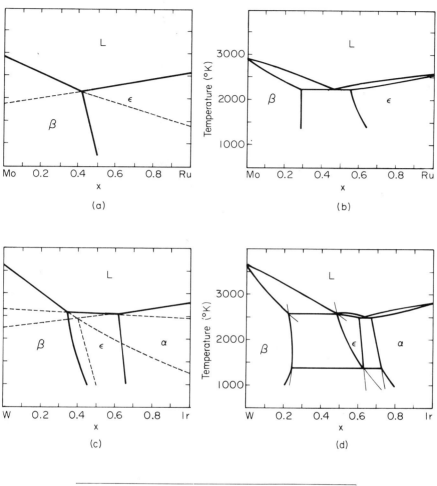

	$\Delta F^{\beta \to L}$	$\Delta F^{\epsilon \to L}$	$\Delta F^{\alpha \to L}$
		(cal/g-atom)	
Mo	$2.0(2900 - T)$	$2.0(1900 - T)$	$2.15(1530 - T)$
Ru	$2.8(1420 - T)$	$2.0(2550 - T)$	$2.80(1780 - T)$
W	$2.0(3650 - T)$	$2.0(2650 - T)$	$2.15(2230 - T)$
Ir	$3.05(1260 - T)$	$2.15(2490 - T)$	$2.0(2750 - T)$

$$L = B = E = A = 0$$

FIG. 33. Illustration of ideal-solution phase diagram calculation. Computations based on Eqs. (81)–(89) are detailed in Table IX.

TABLE IX

ILLUSTRATIVE EXAMPLE OF IDEAL-SOLUTION PHASE DIAGRAM CALCULATION

Case A: Mo–Ru System, Ideal Solution ($B = L = E = 0$)

$$\Delta F_{\text{Mo}}^{\beta \to L} = RT \ln(1 - x_\beta)(1 - x_L)^{-1}, \qquad \Delta F_{\text{Ru}}^{\beta \to L} = RT \ln x_\beta / x_L \qquad (113)$$

Or

$$x_\beta = x_L \exp[\Delta F_{\text{Ru}}^{\beta \to L}/RT] \qquad (114a)$$

$$x_L = (1 - \exp[\Delta F_{\text{Mo}}^{\beta \to L}/RT])/(\exp[\Delta F_{\text{Ru}}^{\beta \to L}/RT] - \exp[\Delta F_{\text{Mo}}^{\beta \to L}/RT]) \qquad (114b)$$

$$\bar{T}_{\text{Mo}}^\beta = 2900°\text{K} \qquad \bar{T}_{\text{Mo}}^\epsilon = 1900°\text{K} \qquad \Delta F_{\text{Mo}}^{\beta \to \epsilon} = \Delta F_{\text{Mo}}^{\beta \to L} + \Delta F_{\text{Mo}}^{L \to \epsilon} = 2000$$

$$\bar{T}_{\text{Ru}}^\beta = 1420°\text{K} \qquad \bar{T}_{\text{Ru}}^\epsilon = 2550°\text{K} \qquad \Delta F_{\text{Ru}}^{\beta \to \epsilon} = -1120 - 0.80T$$

1. At $\dfrac{T = 2400°\text{K}}{\beta/L}$, $\Delta F_{\text{Mo}}^{\beta \to L} = +1000$, $\Delta F_{\text{Ru}}^{\beta \to L} = -2740$

$$\exp[\Delta F_{\text{Mo}}^{\beta \to L}/RT] = 1.234, \qquad \exp[\Delta F_{\text{Ru}}^{\beta \to L}/RT] = 0.562$$
$$x_L = (1 - 1.234)/(0.562 - 1.234) = 0.348$$
$$x_\beta = (0.348)(0.562) = 0.196$$

$\dfrac{T = 2000°\text{K}}{\beta/L}$, $\Delta F_{\text{Mo}}^{\beta \to L} = +1800$, $\Delta F_{\text{Ru}}^{\beta \to L} = -1620$

$$x_L = 0.631, \qquad x_\beta = 0.420$$

2. At $\dfrac{T = 2000°\text{K}}{\epsilon/L}$, $\Delta F_{\text{Mo}}^{\epsilon \to L} = -200$, $\Delta F_{\text{Ru}}^{\epsilon \to L} = +1100$

$$x_L = 0.133, \qquad x_\epsilon = 0.176$$

$\dfrac{T = 2200°\text{K}}{\epsilon/L}$, $\Delta F_{\text{Mo}}^{\epsilon \to L} = -600$, $\Delta F_{\text{Ru}}^{\epsilon \to L} = +700$

$$x_L = 0.425, \qquad x_\epsilon = 0.499$$

3. At $\dfrac{T = 2000°\text{K}}{\epsilon/\beta}$, $\Delta F_{\text{Mo}}^{\epsilon \to \beta} = -2000$, $\Delta F_{\text{Ru}}^{\epsilon \to \beta} = +2720$

$$x_\beta = 0.287, \qquad x_\epsilon = 0.569$$

$\dfrac{T = 1400°\text{K}}{\epsilon/\beta}$, $\Delta F_{\text{Mo}}^{\epsilon \to \beta} = -2000$, $\Delta F_{\text{Ru}}^{\epsilon \to \beta} = +2240$

$$x_\beta = 0.293, \qquad x_\epsilon = 0.655$$

plotted as in Fig. 33a, which shows that the β phase (bcc) is most stable on the Mo side of the diagram and the ϵ phase (hcp) is most stable on the Ru side. The last task is to compute the $T_0[x]$ curve for the β/ϵ case dividing the β and ϵ fields.

The next step is to compute the two-phase boundaries for the β/L, ϵ/L, and β/ϵ equilibria and to construct the eutectic horizontal at the intersection of the $x_L[T]$ curves generated by the β/L and ϵ/L equilibria. The final result is summarized in Fig. 33b. Table IX illustrates these calculations.

The tungsten–iridium case in Figs. 33c and 33d is obtained by the same procedure as shown in Table IX. In this instance, however, the T_0–x curve for the ϵ/L exists at higher temperatures in the center of the phase diagram than the T_0–x curves for the β/L or α/L combinations. Thus, the W–Ir ideal-solution phase diagram computation requires calculation of all six equilibria,

TABLE IX (*continued*)

Case B: W–Ir System—Ideal Solution ($B = L = E = A = 0$)

W: $\quad \bar{T}^\beta = 3650, \qquad \bar{T}^\epsilon = 2650, \qquad \bar{T}^\alpha = 2230, \qquad \Delta F_W^{\alpha \to \epsilon} = 2000$

$\quad \Delta F_W^{\alpha \to \epsilon} = -500 - 0.15T, \qquad \Delta F_W^{\alpha \to \beta} = -2500 - 0.15T$

Ir: $\quad \bar{T}^\beta = 1260, \qquad \bar{T}^\epsilon = 2490, \qquad \bar{T}^\alpha = 2750, \qquad \Delta F_{Ir}^{\beta \to \epsilon} = -1500 - 0.90T$

$\quad \Delta F_{Ir}^{\alpha \to \epsilon} = +150 + 0.15T, \qquad \Delta F_{Ir}^{\alpha \to \beta} = 1650 + 1.05T$

1. At $\dfrac{T = 2800°K}{\beta/L}$, $\quad \Delta F_W^{\beta \to L} = +1700, \qquad \Delta F_{Ir}^{\beta \to L} = -4690$

$\qquad\qquad x_\beta = 0.166, \qquad x_L = 0.386$

2. At $\dfrac{T = 2600°K}{\epsilon/\beta}$, $\quad \Delta F_W^{\beta \to \epsilon} = +2000, \qquad \Delta F_{Ir}^{\beta \to \epsilon} = -3840$

$\qquad\qquad x_\beta = 0.225, \qquad x_i = 0.474$

$\dfrac{T = 2000°K}{\epsilon/\beta}$, $\quad \Delta F_W^{\beta \to \epsilon} = +2000, \qquad \Delta F_{Ir}^{\beta \to \epsilon} = -3300$

$\qquad\qquad x_\beta = 0.234, \qquad x_i = 0.537$

$\dfrac{T = 1000°K}{\epsilon/\beta}$, $\quad \Delta F_W^{\beta \to \epsilon} = +2000, \qquad \Delta F_{Ir}^{\beta \to \epsilon} = -2400$

$\qquad\qquad x_\beta = 0.213, \qquad x_\epsilon = 0.712$

3. At $\dfrac{T = 2567°K}{\epsilon/L}$, $\quad \Delta F_W^{\epsilon \to L} = +167, \qquad \Delta F_{Ir}^{\epsilon \to L} = -167$

$\qquad\qquad x_\epsilon = 0.487, \qquad x = 0.504$

4. At $\dfrac{T = 2000°K}{\epsilon/\alpha}$, $\quad \Delta F_W^{\epsilon \to \alpha} = +800, \qquad \Delta F_{Ir}^{\epsilon \to \alpha} = -450$

$\qquad\qquad x_\epsilon = 0.603, \qquad x_\alpha = 0.676$

$\dfrac{T = 1000°K}{\epsilon/\alpha}$, $\quad \Delta F_W^{\epsilon \to \alpha} = +650, \qquad \Delta F_{Ir}^{\epsilon \to \alpha} = -300$

$\qquad\qquad x_\epsilon = 0.631, \qquad x_\alpha = 0.734$

5. At $\dfrac{T = 2700°K}{\alpha/L}$, $\quad \Delta F_W^{\alpha \to L} = -1005, \qquad \Delta F_{Ir}^{\alpha \to L} = +100$

$\qquad\qquad x_L = 0.901, \qquad x_\alpha = 0.917$

6. At $\dfrac{T = 1000°K}{\beta/\alpha}$, $\quad \Delta F_W^{\alpha \to \beta} = -2650, \qquad \Delta F_{Ir}^{\alpha \to \beta} = +2655$

$\qquad\qquad x_\beta = 0.203, \qquad x_\alpha = 0.790$

β/L, ϵ/L, α/L, β/ϵ, ϵ/α, and α/β. The results are shown in Figure 33d along with the eutectic, peritectic, and eutectoid isotherms.

A selected number of binary systems based upon refractory transition metals were chosen for computation based on the zero-order or ideal-solution calculation (*16*). These cases included 30 binary phase diagrams between partners that did not exhibit large differences in atomic size (maximum difference of 28% by volume or 9% by diameter). Systems dominated by miscibility gaps or compounds were also excluded from consideration. The results obtained on this basis are shown in Figs. 33–41. In each case, the figure shows the free energy differences between the bcc, hcp, fcc and liquid phases of the component pairs employed for the ideal solution phase diagram computations. The *computed phase diagrams are shown by heavy lines super-imposed on the experimental phase diagram.* The source of each experimental phase diagram is referenced in the figure legend, starting from the upper left corner in a clockwise direction.

In this manner, the ideal-solution computed phase diagrams depicted in Figs. 34–41 were obtained. However, the (Nb/Ta) vs. (Ru/Os)* diagrams that are compared with experiment in Fig. 34 show that compounds occur in these systems. The compound fields, which are shown cross hatched, limit the stability of the bcc, hcp, and liquid phases. The existence of compounds indicates that negative values of the interaction parameters (rather than a zero value) are required. Moreover, comparison of the relative positions of the computed and observed β and ϵ fields suggests that B is probably less than E, since the ideal-solution computation ($B = E = 0$) yields an over-estimate of ϵ stability relative to β. Nevertheless, agreement in the melting range and eutectic conditions is to be noted for these systems.

Figure 35 shows the (Nb/Ta) and (Mo/W) vs. Re cases. Each of these systems exhibits compound intrusion, indicating negative values of the inter-action parameters, and as before, we may conclude that B is probably less than E. Nevertheless, the ideal-solution diagrams show reasonable eutectic conditions relative to the observed diagrams. The opposite effect is seen in Fig. 36, which depicts the (Mo/W) vs. (Ru/Os) cases. Again, compound intrusion is evident, and good agreement as regards eutectic conditions results. However, the $B = E = 0$ ideal-solution calculation overestimates the β stability relative to the hcp stability, thus suggesting that $E < B < 0$.

The occurrence of hcp (ϵ) phases intermediate between bcc (β) and fcc (α) phases is evident in both the computed and observed phase diagrams of Fig. 37. The occurrence of compound phases indicates that negative interaction parameters are required. In each case, the melting range for the ϵ phase is too low, suggesting that E must be less than L. Moreover, examina-tion of the observed and computed β/ϵ and ϵ/α equilibria suggests that E is less than B, and E is less than A. Notwithstanding these discrepancies, the features of the experimental diagrams come through remarkably well in the ideal-solution computations.

. The (Mo/W) vs. (Pd/Pt) diagrams in Fig. 38 constitute an especially interesting set, since Mo–Pt and W–Pd exhibit opposite interaction-para-meter tendencies. In the Mo–Pt case a tetragonal modification of the α phase exists (shown cross hatched) which indicates ordering tendencies requiring negative interaction parameters. In the W–Pd case, however, the β/L and β/α two-phase fields are much wider than the ideal-solution calculations, pointing to positive interaction parameters. The ideal-solution computations

---

* The notation (Nb/Ta) vs. (Ru/Os) is used as shorthand notation for the Nb–Ru, Nb–Os, Ta–Ru, and Ta–Os systems.

(text continues on page 60)

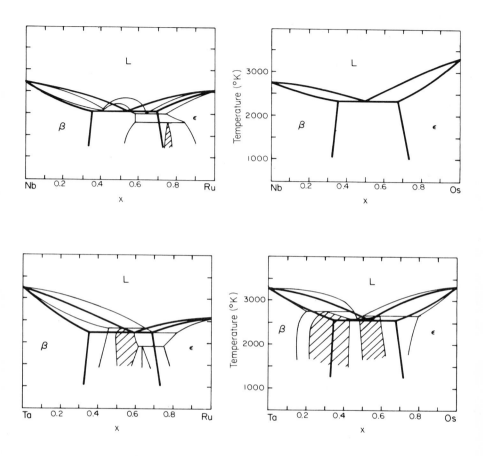

	$\Delta F^{\beta \to L}$	$\Delta F^{\epsilon \to L}$	$\Delta F^{\alpha \to L}$
		(cal/g-atom)	
Nb	2.0(2740 − T)	2.8(1420 − T)	2.85(1170 − T)
Ta	2.0(3270 − T)	2.8(1800 − T)	2.85(1540 − T)
Ru	2.8(1420 − T)	2.0(2550 − T)	2.8(1780 − T)
Os	2.8(1960 − T)	2.0(3300 − T)	2.8(2310 − T)

$$L = B = E = A = 0$$

FIG. 34. Comparison of ideal-solution computed and experimental phase diagrams (57), (3), (3). Thick lines are computed and thin lines are observed.

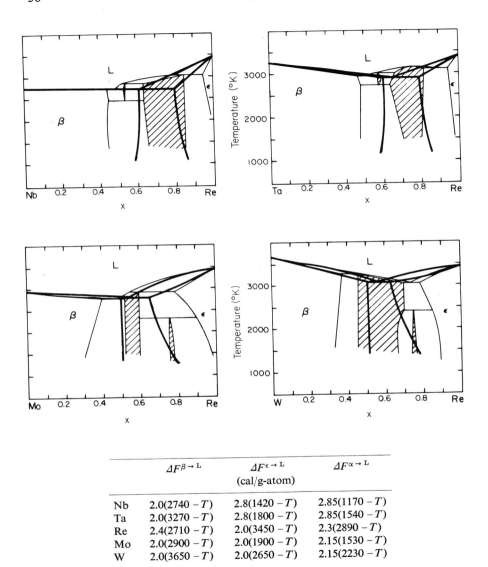

	$\Delta F^{\beta \to \mathrm{L}}$	$\Delta F^{\epsilon \to \mathrm{L}}$	$\Delta F^{\alpha \to \mathrm{L}}$
		(cal/g-atom)	
Nb	$2.0(2740 - T)$	$2.8(1420 - T)$	$2.85(1170 - T)$
Ta	$2.0(3270 - T)$	$2.8(1800 - T)$	$2.85(1540 - T)$
Re	$2.4(2710 - T)$	$2.0(3450 - T)$	$2.3(2890 - T)$
Mo	$2.0(2900 - T)$	$2.0(1900 - T)$	$2.15(1530 - T)$
W	$2.0(3650 - T)$	$2.0(2650 - T)$	$2.15(2230 - T)$

$$L = B = E = A = 0$$

FIG. 35. Comparison of ideal-solution computed and experimental phase diagrams (*3*) (*3*), (*3*), (*58*). Thick lines are computed and thin lines are observed.

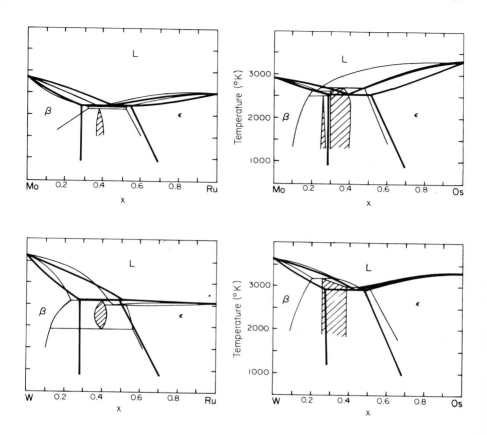

	$\Delta F^{\beta \to L}$	$\Delta F^{\epsilon \to L}$	$\Delta F^{\alpha \to L}$
		(cal/g-atom)	
Mo	2.0(2900 − T)	2.0(1900 − T)	2.15(1530 − T)
W	2.0(3650 − T)	2.0(2650 − T)	2.15(2230 − T)
Ru	2.8(1420 − T)	2.0(2550 − T)	2.8(1780 − T)
Os	2.8(1960 − T)	2.0(3300 − T)	2.8(2310 − T)

$$L = B = E = A = 0$$

FIG. 36. Comparison of ideal-solution computed and experimental phase diagrams (3), (59), (3), (60). Thick lines are computed and thin lines are observed.

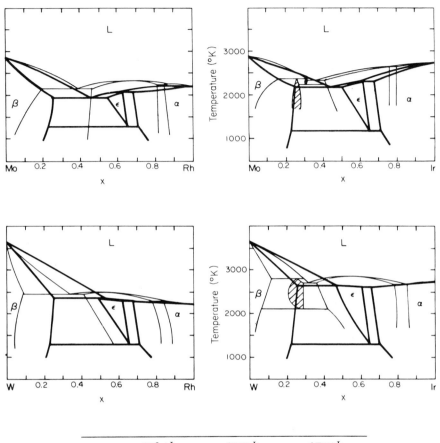

	$\Delta F^{\beta \rightarrow L}$	$\Delta F^{\epsilon \rightarrow L}$	$\Delta F^{\alpha \rightarrow L}$
		(cal/g-atom)	
Mo	2.0(2900 − T)	2.0(1900 − T)	2.15(1530 − T)
W	2.0(3650 − T)	2.0(2650 − T)	2.15(2230 − T)
Rh	3.05(930 − T)	2.15(2010 − T)	2.0(2240 − T)
Ir	3.05(1260 − T)	2.15(2490 − T)	2.0(2750 − T)

$$L = B = E = A = 0$$

FIG. 37. Comparison of ideal-solution computed and experimental phase diagrams (3), (53), (54) (54). Thick lines are computed and thin lines are observed.

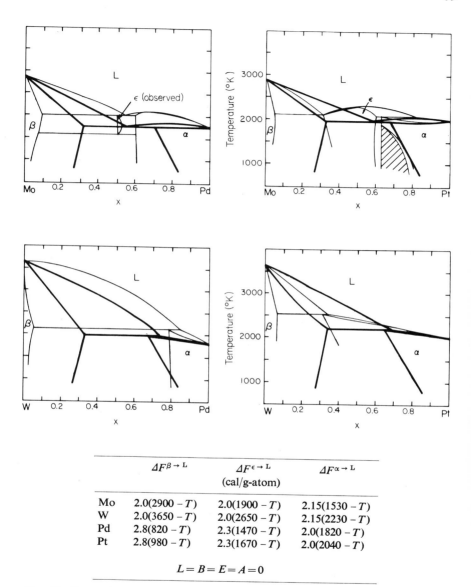

FIG. 38. Comparison of ideal-solution computed and experimental phase diagrams (61), (3), (2), (62). Thick lines are computed and thin lines are observed.

	$\Delta F^{\beta \to L}$	$\Delta F^{\epsilon \to L}$	$\Delta F^{\alpha \to L}$
		(cal/g-atom)	
Mo	2.0(2900 − T)	2.0(1900 − T)	2.15(1530 − T)
W	2.0(3650 − T)	2.0(2650 − T)	2.15(2230 − T)
Pd	2.8(820 − T)	2.3(1470 − T)	2.0(1820 − T)
Pt	2.8(980 − T)	2.3(1670 − T)	2.0(2040 − T)

$$L = B = E = A = 0$$

indicate that the ϵ phase is slightly unstable. The experimental diagrams show a well-developed ϵ field in Mo–Pt, a barely stable ϵ field in Mo–Pd, and no ϵ phase in W–Pt or W–Pd. On the basis of these comparisons, it appears that E must be more negative than A, B, and L and that some of these parameters become positive in the W–Pd system.

The Re vs. (Rh/Ir) and (Pd/Pt) cases shown in Fig. 39 also display evidence for positive interaction parameters in the Re–Rh and Re–Pd systems, as do the Ru–Pd and Os–Pd systems shown in Fig. 40. Nevertheless, the peritectic and eutectic conditions predicted by the ideal-solution computations as well as the T_0–x trends of the ϵ/α equilibria are in good agreement with observations. This is quite satisfying in view of the small energy differences between the fcc and hcp forms of the elements (Figs. 31 and 32 and Table VIII). These small differences are quite reasonable, in view of the structural similarity between the α and ϵ phases.

The Ru/Os vs. Ir comparison in Fig. 41 suggests that A is less than E, since the ideal-solution computation indicates an overestimate of hcp stability relative to α.

In spite of the foregoing differences, the general agreement between the computed and experimental phase diagrams and those observed is quite good. This is rather surprising, in view of the fact that for each group (i.e., Nb/Ta, Mo/W, Rh/Ir, etc.) a single set of stability parameters representing $\Delta H_i^{\beta \to \epsilon}$, $\Delta H_i^{\epsilon \to \alpha}$, $\Delta S_i^{\beta \to \epsilon}$, and $\Delta S_i^{\epsilon \to \alpha}$ has been employed.

An additional check on these values was obtained by computing vanadium, chromium, nickel, and cobalt binaries using the second- and third-row counterpart stability parameters for these elements. Eighteen additional binary diagrams were computed on the basis of ideal solutions, bringing the total to 48 in all. The 18 computed binaries based on V, Cr, Ni, and Co showed similar correlations with observed behavior (16).*

In summary then, 48 ideal-solution binary phase diagrams were computed. The 96 component elements represented included ten cases where Group 5 metals were considered (i.e., V, Nb, Ta), 22 cases where Group 6 metals were represented (Cr, Mo, W), 12 cases where Group 7 elements were included, 18 examples where Group 8 metals were involved, 14 systems where Group 9 elements were considered, and 20 cases encompassing Group 10 metals (i.e., Ni, Pd, Pt). In all of these calculations, a single set of stability parameters was employed to describe the free energy differences between the bcc, hcp, fcc, and liquid phases. Under these circumstances the present values shown in Figs. 31 and 32 and in Table VIII constitute a good working description of the lattice stability of the transition metals.

* The revised, experimental Cr–Ru phase diagram on p. 278 of F. A. Shunk "Constitution of Binary Alloys—Second Supplement," McGraw-Hill, New York, 1969 agrees well with computations (16).

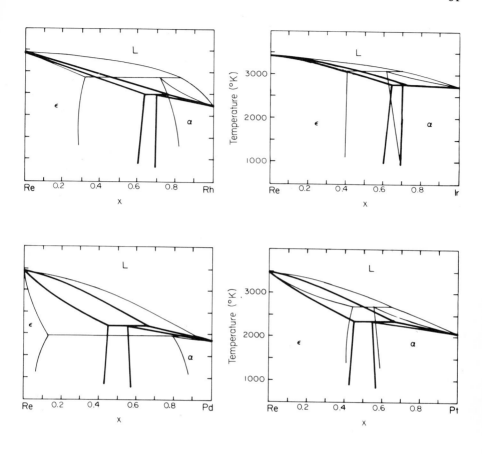

	$\Delta F^{\beta \rightarrow L}$	$\Delta F^{\epsilon \rightarrow L}$	$\Delta F^{\alpha \rightarrow L}$
		(cal/g-atom)	
Re	$2.4(2710 - T)$	$2.0(3450 - T)$	$2.3(2890 - T)$
Rh	$3.05(930 - T)$	$2.15(2010 - T)$	$2.0(2240 - T)$
Ir	$3.05(1260 - T)$	$2.15(2490 - T)$	$2.0(2750 - T)$
Pd	$2.8(820 - T)$	$2.3(1470 - T)$	$2.0(1820 - T)$
Pt	$2.8(980 - T)$	$2.3(1670 - T)$	$2.0(2040 - T)$

$$L = B = E = A = 0$$

FIG. 39. Comparison of ideal-solution computed and experimental phase diagrams (63), (63), (2), (63). Thick lines are computed and thin lines are observed.

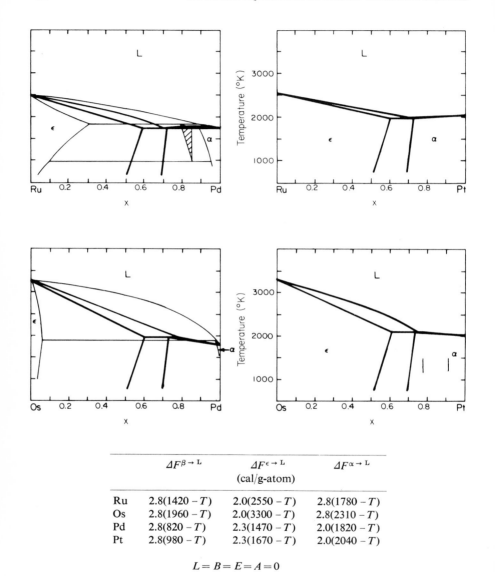

	$\Delta F^{\beta \rightarrow L}$	$\Delta F^{\epsilon \rightarrow L}$	$\Delta F^{\alpha \rightarrow L}$
		(cal/g-atom)	
Ru	$2.8(1420 - T)$	$2.0(2550 - T)$	$2.8(1780 - T)$
Os	$2.8(1960 - T)$	$2.0(3300 - T)$	$2.8(2310 - T)$
Pd	$2.8(820 - T)$	$2.3(1470 - T)$	$2.0(1820 - T)$
Pt	$2.8(980 - T)$	$2.3(1670 - T)$	$2.0(2040 - T)$

$$L = B = E = A = 0$$

FIG. 40. Comparison of ideal-solution computed and experimental phase diagrams (3), (2), (64). Thick lines are computed and thin lines are observed.

3. Procedure Employed for Estimating the Lattice Stability Parameters for Refractory Transition Metals

The earlier discussion of Figs. 23–27 illustrating the influence of lattice stability terms and phase equilibria, as well as the discussion of the ideal-solution or zero-order calculation of phase diagrams presented in Section 2, demonstrates the indispensable requirement for such information in any calculation of phase equilibria.

Unfortunately, first-principle calculations of these terms are beyond current capabilities, as indicated in Section 2 of Chapter II. As a consequence, the only means available for deriving such data is utilization of the experimental techniques described in Section 1 of Chapter II and Section 2 of Chapter III. These techniques were employed in deriving the estimates for

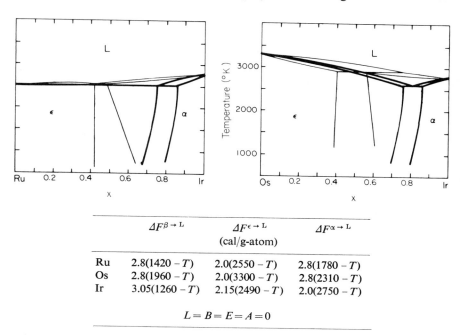

	$\Delta F^{\beta \to L}$	$\Delta F^{\epsilon \to L}$	$\Delta F^{\alpha \to L}$
		(cal/g-atom)	
Ru	2.8(1420 − T)	2.0(2550 − T)	2.8(1780 − T)
Os	2.8(1960 − T)	2.0(3300 − T)	2.8(2310 − T)
Ir	3.05(1260 − T)	2.15(2490 − T)	2.0(2750 − T)

$$L = B = E = A = 0$$

FIG. 41. Comparison of ideal-solution computed and experimental phase diagrams (65), (66). Thick lines are computed and thin lines are observed.

the lattice stability of the bcc, hcp, and fcc forms of the transition metals shown in Figs. 31 and 32 and Table VIII. As indicated earlier, these figures show the free energy difference between the bcc, hcp, and fcc forms of all of the transition metals in a given column of the periodic table. With the exception of the Group 4 elements, where a different value of $\Delta H^{\beta \to \epsilon}$ is indicated for Hf than is shown for Ti and Zr, and except for Co, which has a different value of $\Delta H^{\alpha \rightarrow \epsilon}$ than the other Group 9 metals (Rh, Ir), all metals in a given column are characterized by the same parameters. Thus the parameters shown in Figs. 31 and 32 represent average values. These values were estimated by starting with available data concerning the thermodynamics of phase stability of the pure metals. In order to keep the procedure as simple as possible, free energy differences were described by the approximations of Eqs. (9)–(11) wherein constant enthalpy and entropy differences were considered. On this basis elements such as iron and manganese, in which magnetic contributions are dominant (see Figs. 15 and 20), were excluded from consideration.

With these restrictions it is possible to consider the construction of the lattice stability parameter versus group number curves shown in Figs. 31 and 32. To begin with, values for $\Delta H^{\beta \rightarrow \epsilon}$ and $\Delta S^{\beta \rightarrow \epsilon}$ are available for Ti and Zr (33). These values were employed earlier [Eqs. (21) and (24)] in considering the high-pressure behavior of these metals. The entropy of the ϵ form is about 0.9 cal/g-atom-°K less than that of the β form, while the enthalpy of the ϵ form is about 1030 cal/g-atom less than that of the β form.

At this point in the procedure the values of $\Delta H^{\beta \rightarrow \epsilon}$, $\Delta S^{\beta \rightarrow \epsilon}$, $\Delta H^{\alpha \rightarrow \epsilon}$, and $\Delta S^{\alpha \rightarrow \epsilon}$ are not known for the Group 5 metals (V, Nb, Ta) or the Group 6 metals (Cr, Mo, W). However, since these are all stable in the bcc form, $\Delta H^{\beta \rightarrow \epsilon}$ must be positive in these cases. Thus if the $\Delta H^{\beta \rightarrow \epsilon}$ vs. group-number curve is *assumed* to be a continuous function of group number, this curve must cross the zero axis between Groups 4 and 5. Moreover, since the metals in Groups 7 (Tc, Re) and 8 (Ru, Os) are hcp, the $\Delta H^{\beta \rightarrow \epsilon}$ vs. group number curve must cross the zero axis from the positive to the negative side again between Group 6 and Group 7.

For the Ru/Os case, estimates of the lattice stability are available. These estimates were derived on the basis of an analyses of the Fe–Ru system (36, 43). As indicated earlier (see Fig. 17) in Section 1 of Chapter II, the lattice stabilities of the bcc, hcp, and fcc forms of ruthenium were taken to be identical to that of iron (which occupies the same column in the periodic table), *except* that the magnetic free energy of bcc iron and the magnetic free energy due to the "two-spin-state" population of fcc iron are omitted. Under these conditions, the enthalpy of hcp ruthenium is approximately 1120 cal/g-atom less than that of the bcc form, while the entropy of hcp ruthenium is 0.8 greater than that of the bcc form. These results are plotted on Fig. 31. Moreover, the enthalpy of hcp ruthenium is approximately 120 cal/g-atom less than that of the fcc form, while the entropy of the hcp form is 0.8 cal/g-atom-°K greater than that of the fcc form (see Fig. 17). These values are plotted in Fig. 32.

As we proceed to higher group numbers in Fig. 31, no further quantitative estimates are available until zinc is reached. Here, on the basis of Fig. 30 and Table VII, we can estimate $\Delta H^{\beta \to \epsilon} = -700$ cal/g-atom and $\Delta S^{\beta \to \epsilon} = -0.60$. These points are plotted in Fig. 31. Thus, at this stage in the discussion, specific values of $\Delta H^{\beta \to \epsilon}$ and $\Delta S^{\beta \to \epsilon}$ are available for the (Ti, Zr, Hf) group, the (Ru, Os) group and the (Zn, Cd) group. Moreover, the general shape of the $\Delta H^{\beta \to \epsilon}$ and $\Delta S^{\beta \to \epsilon}$ vs. group number curves are known. Turning to Fig. 32, the situation is more uncertain. As indicated previously, values for $\Delta H^{\alpha \to \epsilon}$ and $\Delta S^{\alpha \to \epsilon}$ are available for ruthenium. In addition, values for zinc are available from Table VII. These results, which suggest that $\Delta H^{\alpha \to \epsilon} = -450$ cal/g-atom and $\Delta S^{\alpha \to \epsilon} = -0.40$ cal/g-atom-°K, are plotted in Fig. 32. Similarly, for cobalt, $\Delta H^{\alpha \to \epsilon} = -110$ cal/g-atom and $\Delta S^{\alpha \to \epsilon} = -0.15$ cal/g-atom-°K. Although the fcc and hcp forms of cobalt are both ferromagnetic, the magnetic properties of both forms are quite similar (in constrast to the case of iron); consequently, the difference in magnetic free energy between α and ϵ should be quite small. In the case of Rh and Ir, $\Delta H^{\alpha \to \epsilon}$ is positive, since the fcc forms are stable. This is also the case for the metals of the (Ni, Pd, Pt) group and the (Cu, Ag, Au) group. At this point quantitative values for $\Delta H^{\alpha \to \epsilon}$ and $\Delta S^{\alpha \to \epsilon}$ are not available.

To begin filling in some values for $\Delta H^{\beta \to \epsilon}$ and $\Delta S^{\beta \to \epsilon}$ in Fig. 31, it is instructive to plot the trajectories of the observed β/L equilibria for rhenium base binary systems and the observed ϵ/L equilibria for tungsten-base binary systems as shown in Fig. 42. As indicated in Fig. 35, the extension of the β/L trajectories for rhenium binaries (shown in Fig. 42) to pure rhenium corresponds to $\overline{T}_{Re}^{\beta}$, which is the melting point of bcc Re. Similarly, extension of the trajectories of the ϵ/L trajectories for tungsten binaries (shown in Fig. 42) to pure tungsten yields the melting point of hcp tungsten, $\overline{T}_{W}^{\epsilon}$. The free energy difference between bcc and liquid tungsten is represented by Eq. 115 (from Table VIII) as

$$\Delta F_{W}^{\beta \to L} = 7300 - 2.0T \quad \text{cal/g-atom} \tag{115}$$

based on the known melting point and an assumed entropy of fusion of 2.0 cal/g-atom-°K. Reference to Fig. 42 indicates a melting point of $\overline{T}_{W}^{\epsilon} = 2650°K$. Hence

$$\Delta F_{W}^{\epsilon \to L} = \Delta F_{W}^{\epsilon \to \beta} + \Delta F_{W}^{\beta \to L} = 0 \quad \text{at} \quad 2650°K \tag{116}$$

or

$$\Delta F_{W}^{\beta \to \epsilon} = 2000 \quad \text{cal/g-atom} \quad \text{at} \quad 2650°K \tag{117}$$

Inspection of Fig. 31 at this point shows a value of $\Delta S^{\beta \to \epsilon} = -0.90$ cal/g-atom-°K for Group 4 (Ti, Zr, Hf) and $\Delta S^{\beta \to \epsilon} = 0.80$ cal/g-atom-°K for Group 8 (Ru, Os). On this basis we assume $\Delta S^{\beta \to \epsilon} = 0.0$ cal/g-atom-°K for Group 6 (Cr, Mo, W). This yields $\Delta H^{\beta \to \epsilon} = 2000$ cal/g-atom for Group

6. This result is checked by performing the ideal-solution calculations of the Mo–Ru, Mo–Os, W–Ru, and W–Os phase diagrams, which are compared with the observed diagrams in Fig. 36.

Returning to Fig. 42, we estimate the melting point of bcc rhenium as 2710°K. Describing the free energy difference between the hcp and liquid forms of rhenium by Eq. (118) (see Table VIII) as

$$\Delta F_{\mathrm{Re}}^{\epsilon \to \mathrm{L}} = 6900 - 2.0T \quad \text{cal/g-atom} \tag{118}$$

Hence

$$\Delta F_{\mathrm{Re}}^{\beta \to \mathrm{L}} = \Delta F_{\mathrm{Re}}^{\beta \to \epsilon} + \Delta F_{\mathrm{Re}}^{\epsilon \to \mathrm{L}} = 0 \qquad \text{at} \quad 2710°\mathrm{K} \tag{119}$$

or

$$\Delta F_{\mathrm{Re}}^{\beta \to \epsilon} = -1480 \qquad\qquad \text{at} \quad 2710°\mathrm{K} \tag{120}$$

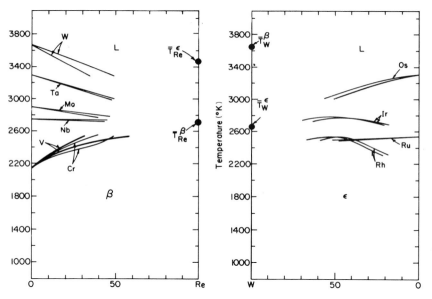

FIG. 42. Comparison of computed melting temperatures of bcc Re and hcp W with β/L and ϵ/L equilibria. In (a) the curves indicate the observed trajectories of the bcc (β)–liquid phase boundaries in the Re–W, Re–Ta, Re–Mo, Re–Nb, Re–V, and Re–Cr systems. In (b) the curves indicate the observed trajectories of hcp (ϵ)–liquid phase boundaries in the W–Os, W–Ir, W–Ru, and W–Rh systems.

Equation (120), along with interpolation between the Group 6 (Cr, Mo, W) and Group 8 (Ru, Os) values in Fig. 31, yields $\Delta H^{\beta \to \epsilon} = -400$ cal/g-atom and $\Delta S^{\beta \to \epsilon} = 0.40$ for the Group 7 metals (Tc, Re). Location of these results permits ideal-solution calculation of the Mo–Re and W–Re diagrams shown in Fig. 35 as a check.

At this stage in the procedure, estimates for $\Delta H^{\beta \to \epsilon}$ and $\Delta S^{\beta \to \epsilon}$ are available for Groups 4, 6, 7, and 8. The $\Delta H^{\beta \to \epsilon}$ and $\Delta S^{\beta \to \epsilon}$ values for Group 5 (V, Nb, Ta) are chosen to be 1500 cal/g-atom and -0.80 cal/g-atom-°K based on interpolation. These results are checked by ideal-solution computation of the Nb–Ru, Ta–Os, Nb–Re, and Ta–Re systems, which are compared with the observed diagrams in Figs. 34 and 35.

The values of $\Delta H^{\beta \to \epsilon} = -1500$ cal/g-atom and $\Delta S^{\beta \to \epsilon} = 0.90$ for the Group 9 metals (Rh, Ir) are chosen by extension of the enthalpy and entropy difference curves and the observed β/ϵ equilibria in the Mo–Rh, Mo–Ir, W–Rh, and W–Ir systems shown in Fig. 37. Once these are assigned, the remaining points for the (Ni, Pd, Pt) and (Cu, Ag, Au) groups are fixed by interpolation in Fig. 31, since the values for Zn are known.

The curves shown in Fig. 31 can be employed in fixing the $\Delta H^{\alpha \to \epsilon}$ and $\Delta S^{\alpha \to \epsilon}$ curves shown in Fig. 32. The values of $\Delta H^{\beta \to \epsilon} = -700$ cal/g-atom and $\Delta S^{\beta \to \epsilon} = -0.50$ cal/g-atom-°K for the (Cu, Ag, Au) group when coupled with the mean free energy difference between the α and β forms of these metals [Eq. (112), Table VII] yields Eq. (121).

$$\Delta F^{\alpha \to \epsilon} = \Delta F^{\alpha \to \beta} + \Delta F^{\beta \to \epsilon}$$
$$= 850 - 0.20T - 700 + 0.50T = 150 + 0.30T \qquad (121)$$

Thus, $\Delta H^{\alpha \to \epsilon} = 150$ cal/g-atom and $\Delta S^{\alpha \to \epsilon} = -0.30$ cal/g-atom-°K for the (Cu, Ag, Au) group, in line with the observed stability of the fcc phase in these metals. These points are shown in Fig. 32, permitting the assignment of $\Delta S^{\alpha \to \epsilon} = -0.30$ cal/g-atom-°K to the Group 10 metals (Ni, Pd, Pt) by interpolation. The value of $\Delta H^{\alpha \to \epsilon} = 150$ cal/g-atom for Rh and Ir stems from a consideration of β/L equilibria in the (Mo/W) vs. (Rh/Ir) set shown in Fig. 37 on the basis of ideal solutions. Assignment of this result for Rh and Ir leads to $\Delta H^{\alpha \to \epsilon} = +250$ cal/g-atom for the Group 10 metals (Ni, Pd, Pt).

At this stage, estimates of $\Delta H^{\alpha \to \epsilon}$ and $\Delta S^{\alpha \to \epsilon}$ are available for all of the groups to the right of (Ru, Os). The values of $\Delta H^{\alpha \to \epsilon}$ and $\Delta S^{\alpha \to \epsilon}$ for rhenium are determined to be -250 cal/g-atom and $+0.30$ cal/g-atom-°K by extension of the $\Delta H^{\alpha \to \epsilon}$ curve and consideration of the Re–Pt system shown in Fig. 39 on the basis of an ideal-solution computation.

The $\Delta H^{\alpha \to \epsilon}$ and $\Delta S^{\alpha \to \epsilon}$ estimates for the Group 6 metals (Cr, Mo, W) are based on extrapolation and consideration of the α/L equilibria in the Mo–Rh, Mo–Ir, W–Rh, and W–Ir systems shown in Fig. 37 .an the basis of an ideal-solution calculation. These considerations yield values of $\Delta H^{\alpha \to \epsilon} = -500$ cal/g-atom and $\Delta S^{\alpha \to \epsilon} = 0.15$ cal/g-atom-°K for Cr, Mo, and W shown in Fig. 32 and Table VIII. Extension of the $\Delta H^{\alpha \to \epsilon}$ and $\Delta S^{\alpha \to \epsilon}$ curves to the Group 5 (V, Nb, Ta) and Group 4 (Ti, Zr, Hf) metals is accomplished by extrapolation.

Recently, Chopra and co-workers (67) have deposited fcc films of a number of transition metals which are not stable in this form. These films ranged in thicknesses between 500 Å and 2 μ. Table X summarizes the results along with estimates of the enthalpy differences between the fcc forms and the stable forms of these metals on the basis of Table VIII. Measurements of the heat of transformation of these structures from the metastable fcc form to the stable form could próvide a direct comparison with the numerical values shown in Table VIII.

TABLE X

VOLUME OF METASTABLE fcc FORMS OF TRANSITION METALS FORMED IN THIN FILMS [a]

Metal	Volume of stable phase [b] (cm^3/g-atom)	Volume of metastable fcc (α) phase (cm^3/g-atom)	Volume difference (cm^3/g-atom)	Predicted enthalpy difference (cal/g-atom)
Zr	$V^\epsilon = 14.04$	$V^\alpha = 14.75$	$\Delta V^{\epsilon \to \alpha} = 0.71$	$\Delta H^{\alpha \to \epsilon} = -800$
Hf	$V^\epsilon = 13.05$	$V^\alpha = 19.05$	$\Delta V^{\epsilon \to \alpha} = 6.00$	$\Delta H^{\alpha \to \epsilon} = -800$
Ta	$V^\beta = 10.91$	$V^\alpha = 12.74$	$\Delta V^{\beta \to \alpha} = 1.83$	$\Delta H^{\alpha \to \beta} = -2150$
Mo	$V^\beta = 9.40$	$V^\alpha = 11.08$	$\Delta V^{\beta \to \alpha} = 1.68$	$\Delta H^{\alpha \to \beta} = -2500$
W	$V^\beta = 9.58$	$V^\alpha = 10.61$	$\Delta V^{\beta \to \alpha} = 1.03$	$\Delta H^{\alpha \to \beta} = -2500$
Re	$V^\epsilon = 8.86$	$V^\alpha = 9.86$	$\Delta V^{\epsilon \to \alpha} = 1.00$	$\Delta H^{\alpha \to \epsilon} = -250$

[a] After Chopra, Randlett, and Duff (67).
[b] At 298°K and 1 atm, β = bcc, ϵ = hcp, and α = fcc.

CHAPTER IV

Application of the Regular-Solution Approximation to Refractory Metal Systems

In spite of the fact that the ideal-solution approach, when coupled with appropriate lattice stability parameters for the pure metals, permits computation of the relevant features of a significant number of binary phase diagrams, several deficiences are apparent. The first is the absence of miscibility gaps. As indicated earlier, miscibility gaps require relatively large positive interaction parameters. For regular solutions, the interaction parameter is $2R$ times the gap critical temperature T_c as shown previously.

In order to illustrate the appearance of phase-boundary curves when large positive interaction parameters characterize the phases in equilibrium, it is instructive to consider a hypothetical case of a binary $i–j$ system at 2500°K, where the melting point of the ϵ form of i is 3000°K and that of the ϵ form of j is 2000°K. As examples, we fix the difference between E and L at 2000 cal/g-atom so that

$$E - L = +2000 \quad \text{cal/g-atom} \tag{122}$$

We now consider three cases, where $E = +1000$ and $L = -1000$ cal/g-atom in the first instance, $E = -6000$ and $L = -8000$ cal/g-atom in the second, and $E = +8000$ and $L = +6000$ cal/g-atom in the third. Figures 43–45 show the free energy of the ϵ and liquid phases as a function of composition at 2500°K. Since $E - L = 2000$ cal/g-atom in each case, the value of x_0 is the same in all cases [see Eqs. (86) and (87)], $x_0 = 0.293$. However, the nearly ideal case (Fig. 43) has a somewhat wider two-phase field than the case where E and L are both negative (Fig. 44). This difference is slight, however, being 11.7 at. % in the nearly ideal case ($E = +1000$, $L = -1000$) and 7.4 at. % in the negative case ($E = -6000$ and $L = -8000$). On the other hand, the case where positive deviations are present ($E = +8000$, $L = +6000$) yields the

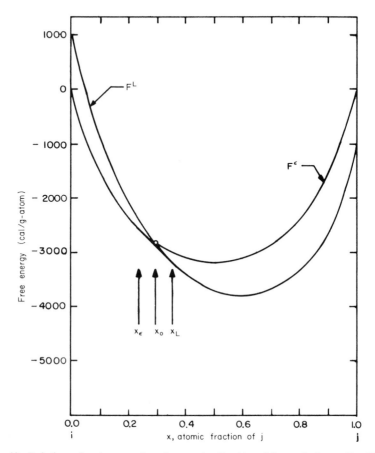

FIG. 43. Relation of x_0 to x_ϵ and x_L for regular liquid and hcp solutions: $T = 2500\,^\circ\mathrm{K}$, $\Delta F_i^{L \to \epsilon} = -1000$ cal/g-atom, $\Delta F_j^{L \to \epsilon} = +1000$ cal/g-atom, $E = +1000$ cal/g-atom, $L = -1000$ cal/g-atom, $x_\epsilon = 0.234$, $x_0 = 0.293$, $x_L = 0.351$.

widest two-phase field, 26.2 at. %. In all of these situations the approximation $(x_\epsilon + x_L)/2 = x_0$ works reasonably well. Thus, large positive interaction
parameters spread the two-phase field more drastically than the contraction
due to negative interaction parameters. As mentioned earlier, some of the
systems discussed in terms of ideal solutions exhibited definite evidence for
positive interaction parameters. In addition to the W–Pd system (Fig. 38)
the Re–Rh and Re–Pd systems (Fig. 39) and the Ru–Pd and Os–Pd cases
(Fig. 40) discussed above, there are a number of additional cases among the
binary systems of interest which require positive interaction parameters.
These systems include the (Zr/Hf) vs. (Nb/Ta) set of four diagrams, the

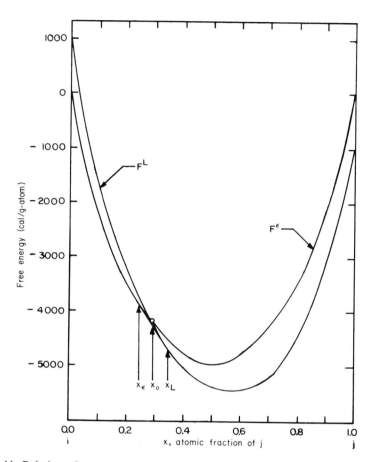

FIG. 44. Relation of x_0 to x_ϵ and x_L for regular liquid and hcp solutions. $T = 2500\,°K$, $F_i^{L \to \epsilon} = -1000\ cal/g\text{-}atom$, $F_j^{L \to \epsilon} = +1000\ cal/g\text{-}atom$, $E = -6000\ cal/g\text{-}atom$, $L = -8000$ cal/g-atom, $x_\epsilon = 0.256$, $x_0 = 0.293$, $x_L = 0.330$.

(Zr/Hf) vs. (Mo/W) set of four diagrams, and the (Rh/Ir) vs. (Pd/Pt) quartet. These systems were deliberately excluded from consideration in the previous ideal-solution calculations. An additional feature that has been mentioned previously is the case where compounds form. Although there are presently no experimental measurements of the heats of formation of binary transition metal compounds of the elements under consideration here, indirect evidence that will be discussed below (25, 68) suggests that compounds in the (Zr/Hf) vs. (Ru/Os), (Rh/Ir), and (Pd/Pt) binaries exhibit large negative heats and free energies of formation.

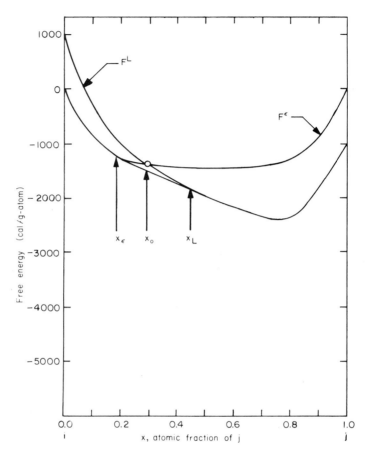

FIG. 45. Relation of x_0 to x_ϵ and x_L for regular liquid and hcp solutions. $T = 2500\,°K$, $F_i^{L \to \epsilon} = -1000\,\text{cal/g-atom}$, $F_j^{L \to \epsilon} = +1000\,\text{cal/g-atom}$, $E = +8000\,\text{cal/g-atom}$, $L = +6000$ cal/g-atom, $x_\epsilon = 0.176$, $x_0 = 0.293$, $x_L = 0.438$.

Recent experiments in our own laboratory indicate that a similar situation exists for compounds in the (Nb/Ta) vs. (Rh/Ir) and (Pd/Pt) systems. Equilibria between liquid, β, ϵ, and/or α phases with such compounds demand large negative interaction parameters for these systems. Again it should be noted that none of the aforementioned cases was included in the foregoing ideal-solution analysis.

A third area of difficulty that the ideal case encounters is proper prediction of the effect of alloying on the $\beta \to \epsilon$ transition in Zr and Hf. If the Zr/Hf binary diagrams were considered on the basis of ideal solutions (i.e.,

$B = E = 0$), then reference to Figs. 31 and 32 shows that we would predict that additions of Nb/Ta or Mo/W stabilize the β phase in agreement with experiment, but that additions of Re, Ru/Os, Rh/Ir, or Pd/Pt stabilize the ϵ phase at the expense of the β. By contrast, it is found that Re stabilizes the β phase, and no transition metals are known to stabilize the ϵ phase at the expense of the β phase in titanium or zirconium. For these reasons Zr/Hf binaries were not included in the foregoing discussion.

A final area of difficulty which the ideal solution approach must encounter is the compositional trajectory of the melting range. If the β/L, ϵ/L, or α/L equilibria in a system i–j is computed for ideal solutions ($L = B = E = A = 0$), then the melting range will cover the temperature range between $\overline{T}_i{}^\beta$ and $\overline{T}_j{}^\beta$, or $\overline{T}_i{}^\epsilon$ and $\overline{T}_j{}^\epsilon$, or $\overline{T}_i{}^\alpha$ and $\overline{T}_j{}^\alpha$ in approximately linear fashion. The width of the two-phase field will pe proportional to the difference between the melting points $(\overline{T}_i{}^\beta - \overline{T}_j{}^\beta,\ \overline{T}_i{}^\epsilon - \overline{T}_j{}^\epsilon$, or $\overline{T}_i{}^\alpha - \overline{T}_j{}^\alpha)$. However, there are many systems in which the liquid/solid range does not run linearly or even nearly linearly between the melting points of bcc, fcc, or hcp phases. In many cases clear minima exist. Such behavior has been attributed to strain energy effects due to differences in size between i and j by Hume-Rothery and co-workers (69).

In the cases already discussed, the ϵ/L equilibria in the (Mo/W) vs. (Rh/Ir) quartet (see Fig. 37) show the unusual occurrence of a clear maximum. This is also evident in the α/L equilibrium of Mo–Pd, and the ϵ/L equilibria of Mo/Pt of Fig. 38. Systems showing pronounced minima have not been discussed as yet, since the systems exhibiting large size differences between components were excluded from ideal-solution consideration. However, the (Zr/Hf) vs. (Nb/Ta) series and the (Zr/Hf) vs. (Mo/W) quartet all exhibit minima in the course of the β/L equilibria curves. Similar minima occur in the ϵ/L equilibria of the (Zr/Hf) vs. Re systems. All ten of the foregoing examples are systems with very large volume differences (ranging from 20 to 45% by volume). Needless to say, these cases would not compare well with computations based on the assumption that $L = B = E = A = 0$.

The regular-solution approximation constitutes the simplest description of the departure from ideality, since only one parameter is required to describe F_E, the excess free energy of mixing of each phase. Nevertheless, if we are to describe the equilibria between the liquid, bcc, hcp, and fcc phases in each of 72 systems, a total of $4 \times 72 = 288$ such parameters must be determined! One might consider extracting each of these parameters individually from the observed two-phase equilibria in the 72 binary diagrams. However, such a procedure would not be possible because all of the phases do not appear on each diagram and the result would be little more than an exercise in curve fitting. Accordingly, we have employed the following method of estimating the various components of L, B, E, and A.

1. Estimation of the Interaction Parameter for the Liquid Phase

Kleppa (70) has reviewed various methods for estimating the interaction parameter for liquid phases and has discussed the method developed by Hildebrand and Scott (14) and extended by Mott (71). This method, which has been employed by Furakawa (72) and Wada (73) along similar lines, consists of definining L as the sum of two terms.

$$L = e_0 + e_p \tag{123}$$

The second of these terms e_p, has been called the internal pressure contribution by Brewer (24) and is described in terms of the " solubility parameters " for the binary partners. These solubility parameters δ_i and δ_j are defined as

$$\delta_i = (-H_i/V_i)^{1/2} \quad (\text{cal/cm}^3)^{1/2} \tag{124}$$

$$\delta_j = (-H_j/V_j)^{1/2} \quad (\text{cal/cm}^3)^{1/2} \tag{125}$$

where H_i and H_j are the enthalpies of vaporization and V_i and V_j the volumes of the binary partners per gram atom. The internal pressure term e_p is evaluated by squaring the difference between δ_i and δ_j and multiplying by the average volume. However, as Brewer has noted, this procedure would require accurate calculations of δ_i at a variety of temperatures, particularly in the liquid phase, since H_i and V_i are temperature dependent. Hildebrand and Scott (14) and Mott (71) evaluated δ_i at 2000°K. Brewer (24), on the other hand, reexamined the enthalpy of sublimation data for all of the metals, but chose to compute e_p based on room temperature values. However, Brewer noted that δ_i should be computed for the liquid at high temperatures and that the more appropriate values would be lower than those computed at room temperature. Because of this situation, a multiplicity of values has been tabulated. For example, Brewer (24) gives solubility parameters of 102 and 130 for Zr and Mo respectively at 298°K. On the other hand, if one computes δ_i for Zr above its melting point at 2200°K using available thermodynamic (74) and expansion data (75) and allowing for a 4% volume expansion on melting, a value of 96 (cal/cm³)^{1/2} is obtained. A similar calculation for Mo at 3000°K leads to a solubility parameter of 118 (cal/cm³)^{1/2}. These smaller values of δ_i and δ_j would lead to smaller values of e_p. In order to avoid the difficulty of evaluating δ_i at many temperatures and/or choosing an arbitrary temperature for evaluating the solubility parameter, we have chosen to adopt Brewer's procedure of evaluating δ_i at 298°K, and have reduced the Hildebrand and Scott (14) definition of the internal pressure by 40%. The results of the approximation will be discussed in Chapter V when the observed and computed phase equilibria are discussed. Thus, the present definition of the internal pressure term is

$$e_p = 0.3(V_i + V_j)[(-H_i/V_i)^{1/2} - (-H_j/V_j)^{1/2}]^2 \quad \text{cal/g-atom} \tag{126}$$

where the factor 0.3 reflects the above mentioned 40% reduction [the Hildebrand–Scott coefficient (14) is 0.5] and the values of H and V refer to room temperature.

According to Eq. (123), the parameter L consists of two terms e_0 and e_p. The latter is always positive, and if e_0 were always zero, only positive deviations from ideality would be predicted, thus causing obviously incorrect results in certain cases. In the systems of interest here some notable examples are worth mentioning. Criscione and co-workers (76) have reacted ZrC and HfC with Ir at temperatures between 1500 and 2500°K and obtained reactions of the type

$$ZrC + 3Ir \rightarrow ZrIr_3 + C \qquad (127)$$

for both cases. Since the free energy of formation per mole (2 g-atoms) is -38 kcal for ZrC (77) and -55 kcal for HfC (78) at 2000°K, the observation of such a reaction implies that $ZrIr_3$ has a free energy of formation per mole that is more negative than -38 kcal and $HfIr_3$ has a free energy of formation that is more negative than -55 kcal/mole. Brewer and co-workers (25, 68) have repeated and extended these observations to reactions of Pt and Ir with ZrC. At temperatures above 1500°K, $ZrPt_3$ and $ZrIr_{3-x}$ were observed, indicating that $ZrIr_3$, $ZrPt_3$, $HfIr_3$, and $HfPt_3$ all have free energies of formation at 2000°K that are less (more negative) than -40 to -60 kcal/mole or -10 to -15 kcal/g-atom (since 1 mole of $HfIr_3$ contains 4 g-atom). Brewer (25, 68) has estimated heats of formation near -80 kcal/mole or -20 kcal/g-atom for these compounds.

Recent experiments in our laboratory with NbC–Pt, NbC–Ir, and NbC–Re mixtures in graphite crucibles at 2000°K demonstrated the formation of the α Mn form of $NbRe_4$ ($a_0 = 9.65$Å) (3), the $AuCu_3$ type of $NbIr_3$, and the α and β forms of $NbPt_3$ (79). Since the free energy of formation of NbC at 2000°K is about -29 kcal/mole, the free energy of formation of $NbIr_3$ must be less (more negative) than -7250 cal/g-atom near 2000°K. The free energy of formation of $NbRe_4$ must be less (more negative) than -5800 cal/g-atom at 2000°K.*

These observations are related to the present considerations as follows. Although the melting temperatures of the $ZrIr_3$, $HfIr_3$, $ZrPt_3$, $HfPt_3$, $NbIr_3$, $NbPt_3$, etc., compounds have not all been measured, available data indicate that these compounds melt near 2500°K (see Chapter V). At this temperature, Eqs. (65) and (66) define the free energy of formation of the liquid phase ΔF_f^L as

$$\Delta F_f^L \cong Lx(1-x) + RT[x \ln x + (1-x) \ln(1-x)] \quad \text{cal/g-atom} \qquad (128)$$

which becomes

$$\Delta F_f^L \cong 0.1875L - 1.118T \quad \text{cal/g-atom} \qquad (129)$$

* E. Raub and G. Faulkenberg [Z. Metallk. 55, 190 (1964)] have decomposed TaC, WC, and NbC with Pt and Pd at high temperatures.

for $x = 0.25$.   Letting $T = 2500°K$ yields

$$\Delta F_{\mathrm{f}}{}^{\mathrm{L}} \cong 0.1875L - 2795 \quad \text{cal/g-atom} \tag{130}$$

Inasmuch as the free energies of the $ZrIr_3$, $HfIr_3$, $\ldots$, etc., compounds must be equal to those of their respective liquid phases at the compounds' melting temperature, then the L parameters for the systems in question must be more negative than $-38,400$ to $-65,000$ cal/g-atom for the Zr/Hf cases and more negative than $-23,700$ cal/g-atom in the $NbIr_3$ and $NbPt_3$ cases. Since L is defined as the sum of e_0 and e_p, and e_p is about $+7000$ cal/g-atom for Zr–Ir and Hf–Ir, the e_0 values for these systems must be more negative than $-45,400$ to $-72,000$ cal/g-atom. In the Nb–Ir and Nb–Pt cases, e_p is only 500 cal/g-atom, so that e_0 must be more negative than $-24,200$ for these situations.

These examples clearly indicate that the e_0 term must take on large negative values in specific instances. One of the methods employed for estimating e_0 is to use Pauling's electronegativity values and setting

$$e_0 = -23,060\bar{n}(X_i - X_j)^2 \quad \text{cal/g-atom} \tag{131}$$

where $\bar{n} = 5$ and X_i and X_j are the electronegativity parameters (70–73). Although this relation yields a negative value for e_0 when suitable electronegatives are employed, it does not provide uniformly suitable results for the systems in question. Section 1 of Chapter VI discusses the difficulties that arise when Eq. (131) is employed to compute e_0.

Recently, Rudman (80) has suggested an alternate method for estimating the heats of formation of transition metal alloy systems. The following is an application of his method to the problem at hand.

If we consider a liquid solution with coordination number Z, in the i–j system, then following the regular-solution interaction parameter description presented by Cottrell (81) we define

$$
\begin{aligned}
Zx &= \text{number of } j \text{ atoms next to any atom} \\
Z(1 - x) &= \text{number of } i \text{ atoms next to any atom} \\
Nx &= \text{number of } j \text{ atoms per gram atom} \\
N(1 - x) &= \text{number of } i \text{ atoms per gram atom}
\end{aligned}
$$

where N is Avogadro's number, and

$$
\begin{aligned}
N_{jj} &= \text{number of } jj \text{ bonds} \\
N_{ii} &= \text{number of } ii \text{ bonds} \\
N_{ij} &= \text{number of } ij \text{ bonds} \\
N_{ij} &= (Nx)Z(1 - x)
\end{aligned}
$$

In determining the number of N_{ii} and N_{jj} bonds we must introduce a factor of 0.5 to avoid duplication in counting; thus

$$N_{jj} = 0.5(Nx)Zx$$
$$N_{ii} = 0.5N(1-x)Z(1-x)$$

Consequently, the enthalpy of the liquid phase L is given as

$$H^L = N_{ii}H_{ii}^L + N_{jj}H_{jj}^L + N_{ij}H_{ij}^L$$
$$H^L = 0.5NZ(1-x)^2H_{ii}^L + 0.5NZx^2H_{jj}^L + NZx(1-x)H_{ij}^L \qquad (132)$$

Rearranging Eq. (132) yields

$$H^L = (1-x)0.5NZH_{ii}^L + x0.5NZH_{jj}^L$$
$$+ 2x(1-x)\,(0.5NZH_{ij}^L - 0.25NZH_{ii}^L - 0.25NZH_{jj}^L) \qquad (133)$$

where H_{ii}^L, H_{jj}^L, and H_{ij}^L are the enthalpy per ii, jj, and ij bond. Since the enthalpies per gram atom of the pure components i and j are defined as

$$H_i{}^L = 0.5NZH_{ii}^L \quad \text{and} \quad H_j{}^L = 0.5NZH_{jj}^L$$

we let

$$H^L[(i+j)/2] = 0.5NZH_{ij}^L$$

and obtain

$$H^L = (1-x)H_i{}^L + xH_j{}^L$$
$$+ x(1-x)2\{H^L[(i+j)/2] - 0.5H_i{}^L - 0.5H_j{}^L\} \quad \text{cal/g-atom} \qquad (134)$$

where

$$e_0 = 2\{H^L[(i+j)/2] - 0.5H_i{}^L + 0.5H_j{}^L\} \quad \text{cal/g-atom} \qquad (135)$$

so that

$$H^L = (1-x)H_i{}^L + xH_j{}^L + x(1-x)e_0 \quad \text{cal/g-atom} \qquad (136)$$

Equations (134) and (135) describe e_0 the "electronic component" of L as a function of group number. In order to illustrate the procedure, Fig. 46 shows the enthalpy of vaporization of the second- and third-row transition metals as a function of group number. The open circles are H_i at 298°K (24) averaged between partners. Thus, when $i = 5$ (Nb/Ta), a value of -180 kcal, which is the mean of H_{Nb} (-173 kcal) and H_{Ta} (-187 kcal), is shown. Following Eq. (135) the value of e_0 for 4–10 systems (i.e., Zr/Hf–Pd/Pt) would be

$$e_0 = 2\{H^L[(i+j)/2] - 0.5(-148) - 0.5(-114)\} \quad \text{kcal/g-atom} \qquad (137)$$

with $(i+j)/2 = 7 = $ (Tc/Re), giving

$$e_0 = 2(-170 + 131) \quad \text{kcal/g-atom}$$
$$= -68,000 \quad \text{cal/g-atom} \qquad (138)$$

Thus, this procedure yields a suitably large negative value of e_0 for the Zr–Pd, Zr–Pt, Hf–Pd, and Hf-Pt cases. The variation of the enthalpy of vaporization with group number leads to large negative values of e_0 for the (Zr/Hf) vs. (Tc/Re), (Ru/Os), and (Rh/Ir) cases as well.

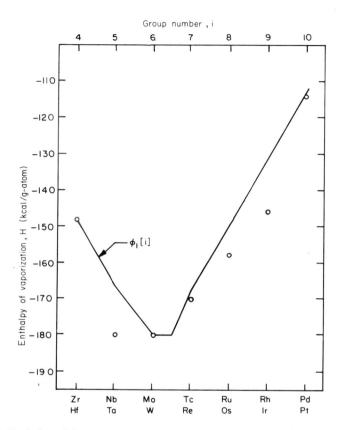

FIG. 46. Variation of the averaged and adjusted enthalpy of vaporization of the transition metals with group number. The $\phi_1[i]$ curve is used in the computation of the interaction parameter L for the liquid as shown in Table XI. Points represent averaged experimental values at $298\,°K$; the solid line is the adjusted curve for interaction parameter computation.

The H_i values have been adjusted in order to obtain favorable numerical values for all of the systems in question. The result is shown in Fig. 46 as the $\phi_1[i]$ curve, which distorts the observed H_i values at $i = 5, 8$, and 9 but maintains the overall shape and magnitude. Thus, for computational purposes,

$$e_0 = 2\{\phi_1[i+j)/2] - \tfrac{1}{2}\phi_1[i] - \tfrac{1}{2}\phi_1[j]\} \quad \text{cal/g-atom} \tag{139}$$

and

$$L = e_0 + e_\mathrm{p}$$

based on Eq. (123), where e_p is given by Eq. (126).

2. ESTIMATION OF THE INTERACTION PARAMETERS FOR THE BCC, HCP, AND FCC PHASES

In order to estimate the regular-solution interaction parameter B for the bcc β phase, it is instructive to consider Fig. 47, which illustrates three types of β/L trajectories in an i–j system. In case 1 the liquidus and solidus run in nearly linear fashion between $\overline{T}_i{}^\beta$ and $\overline{T}_j{}^\beta$. If the T_0–x curve for β/L defined by Eqs. (86) and (87) is taken to lie midway between x_β and x_L as shown by the dashed lines in Fig. 47, then for case 1, $B - L \approx 0$.

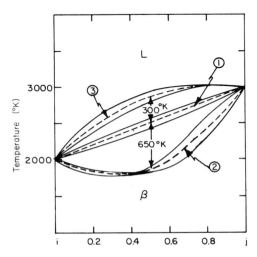

FIG. 47. Illustration of factors controlling β/L equilibria. Melting trajectories showing linear, minimum, and maximum behavior are shown. Case 1, $B - L \approx 0$; case 2, $B - L \approx 5200$ cal/g-atom; case 3, $B - L \approx -2400$ cal/g-atom. $B - L = e_1 + e_2$ where e_1 is the strain energy term

$$e_1 \approx -0.5(H_i + H_j)(V_i - V_j)^2(V_i + V_j)^{-2} \quad \text{cal/g-atom} \quad \propto VG(\Delta V/V)^2$$
$$G \propto H/V$$

The second case in Fig. 47 illustrates a depression in the β/L equilibrium range. This behavior usually occurs when the components i and j have different sizes (69). Under these conditions, Eq. (86) implies that $B > L$. The third case in Fig. 47 illustrates a β/L range lying above the linear T_0–x curve. Here Eq. (86) required $B < L$. As pointed out earlier, case 3 is rare. However, the (Mo/W) vs. (Rh/Ir) quartet shown in Fig. 37 clearly indicates this behavior. Figure 47 shows how the β/L trajectories can be employed to estimate $B - L$. If the entropy of fusion of the β phase is approximately 2.0 cal/g-atom-°K, then Eq. (86) suggests that

$$(B - L) \approx 4(\overline{T}_i{}^\beta + \overline{T}_j{}^\beta - 2T_0{}^{\beta L}[\text{at } x = 0.5]) \quad \text{cal/g-atom}$$

In order to estimate B quantitatively, we consider that the difference $B - L$ is composed of two terms

$$B = L + e_1 + e_2 \qquad (140)$$

The first of these, e_1, being a strain energy term depending on the size or volume difference between i and j. This term will always be positive, and when it is dominant, case 2 behavior (Fig. 47) results. When the size differences are small and e_1 is small, negative values of e_2 can lead to case 3 behavior. In order to estimate e_1, we consider it to be proportional to a stress–strain term, replacing the stress by the product of shear modulus G times the fractional difference in volume.*

$$e_1 \propto VG(\Delta V/V)^2 \qquad (141)$$

Taking the shear modulus to be proportional to H/V, we obtain

$$e_1 = -0.5(H_i + H_j)(V_i - V_j)^2(V_i + V_j)^{-2} \quad \text{cal/g-atom} \qquad (142)$$

where the coefficient 0.5 is obtained by examining β/L equilibria in (Zr/Hf) vs. (Nb/Ta) and (Mo/W) systems. It is of interest to note that the simple computation of e_1 given by Eq. (142) yields the same magnitude of the maximum size effect energy afforded by more complex computations. Knapton (82) has reported calculations due to Howlett employing Lawson's size effect energy equation. Howlett computed a maximum strain energy of 500 cal/g-atom for Zr–Nb at 57% Nb, and 1300 cal/g-atom for Zr–Ta at 56% Ta. Equation (142) yields 659 cal/g-atom for Zr–Nb, and 655 cal/g-atom for Zr–Ta at the 50/50 composition.

In order to compute the e_2 component, a development similar to that employed to obtain e_0 is used [see Eqs. (132)–(137)]. To obtain the " electronic component " for the β phase e_2, we let

$$H^\beta = (1 - x)H_i^\beta + xH_j^\beta + x(1 - x)B \quad \text{cal/g-atom} \qquad (143)$$

$$H^L = (1 - x)H_i^L + xH_j^L + x(1 - x)L \qquad (144)$$

so that

$$H^\beta - H^L = \Delta H^{L \to \beta} = (1 - x)\Delta H_i^{L \to \beta} + x\,\Delta H_j^{L \to \beta} + x(1 - x)(B - L) \quad (145)$$

where the e_2 component of $B - L$ is given by

$$e_2 = 2\{\Delta H^{L \to \beta}\,[(i + j)/2] - 0.5\,\Delta H_i^{L \to \beta} - 0.5\,\Delta H_j^{L \to \beta}\} \quad \text{cal/g-atom} \qquad (146)$$

in direct analogy to Eq. (137). Figure 48 shows the variation of $\Delta H_i^{L \to \beta}$ vs. group number. The points are obtained from Table VIII by averaging the enthalpy difference between elements in a given group. For example, $\Delta H^{L \to \beta}$

---

* The shear modulus is appropriate in this case, since the strain-energy term is added to the solid phase and not to the liquid where the shear modulus is zero.

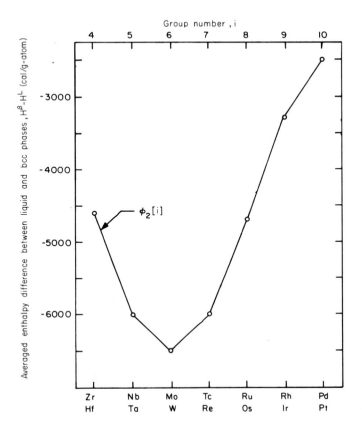

FIG. 48. Variation of the averaged enthalpy difference between the bcc and liquid forms of the transition metals with group number. The $\phi_2[i]$ curve is used in the computation of the interaction parameter B for the bcc phase as shown in Table XI.

is -5800 cal/g-atom for Mo and -7300 cal/g-atom for W. The value of $\phi_2[i = 6]$ shown in Fig. 48 is -6500 cal/g-atom.

Thus, to obtain B from Eq. (140) one computes e_1, which is a positive term reflecting the size difference, and a second term e_2, which depends upon the variation of enthalpy difference between the bcc and liquid phase with group number. Then e_1 and e_2 are added to L to determine B.

The computation of E, which is the interaction parameter for the hcp (ϵ) phase follows the same lines. We define

$$E = B + e_3 \quad \text{cal/g-atom} \qquad (147)$$

Hence

$$e_3 = 2\{\Delta H^{\beta \rightarrow \epsilon} [(i+j)/2] - 0.5\,\Delta H_i^{\beta \rightarrow \epsilon} - 0.5\,\Delta H_j^{\beta \rightarrow \epsilon}\} \quad \text{cal/g-atom} \qquad (148)$$

The variation of the enthalpy difference between ϵ and β phases is shown in Fig. 49. These values are taken directly from Table VIII, except for the (Zr/Hf) point, which is the average of the zirconium and hafnium values.

The final step is to compute A, the interaction parameter for the fcc (α) phase. This parameter is defined by Eq. (149) as

$$A = E + e_4 \qquad (149)$$

where

$$e_4 = 2\{\Delta H^{\epsilon \to \alpha} \,[(i+j)/2] - 0.5\, \Delta H_i^{\epsilon \to \alpha} - 0.5\, \Delta H_j^{\epsilon \to \alpha}\} \quad \text{cal/g-atom} \qquad (150)$$

Figure 50 illustrates the variation of the enthalpy difference between the fcc and hcp forms with group number. The values shown are taken directly from Table VIII.

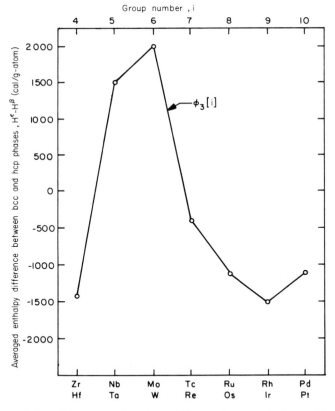

FIG. 49. Variation of the averaged enthalpy difference between the bcc and hcp forms of the transition metals with group number. The $\phi_3\,[i]$ curve is used in the computation of the interaction parameter E for the hcp phase as shown in Table XI.

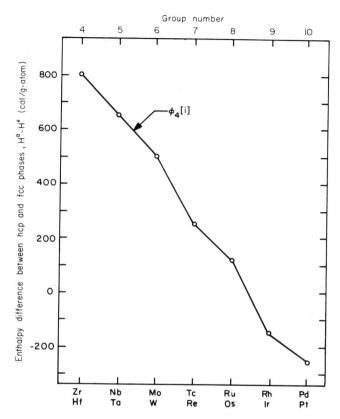

FIG. 50. Variation of the enthalpy difference between the hcp and fcc forms of the transition metals with group number. The ϕ_4 [i] curve is used in the computation of the interaction parameter A for the fcc phase as shown in Table XI.

The computation of L, B, E, and A are detailed in Tables XI–XIII so that all of the individual components can be identified. Table XI contains the enthalpies of vaporization (7, 24, 74) and the volumes (83) employed in computing e_p and e_1. In addition, Table XI contains the $\phi_1[i]$, $\phi_2[i]$, $\phi_3[i]$, and $\phi_4[i]$ functions employed for computing e_0, e_2, e_3, and e_4. Table XII summarizes the numerical values of e_0, e_p, e_1, e_2, e_3, and e_4, while Table XIII contains the resultant L, B, E, and A parameters for all of the binary systems under consideration.

Before proceeding to employ these parameters to compute the regular-solution diagrams, it is worthwhile to reconsider the values in Tables XI–XIII. Reference to Table XIII shows that the differences between L, B, E, and A for any given system is small. This is to be expected of phases that are

TABLE XI

ESTIMATION PROCEDURE FOR COMPUTING THE PARAMETERS L, B, E, AND A

1. Computation of L

Metal (Me)	Index (i)	$\phi_1[i]$ (cal/g-atom)
Zr/Hf	4.0	$-148{,}000$
	4.5	$-157{,}000$
Nb/Ta	5.0	$-166{,}000$
	5.5	$-173{,}000$
Mo/W	6.0	$-180{,}000$
	6.5	$-180{,}000$
Re	7.0	$-168{,}000$
	7.5	$-159{,}000$
Ru/Os	8.0	$-150{,}000$
	8.5	$-141{,}500$
Rh/Ir	9.0	$-132{,}000$
	9.5	$-123{,}000$
Pd/Pt	10.0	$-114{,}000$

Procedure:

The parameter L for the system Me_i–Me_j is given by $L = e_0 + e_p$ where e_0 is defined as follows:

$$2(\phi_1[(i+j)/2] - \tfrac{1}{2}\phi_1[i] - \tfrac{1}{2}\phi_1[j]) = e_0$$

The function $\phi_1[i]$ is the enthalpy of vaporization curve averaged between second- and third-row elements.

Thus, for Nb/Ta–Re

$$i = 5.0, \qquad j = 7.0, \qquad (i+j)/2 = 6.0$$
$$\phi_1[6.0] = -180{,}000$$
$$\phi_1[5.0] = -166{,}000, \qquad \phi_1[7.0] = -168{,}000$$

Hence, $e_0 = -26{,}000$ cal/g-atom.

The second component of L is the internal pressure parameter given by

$$e_p = 0.3(V_1 + V_2)((-H_1/V_1)^{1/2} - (-H_2/V_2)^{1/2})^2$$

where H_1, V_1, H_2, and V_2 are the enthalpies of vaporization and the volumes of the metals.

2. Computation of B

There are two components in the calculation of B. The first is a strain energy e_1 defined as follows:

$$e_1 = -0.5(H_1 + H_2)(V_1 - V_2)^2(V_1 + V_2)^{-2} \quad \text{cal/g-atom}$$

where H_1 and H_2 are the enthalpies of vaporization of the binary partners at $298°K$ and V_1 and V_2 are the atomic volumes of the partners. The following have been used:

Metal	H (cal/ g-atom)	V (cm³/ g-atom)	$(-H/V)^{1/2}$	Metal	H (cal/ g-atom)	V (cm³/ g-atom)	$(-H/V)^{1/2}$
Zr	$-146{,}000$	14.04	102	Ru	$-155{,}000$	8.19	138
Hf	$-150{,}000$	13.50	105	Os	$-162{,}000$	8.43	139
Nb	$-173{,}000$	10.84	126	Rh	$-133{,}000$	8.31	126
Ta	$-187{,}000$	10.91	131	Ir	$-160{,}000$	8.56	136
Mo	$-157{,}000$	9.40	130	Pd	$-91{,}000$	8.86	101
W	$-202{,}000$	9.58	145	Pt	$-135{,}000$	9.10	122
Re	$-186{,}000$	8.86	145				

The parameter B is defined as $B = L + e_1 + e_2$. The first and second terms are defined above. The third term is defined on the following page:

TABLE XI (continued)

Metal (Me)	Index (i)	$\phi_2[i]$ (cal/g-atom)
Zr/Hf	4.0	−4600
	4.5	−5300
Nb/Ta	5.0	−6000
	5.5	−6250
Mo/W	6.0	−6500
	6.5	−6250
Re	7.0	−6000
	7.5	−5350
Ru/Os	8.0	−4700
	8.5	−4000
Rh/Ir	9.0	−3300
	9.5	−2900
Pd/Pt	10.0	−2500

Procedure:

The function $\phi_2[i]$ represents the average enthalpy difference between liquid and bcc phases for each of the groups $i = 4, 5, 6$, etc. The parameter e_2 for the system Me_i–Me_j is given by

$$2(\phi_2[(i+j)/2] - \tfrac{1}{2}\phi_2[i] - \tfrac{1}{2}\phi_2[j]) = +e_2$$

Thus, for Mo/W–Rh/Ir

$$i = 6.0, \qquad j = 9.0, \qquad (i+j)/2 = 7.5$$
$$\phi_2[7.5] = -5350$$
$$\phi_2[6.0] = -6500$$
$$\phi_2[9.0] = -3300$$
$$e_2 = -900 \quad \text{cal/g-atom}$$

3. Computation of E

Metal (Me)	Index (i)	$\phi_3[i]$ (cal/g-atom)
Zr/Hf	4.0	−1430
	4.5	+35
Nb/Ta	5.0	+1500
	5.5	+1750
Mo/W	6.0	+2000
	6.5	+800
Re	7.0	−400
	7.5	−760
Ru/Os	8.0	−1120
	8.5	−1310
Rh/Ir	9.0	−1500
	9.5	−1300
Pd/Pt	10.0	−1100

Procedure:

The function $\phi_3[i]$ represents the difference in enthalpy between the bcc and hcp phases for each of the groups $i = 5, 6, 7, 8$, etc. For $i = 4$ the values for Zr and Hf have been averaged. The parameter E is defined as

$$E = B + e_3$$

where e_3 for the system Me_i–Me_j is given by

$$2(\phi_3[(i+j)/2] - \tfrac{1}{2}\phi_3[i] - \tfrac{1}{2}\phi_3[j]) = e_3$$

Thus, for Zr/Hf–Rh/Ir

$$i = 4.0, \qquad j = 9.0, \qquad (i+j)/2 = 6.5$$
$$\phi_3[6.5] = +800$$
$$\phi_3[4.0] = -1430$$
$$\phi_3[9.0] = -1500$$
$$e_3 = +4530 \quad \text{cal/g-atom}$$

in equilibrium with one another. Examination of Table XII discloses that the largest numerical contributions to the interaction parameters are already present in the liquid phase by virtue of the e_0 and internal pressure (e_p) terms. For the (Zr/Hf) vs. (Ru/Os), (Rh/Ir), and (Pd/Pt) cases, large negative values of e_0 dominate the final values of all of the interaction parameters. A similar result is obtained in the (Nb/Ta) vs. (Ru/Os), (Rh/Ir), and (Pd/Pt) cases, but to a lesser degree. The internal pressure term is quite substantial in all of the (Zr/Hf) systems except when Pd is one of the components. It diminishes in size for the Nb/Ta cases and remains small except for Mo–Pd, W–Pd, W–Pt,

TABLE XI (*continued*)

4. Computation of A

Metal (Me)	Index (i)	$\phi_4[i]$ (cal/g-atom)
Zr/Hf	4.0	$+ 800$
	4.5	$+ 725$
Nb/Ta	5.0	$+ 650$
	5.5	$+ 575$
Mo/W	6.0	$+ 500$
	6.5	$+ 375$
Re	7.0	$+ 250$
	7.5	$+ 185$
Ru/Os	8.0	$+ 120$
	8.5	$- 15$
Rh/Ir	9.0	$- 150$
	9.5	$- 200$
Pd/Pt	10.0	$- 250$

Procedure:

The function $\phi_4[i]$ represents the difference in enthalpy between the hcp and fcc phases for each of the groups $i = 4, 5, 6, 7$, etc. The parameter A is defined as

$$A = E + e_4$$

where e_4 for the system Me_i–Me_j is given by

$$2(\phi_4[(i+j)/2] - \tfrac{1}{2}\phi_4[i] - \tfrac{1}{2}\phi_4[j]) = e_4$$

Thus, for Nb/Ta–Ru/Os

$$i = 5.0, \qquad j = 8.0, \qquad (i+j)/2 = 6.5$$

$$\phi_4[6.5] = + 375$$

$$\phi_4[5.0] = + 650$$

$$\phi_4[8.0] = + 120$$

$$e_4 = - 20 \quad \text{cal/g-atom}$$

Re–Pd, Ru–Pd, Os–Pd, Rh–Pd, and Ir–Pd, where the discussion in Section 2 of Chapter III indicated positive interaction parameters. Examination of the size effect terms, which depend on size or volume differences, shows that e_1 is a large term in all of the (Zr/Hf) binary systems. However, as one shifts attention to the (Nb/Ta) systems, smaller values of e_1 result because of smaller differences in size. In the remaining cases the size factor has little significance.

Finally, it is instructive to compare the relative values of L, B, E, and A given in Table XIII. In all of the (Zr/Hf) cases, B is more negative than E. Thus, the inclusion of interaction parameters will help to stabilize the bcc phase at the expense of the hcp phase, even for systems where the pure alloying element is more stable in the ϵ phase than in the β phase. In the (Zr/Hf) vs. (Nb/Ta) and (Mo/W) cases, B is greater than L. Thus, these parameters will combine to depress the β/L melting range. A similar case obtains in the Zr–Re and Hf–Re cases. On the other hand, in the (Nb/Ta) vs. (Rh/Ir) and (Pd/Pt) cases, the interaction parameters for the solid phases are more negative than for the liquid, leading to elevation of the melting range. This occurrence reappears in a pronounced manner for the (Mo/W) vs. (Rh/Ir) and (Pd/Pt) cases. In all remaining cases the differences between L, B, E, and A are quite small.

In summary then, 288 interaction parameters have been calculated for the 72 binary systems of interest. The interaction parameter L for the liquid was expressed as the sum of a positive internal pressure term e_p and an interaction parameter e_0. The former term, e_p, was computed from known

TABLE XII

COMPUTATION OF REGULAR-SOLUTION PARAMETERS L, B, E, AND A [a]

	e_0	e_p	e_1	e_2	e_3	e_4
Zr–Nb	0	4,297	2,637	0	0	0
Zr–Ta	0	6,299	2,621	0	0	0
Zr–Mo	−4,000	5,512	5,939	−900	2430	0
Zr–W	−4,000	13,109	6,216	−900	2430	0
Zr–Re	−30,000	12,702	8,466	−1900	5330	100
Zr–Ru	−62,000	8,644	10,414	−3700	6550	80
Zr–Os	−62,000	9,227	9,586	−3700	6550	80
Zr–Rh	−80,000	3,865	9,175	−4600	4530	100
Zr–Ir	−80,000	7,838	8,986	−4600	4530	100
Zr–Pd	−74,000	7	6,053	−4900	1730	−50
Zr–Pt	−74,000	2,776	6,405	−4900	1730	−50
Hf–Nb	0	3,219	1,918	0	0	0
Hf–Ta	0	4,948	1,891	0	0	0
Hf–Mo	−4,000	4,300	4,919	−900	2430	0
Hf–W	−4,000	11,072	5,086	−900	2430	0
Hf–Re	−30,000	10,736	7,233	−1900	5330	100
Hf–Ru	−62,000	7,089	9,154	−3700	6550	80
Hf–Os	−62,000	7,606	8,323	−3700	6550	80
Hf–Rh	−80,000	2,889	7,980	−4600	4530	100
Hf–Ir	−80,000	6,362	7,776	−4600	4530	100
Hf–Pd	−74,000	107	5,188	−4900	1730	−50
Hf–Pt	−74,000	1,959	5,390	−4900	1730	−50
Nb–Mo	0	97	834	0	0	0
Nb–W	0	2,213	707	0	0	0
Nb–Re	−26,000	2,133	1,813	−1000	2900	100
Nb–Ru	−44,000	822	3,190	−1800	1220	−20
Nb–Os	−44,000	977	2,618	−1800	1220	−20
Nb–Rh	−38,000	0	2,698	−2700	−800	0
Nb–Ir	−38,000	582	2,298	−2700	−800	0
Nb–Pd	−38,000	3,694	1,334	−2200	−1920	−30
Nb–Pt	−38,000	96	1,178	−2200	−1920	−30
Ta–Mo	0	6	951	0	0	0
Ta–W	0	1,205	821	0	0	0
Ta–Re	−26,000	1,162	1,995	−1000	2900	100
Ta–Ru	−44,000	280	3,446	−1800	1220	−20
Ta–Os	−44,000	371	2,858	−1800	1220	−20
Ta–Rh	−38,000	144	2,916	−2700	−800	0
Ta–Ir	−38,000	146	2,520	−2700	−800	0
Ta–Pd	−38,000	534	1,488	−2200	−1920	−30
Ta–Pt	−38,000	486	1,320	−2200	−1920	−30

TABLE XII (*continued*)

	e_0	e_p	e_1	e_2	e_3	e_4
Mo–Re	– 12,000	1,233	150	0	0	0
Mo–Ru	– 6,000	338	741	– 800	– 1680	– 120
Mo–Os	– 6,000	433	463	– 800	– 1680	– 120
Mo–Rh	– 6,000	85	548	– 900	– 2020	20
Mo–Ir	– 6,000	194	349	– 900	– 2020	20
Mo–Pd	– 6,000	4,609	109	– 400	– 3140	– 10
Mo–Pt	– 6,000	355	37	– 400	– 3140	– 10
W–Re	– 12,000	0	296	0	0	0
W–Ru	– 6,000	261	1,085	– 800	– 1680	– 120
W–Os	– 6,000	194	746	– 800	– 1680	– 120
W–Rh	– 6,000	1,939	846	– 900	– 2020	20
W–Ir	– 6,000	440	566	– 900	– 2020	20
W–Pd	– 6,000	10,706	224	– 400	– 3140	– 10
W–Pt	– 6,000	2,962	110	– 400	– 3140	– 10
Re–Ru	0	250	263	0	0	0
Re–Os	0	187	109	0	0	0
Re–Rh	0	1,859	164	– 100	– 340	140
Re–Ir	0	424	48	– 100	– 340	140
Re–Pd	0	10,300	0	500	– 1120	– 30
Re–Pt	0	2,851	28	500	– 1120	– 30
Ru–Rh	0	713	6	0	0	0
Ru–Ir	0	20	75	0	0	0
Ru–Pd	0	7,009	191	600	– 780	– 170
Ru–Pt	0	1,330	397	600	– 780	– 170
Os–Rh	0	848	8	0	0	0
Os–Ir	0	46	9	0	0	0
Os–Pd	0	7,494	79	600	– 780	– 170
Os–Pt	0	1,520	215	600	– 780	– 170
Rh–Pd	0	3,219	111	0	0	0
Rh–Pt	0	84	271	0	0	0
Ir–Pd	0	6,406	34	0	0	0
Ir–Pt	0	1,039	140	0	0	0

[a] In calories per gram atom.

values of the enthalpy of vaporization and the volume, modified slightly from the original method (*14, 71*) to reflect high-temperature effects. The second term, e_0, is computed on an empirical basis, which is quite explicit nevertheless. As a maximum, Fig. 46, which depicts $\phi_1[i]$ vs. i, contains four

TABLE XIII

REGULAR-SOLUTION PARAMETERS [a]

	L	B	E	A
Zr–Nb	+ 4,297	+ 6,934	+ 6,934	+ 6,934
Zr–Ta	+ 6,299	+ 8,920	+ 8,920	+ 8,920
Zr–Mo	+ 1,512	+ 6,551	+ 8,981	+ 8,981
Zr–W	+ 9,109	+ 14,425	+ 16,855	+ 16,855
Zr–Re	− 17,298	− 10,732	− 5,402	− 5,302
Zr–Ru	− 53,356	− 46,642	− 40,092	− 40,012
Zr–Os	− 52,773	− 46,887	− 40,337	− 40,257
Zr–Rh	− 76,135	− 71,560	− 67,030	− 66,930
Zr–Ir	− 72,162	− 67,776	− 63,246	− 63,146
Zr–Pd	− 73,993	− 72,840	− 71,110	− 71,160
Zr–Pt	− 71,224	− 69,719	− 67,989	− 68,039
Hf–Nb	+ 3,219	+ 5,137	+ 5,137	+ 5,137
Hf–Ta	+ 4,948	+ 6,839	+ 6,839	+ 6,839
Hf–Mo	+ 300	+ 4,319	+ 6,749	+ 6,749
Hf–W	+ 7,072	+ 11,258	+ 13,688	+ 13,688
Hf–Re	− 19,264	− 13,931	− 8,601	− 8,501
Hf–Ru	− 54,911	− 49,457	− 42,907	− 42,827
Hf–Os	− 54,394	− 49,771	− 43,221	− 43,141
Hf–Rh	− 77,111	− 73,731	− 69,201	− 69,101
Hf–Ir	− 73,638	− 70,462	− 65,932	− 65,832
Hf–Pd	− 73,893	− 73,605	− 71,875	− 71,925
Hf–Pt	− 72,041	− 71,551	− 69,821	− 69,871
Nb–Mo	+ 97	+ 931	+ 931	+ 931
Nb–W	+ 2,213	+ 2,920	+ 2,920	+ 2,920
Nb–Re	− 23,867	− 23,054	− 20,154	− 20,054
Nb–Ru	− 43,178	− 41,788	− 40,568	− 40,588
Nb–Os	− 43,023	− 42,205	− 40,985	− 41,005
Nb–Rh	− 38,000	− 40,002	− 38,802	− 38,802
Nb–Ir	− 37,418	− 37,820	− 38,620	− 38,620
Nb–Pd	− 34,306	− 35,172	− 37,092	− 37,122
Nb–Pt	− 37,904	− 38,926	− 40,846	− 40,876
Ta–Mo	+ 6	+ 957	+ 957	+ 957
Ta–W	+ 1,205	+ 2,026	+ 2,026	+ 2,026
Ta–Re	− 24,838	− 23,843	− 20,943	− 20,843
Ta–Ru	− 43,720	− 42,074	− 40,854	− 40,874
Ta–Os	− 43,629	− 42,571	− 41,351	− 41,371
Ta–Rh	− 37,856	− 37,640	− 38,440	− 38,440
Ta–Ir	− 37,854	− 38,034	− 38,834	− 38,834
Ta–Pd	− 37,466	− 38,178	− 40,098	− 40,128
Ta–Pt	− 37,514	− 38,394	− 40,314	− 40,344

TABLE XIII (*continued*)

	L	B	E	A
Mo–Re	− 10,767	− 10,617	− 10,617	− 10,617
Mo–Ru	− 5,662	− 5,721	− 7,401	− 7,521
Mo–Os	− 5,567	− 5,904	− 7,584	− 7,704
Mo–Rh	− 5,915	− 6,267	− 8,287	− 8,267
Mo–Ir	− 5,806	− 6,357	− 8,377	− 8,357
Mo–Pd	− 1,391	− 1,682	− 4,822	− 4,832
Mo–Pt	− 5,645	− 6,008	− 9,148	− 9,158
W–Re	− 12,000	− 11,704	− 11,704	− 11,704
W–Ru	− 5,739	− 5,454	− 7,134	− 7,254
W–Os	− 5,806	− 5,860	− 7,540	− 7,660
W–Rh	− 4,061	− 4,115	− 6,135	− 6,115
W–Ir	− 5,560	− 5,894	− 7,914	− 7,894
W–Pd	+ 4,706	+ 4,530	+ 1,390	+ 1,380
W–Pt	− 3,038	− 3,328	− 6,468	− 6,478
Re–Ru	+ 250	+ 513	+ 513	+ 513
Re–Os	+ 187	+ 296	+ 296	+ 296
Re–Rh	+ 1,859	+ 1,923	+ 1,583	+ 1,723
Re–Ir	+ 424	+ 372	+ 32	+ 172
Re–Pd	+ 10,300	+ 10,800	+ 9,680	+ 9,650
Re–Pt	+ 2,851	+ 3,379	+ 2,259	+ 2,229
Ru–Rh	+ 713	+ 719	+ 719	+ 719
Ru–Ir	+ 20	+ 95	+ 95	+ 95
Ru–Pd	+ 7,009	+ 7,800	+ 7,020	+ 6,850
Ru–Pt	+ 1,330	+ 2,327	+ 1,547	+ 1,377
Os–Rh	+ 848	+ 856	+ 856	+ 856
Os–Ir	+ 46	+ 55	+ 55	+ 55
Os–Pd	+ 7,494	+ 8,173	+ 7,393	+ 7,223
Os–Pt	+ 1,520	+ 2,335	+ 1,555	+ 1,385
Rh–Pd	+ 3,219	+ 3,330	+ 3,330	+ 3,330
Rh–Pt	+ 84	+ 355	+ 355	+ 355
Ir–Pd	+ 6,406	+ 6,440	+ 6,440	+ 6,440
Ir–Pt	+ 1,039	+ 1,179	+ 1,179	+ 1,179

[a] In calories per gram atom.

adjustable parameters corresponding to $\phi_1[i]$ at $i = 5.0$, 6.5, 8.0, and 9.0. With these points, 72 values of L are computed.

The interaction parameters for the solid phases were estimated by computing a size effect term e_1 involving one adjustable parameter and adding a second component e_2 that involved no adjustments. Thus, one additional adjustable parameter [the numerical coefficient of Eq. (142)] generated 72 additional values of B. The remaining 144 values of E and A required no additional adjustable parameters.

CHAPTER V

Computation of Binary Phase Diagrams of Refractory Metals

The regular-solution phase diagrams for the 72 binary systems of interest were computed using the lattice stability values given in Table VIII and the interaction parameters contained in Table XIII. As indicated earlier, hand calculations of all the relevant equilibria would be nearly impossible. However, development of suitable computer programs for simultaneous solution of " two regular-solution equilibria " [i.e., Eqs. (82) and (83)] called TRSE permits these computations to be performed quite readily. Appendix 1 contains the details of the TRSE program.* The T_0–x curves [defined by Eqs. (86) and (87)], two-phase equilibria compositional limits [defined by Eqs. (82) and (83)], and the miscibility-gap conditions [defined by Eqs. (74) and (75)] were all computed numerically on an IBM-7094. This output was fed on tape into an X–Y plotter, which graphically displayed all of the required phase boundaries. Only the eutectic (-oid), peritectic (-oid), and monotectoid isotherms were inserted manually at appropriate intersections. The results are given in Figs. 51–89, which show the computed phase diagrams (thick lines) and their experimental counterparts (thin lines). As before, compounds occurring in the experimental diagrams are cross hatched, and the reference sources for the experimental diagrams are noted in the figure legends; the referencing sequence is clockwise, starting at the upper left corner. In addition, the computed T_0–x diagram, which defines the stability range of L, β, ϵ, and α phases, was computed as the first step. These diagrams are shown opposite the phase diagrams along with the input parameters.

* The authors are indebted to R. Serbagi of Digital Programming Service, Inc., Waltham, Massachusetts for his assistance in these computations. Appendix 1 contains a statement of the program.

1. COMPUTATION OF REGULAR-SOLUTION T_0–x DIAGRAMS

As a start, it is of interest to focus attention on the T_0–x diagrams, since they reflect the values of the lattice stability parameters and the differences $(L - B), (L - E), (L - A), (B - E), (E - A)$, and $(A - B)$ [see Eqs. (86) and (87)]. Thus, the T_0–x curves do not depend at all on e_0, or the internal pressure term e_p: rather, the T_0–x curves depend only on the " strain energy " or size effect term, e_1 and the energies, e_2, e_3, and e_4, derived from the enthalpy differences between the L, β, ϵ, and α forms of the pure metals.

Figure 51 shows the T_0–x diagram for the (Zr/Hf) vs. (Nb/Ta) group. The β/L curve is depressed below linear behavior and the β phase is stabilized with respect to the ϵ phase in agreement with experiment. Similar results are seen in Fig. 53 for the (Zr/Hf) vs. (Mo/W) cases. Here the β/L depression is even more severe owing to a larger value of the strain energy parameter e_1 due to a larger size difference. These results agree with the experimental phase diagrams shown in Fig. 54. The T_0–x curves for (Zr/Hf) vs. (Ru/Os) cases (Fig. 57) show that β is stabilized at the expense of ϵ as (Ru/Os) is added to (Zr/Hf). Very large depressions of the β/L and ϵ/L curves occur due to the large size differences between partners. Figure 59, which illustrates the T_0–x behavior for the (Zr/Hf) vs. (Rh/Ir) quartet shows behavior similar to the (Zr/Hf) vs. (Ru/Os) set, except that the e_2 term (see Table XII) has grown to a size where it partially offsets the large strain energy value. Finally, in the (Zr/Hf) vs. (Pd/Pt) cases, Fig. 61, smaller β/L, ϵ/L, and α/L depressions (relative to linear behavior) are noted and the $B - E$ term is no longer sufficient to stabilize the β phase at the expense of the ϵ phase. *This is the only deficiency encountered in the entire sequence.*

The (Nb/Ta) sequence starts with the (Mo/W) quartet in Fig. 63, which is almost trivial. Figure 65, showing (Nb/Ta) vs. (Ru/Os), exhibits β/L and ϵ/L depressions (relative to linear behavior) that are much smaller than the (Zr/Hf) depressions noted earlier. This is simply due to the fact that size differences are much smaller. In addition, B is less than E as anticipated from the earlier comparison of the ideal-solution computation with the observed phase diagrams shown in Fig. 34.

The T_0–x curves for the (Nb/Ta) vs. (Rh/Ir) set shown in Fig. 67 illustrate almost linear solid/liquid T_0–x curves due to the fact that the strain energy term is compensated for by the e_2 term (see Table XII) and $(L - E)$ and $(L - A)$ actually become slightly negative. Finally, the (Nb/Ta) vs. (Pd/Pt) set (Fig. 69) actually show maxima and near maxima in the α/L trajectories. This is in agreement with the observations as shown in Fig. 70.

The (Mo/W) sequence begins with Fig. 71 showing the T_0–x behavior for the (Ru/Os) quartet. Size differences and strain energies are much smaller than in the (Zr/Hf) and (Nb/Ta) series discussed earlier (see Tables XI and

(text continues on page 122)

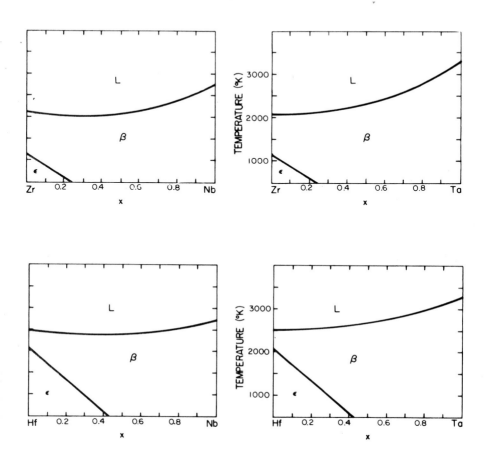

	$\Delta F^{\beta \to L}$	$\Delta F^{\epsilon \to L}$ (cal/g-atom)	$\Delta F^{\alpha \to L}$		L	B	E (cal/g-atom)	A
Zr	2.0(2125 − T)	2.9(1820 − T)	2.9(1544 − T)	Zr–Nb	4297	6934	6934	6934
Hf	2.0(2495 − T)	2.9(2351 − T)	2.9(2076 − T)	Zr–Ta	6299	8920	8920	8920
Nb	2.0(2740 − T)	2.8(1420 − T)	2.85(1170 − T)	Hf–Nb	3219	5137	5137	5137
Ta	2.0(3270 − T)	2.8(1800 − T)	2.85(1540 − T)	Hf–Ta	4948	6839	6839	6839

FIG. 51. Regular-solution computed T_0 vs. x diagrams.

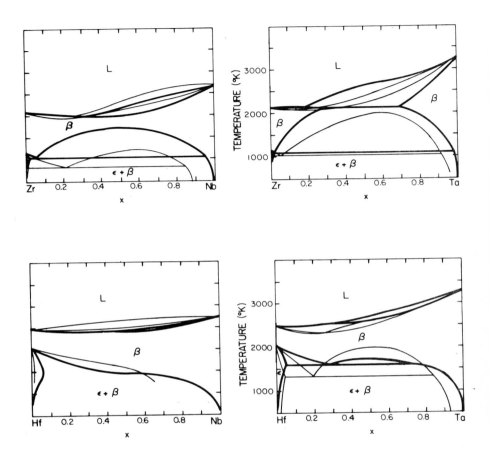

FIG. 52. Comparison of regular-solution computed and experimental phase diagrams (3), (3), (3), (52). Thick lines are computed and thin lines are observed.

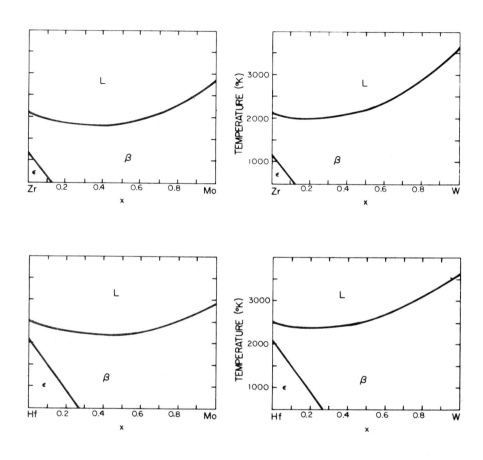

	$\Delta F^{\beta \to L}$	$\Delta F^{\epsilon \to L}$ (cal/g-atom)	$\Delta F^{\alpha \to L}$		L	B	E (cal/g-atom)	A
Zr	2.0(2125 − T)	2.9(1820 − T)	2.9(1544 − T)	Zr–Mo	1,512	6,551	8,981	8,981
Hf	2.0(2495 − T)	2.9(2351 − T)	2.9(2076 − T)	Zr–W	9,109	14,425	16,855	16,855
Mo	2.0(2900 − T)	2.0(1900 − T)	2.15(1530 − T)	Hf–Mo	300	4,319	6,749	6,749
W	2.0(3650 − T)	2.0(1900 − T)	2.15(2230 − T)	Hf–W	7,072	11,258	13,688	13,688

FIG. 53. Regular-solution computed T_0 vs. x diagrams.

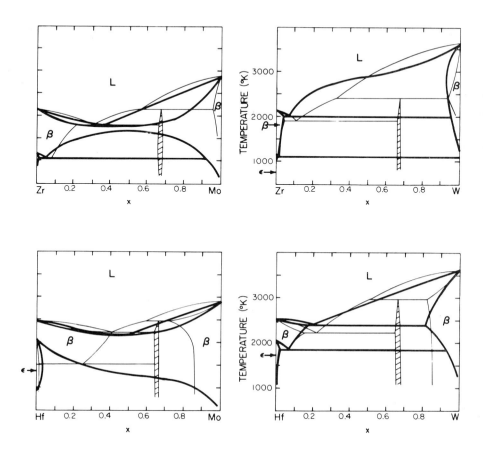

FIG. 54. Comparison of regular-solution computed and experimental phase diagrams (2), (2), (3), (3). Thick lines are computed and thin lines are observed.

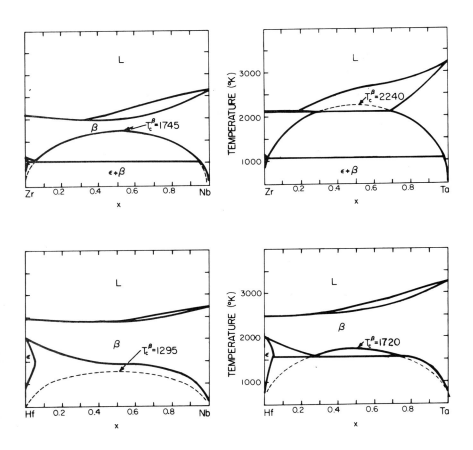

	$\Delta F^{\beta \to L}$	$\Delta F^{\epsilon \to L}$	$\Delta F^{\alpha \to L}$		L	B	E	A
		(cal/g-atom)				(cal/g-atom)		
Zr	2.0(2125 − T)	2.9(1820 − T)	2.9(1544 − T)	Zr–Nb	4297	6934	6934	6934
Hf	2.0(2495 − T)	2.9(2351 − T)	2.9(2076 − T)	Zr–Ta	6299	8920	8920	8920
Nb	2.0(2740 − T)	2.8(1420 − T)	2.85(1170 − T)	Hf–Nb	3219	5137	5137	5137
Ta	2.0(3270 − T)	2.8(1800 − T)	2.85(1540 − T)	Hf–Ta	4948	6839	6839	6839

FIG. 55. Illustration of regular-solution phase diagram computation and the intrusion of miscibility gaps.

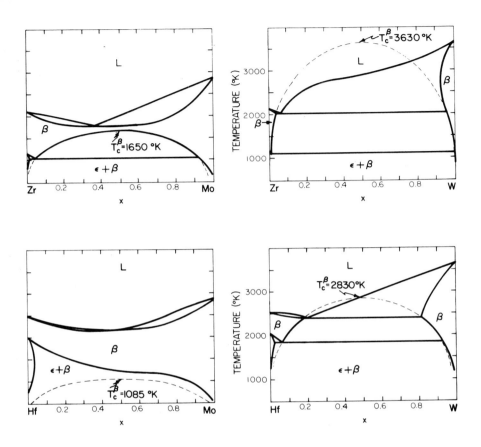

	$\Delta F^{\beta \to L}$	$\Delta F^{\epsilon \to L}$	$\Delta F^{\alpha \to L}$			L	B	E	A
		(cal/g-atom)					(cal/g-atom)		
Zr	2.0(2125 − T)	2.9(1820 − T)	2.9(1544 − T)		Zr–Mo	1,512	6,551	8,981	8,981
Hf	2.0(2495 − T)	2.9(2351 − T)	2.9(2076 − T)		Zr–W	9,109	14,425	16,855	16,855
Mo	2.0(2900 − T)	2.0(1900 − T)	2.15(1530 − T)		Hf–Mo	300	4,319	6,749	6,749
W	2.0(3650 − T)	2.0(1900 − T)	2.15(2230 − T)		Hf–W	7,072	11,258	13,688	13,688

FIG. 56. Illustration of regular-solution phase diagram computation and intrusion of miscibility gaps.

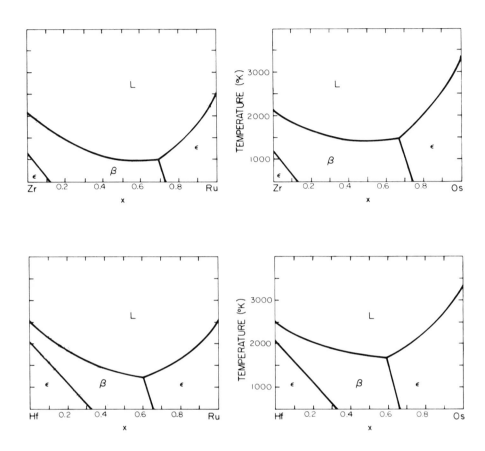

	$\Delta F^{\beta \to L}$	$\Delta F^{\epsilon \to L}$	$\Delta F^{\alpha \to L}$			L	B	E	A
		(cal/g-atom)						(cal/g-atom)	
Zr	2.0(2125−T)	2.9(1820−T)	2.9(1544−T)		Zr–Ru	−53,356	−46,642	−40,092	−40,012
Hf	2.0(2495−T)	2.9(2351−T)	2.9(2076−T)		Zr–Os	−52,773	−46,887	−40,337	−40,257
Ru	2.8(1420−T)	2.0(2550−T)	2.8(1780−T)		Hf–Ru	−54,911	−49,457	−42,907	−42,827
Os	2.8(1960−T)	2.0(3300−T)	2.8(2310−T)		Hf–Os	−54,394	−49,771	−43,221	−43,141

FIG. 57. Regular-solution computed T_0 vs. x diagrams.

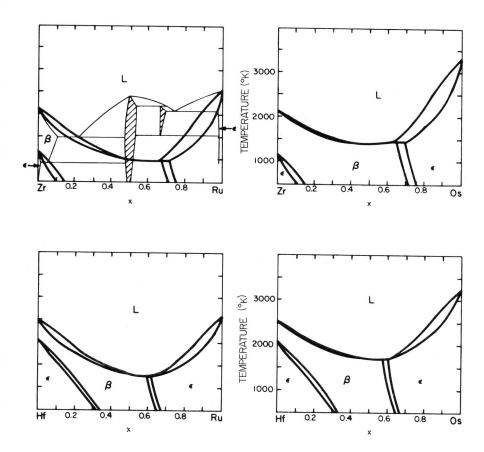

FIG. 58. Regular-solution computed phase diagrams (*84*). Thick lines are computed and thin lines are observed.

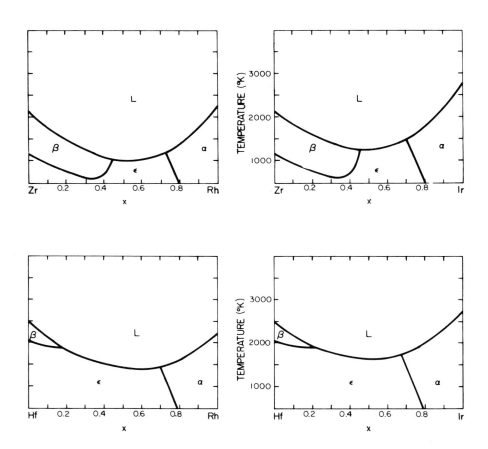

| | $\Delta F^{\beta \to L}$ | $\Delta F^{\epsilon \to L}$ | $\Delta F^{\alpha \to L}$ | | L | B | E | A |
		(cal/g-atom)					(cal/g-atom)	
Zr	$2.0(2125-T)$	$2.9(1820-T)$	$2.9(1544-T)$	Zr–Rh	$-76,135$	$-71,560$	$-67,030$	$-66,930$
Hf	$2.0(2495-T)$	$2.9(2351-T)$	$2.9(2076-T)$	Zr–Ir	$-72,162$	$-67,776$	$-63,246$	$-63,146$
Rh	$3.05(930-T)$	$2.15(2010-T)$	$2.0(2240-T)$	Hf–Rh	$-77,111$	$-73,731$	$-69,201$	$-69,101$
Ir	$3.05(1260-T)$	$2.15(2490-T)$	$2.0(2740-T)$	Hf–Ir	$-73,638$	$-70,462$	$-65,932$	$-65,832$

FIG. 59. Regular-solution computed T_0 vs. x diagrams.

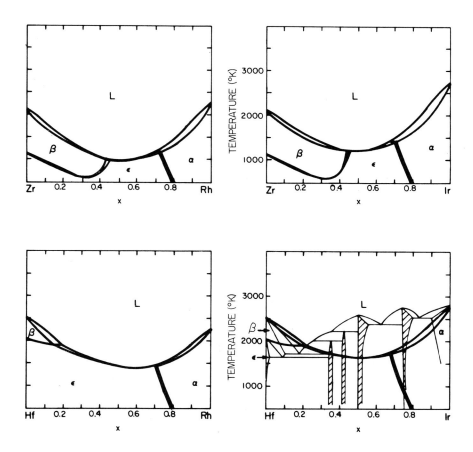

FIG. 60. Regular-solution computed phase diagrams (85). Thick lines are computed and thin lines are observed.

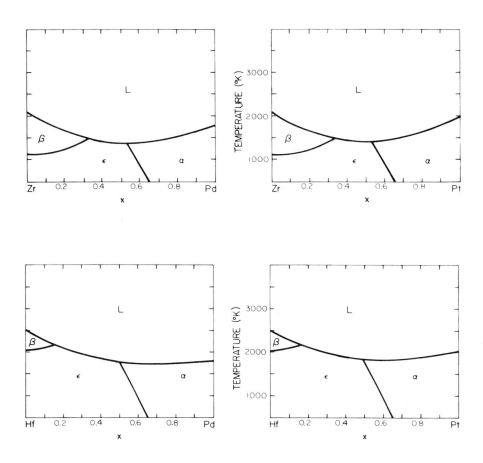

| | $\Delta F^{\beta \to L}$ | $\Delta F^{\epsilon \to L}$ | $\Delta F^{\alpha \to L}$ | | L | B | E | A |
		(cal/g-atom)					(cal/g-atom)	
Zr	2.0(2125−T)	2.9(1820−T)	2.9(1544−T)	Zr–Pd	−73,993	−72,840	−71,110	−71,160
Hf	2.0(2495−T)	2.9(2351−T)	2.9(2076−T)	Zr–Pt	−71,224	−69,719	−67,989	−68,039
Pd	2.8(820−T)	2.3(1470−T)	2.0(1820−T)	Hf–Pd	−73,893	−73,605	−71,875	−71,925
Pt	2.8(980−T)	2.3(1670−T)	2.0(2040−T)	Hf–Pt	−72,041	−71,551	−69,821	−69,871

FIG. 61. Regular-solution computed T_0 vs. x diagrams.

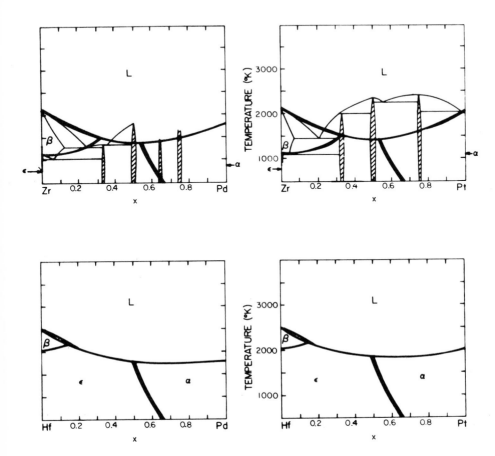

FIG. 62. Comparison of regular-solution computed and experimental phase diagrams
(3), (3). Thick lines are computed and thin lines are observed.

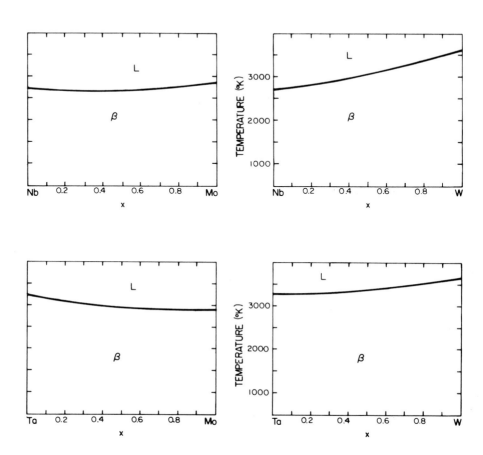

	$\Delta F^{\beta \to L}$	$\Delta F^{\epsilon \to L}$ (cal/g-atom)	$\Delta F^{\alpha \to L}$		L	B	E (cal/g-atom)	A
Nb	2.0(2740 − T)	2.8(1420 − T)	2.85(1170 − T)	Nb–Mo	97	931	931	931
Ta	2.0(3270 − T)	2.8(1800 − T)	2.85(1540 − T)	Nb–W	2213	2920	2920	2920
Mo	2.0(2900 − T)	2.0(1900 − T)	2.15(1530 − T)	Ta–Mo	6	957	957	957
W	2.0(3650 − T)	2.0(2650 − T)	2.15(2230 − T)	Ta–W	1205	2026	2026	2026

FIG. 63. Regular-solution computed T_0 vs. x diagrams.

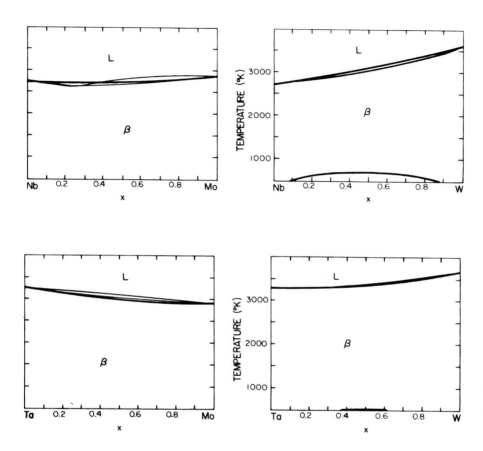

FIG. 64. Comparison of regular-solution computed and experimental phase diagrams (3), (2). Thick lines are computed and thin lines are observed.

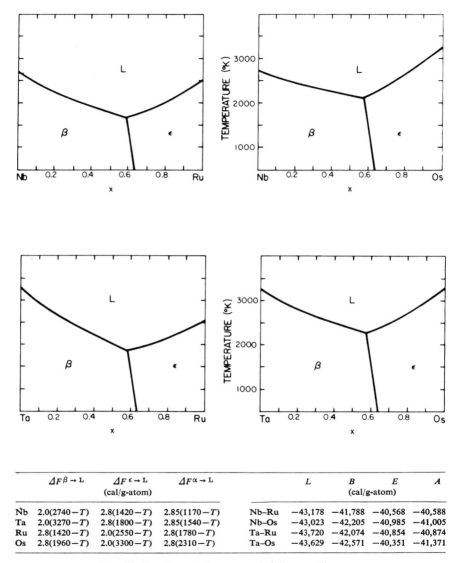

	$\Delta F^{\beta \to L}$	$\Delta F^{\epsilon \to L}$	$\Delta F^{\alpha \to L}$			L	B	E	A
		(cal/g-atom)						(cal/g-atom)	
Nb	2.0(2740−T)	2.8(1420−T)	2.85(1170−T)		Nb–Ru	−43,178	−41,788	−40,568	−40,588
Ta	2.0(3270−T)	2.8(1800−T)	2.85(1540−T)		Nb–Os	−43,023	−42,205	−40,985	−41,005
Ru	2.8(1420−T)	2.0(2550−T)	2.8(1780−T)		Ta–Ru	−43,720	−42,074	−40,854	−40,874
Os	2.8(1960−T)	2.0(3300−T)	2.8(2310−T)		Ta–Os	−43,629	−42,571	−40,351	−41,371

FIG. 65. Regular-solution computed T_0 vs. x diagrams.

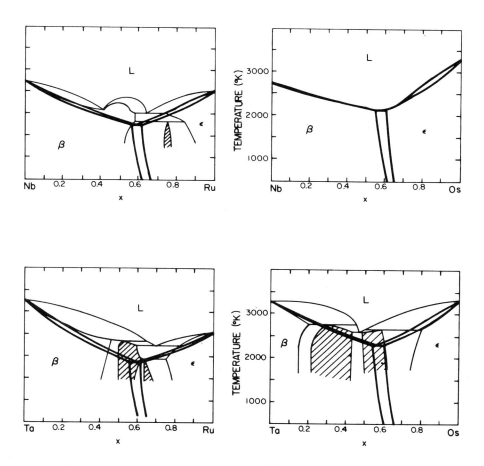

FIG. 66. Comparison of regular-solution computed and experimental phase diagrams (57), (3), (3). Thick lines are computed and thin lines are observed.

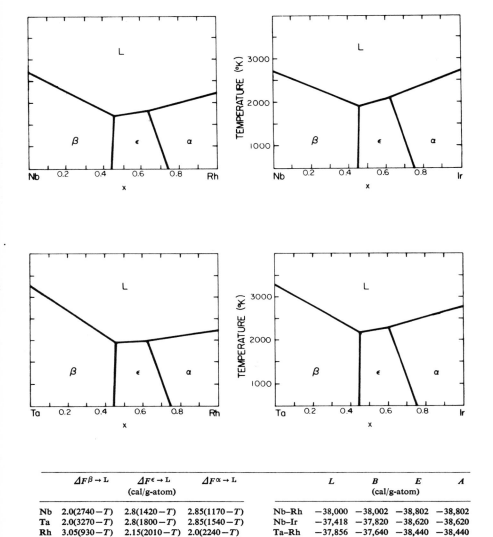

	$\Delta F^{\beta \to L}$	$\Delta F^{\epsilon \to L}$	$\Delta F^{\alpha \to L}$			L	B	E	A
		(cal/g-atom)					(cal/g-atom)		
Nb	2.0(2740−T)	2.8(1420−T)	2.85(1170−T)		Nb–Rh	−38,000	−38,002	−38,802	−38,802
Ta	2.0(3270−T)	2.8(1800−T)	2.85(1540−T)		Nb–Ir	−37,418	−37,820	−38,620	−38,620
Rh	3.05(930−T)	2.15(2010−T)	2.0(2240−T)		Ta–Rh	−37,856	−37,640	−38,440	−38,440
Ir	3.05(1260−T)	2.15(2490−T)	2.0(2750−T)		Ta–Ir	−37,954	−38,034	−38,834	−38,834

FIG. 67. Regular-solution computed T_0 vs. x diagrams.

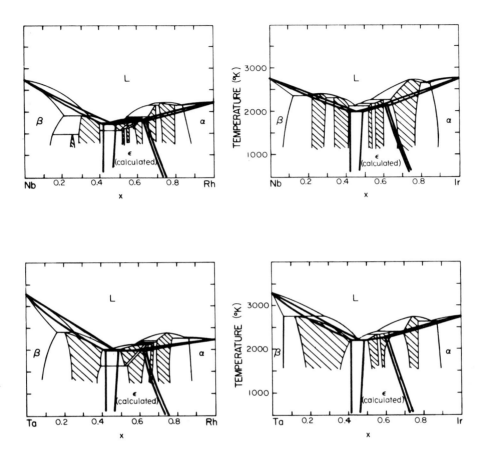

FIG. 68. Comparison of regular-solution computed and experimental phase diagrams (86), (87), (88), (89). Thick lines are computed and thin lines are observed.

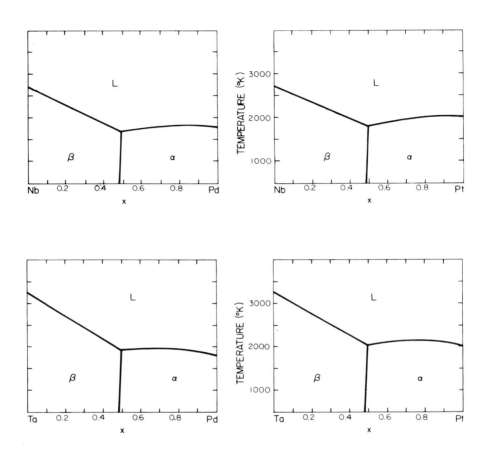

	$\Delta F^{\beta \rightarrow L}$	$\Delta F^{\epsilon \rightarrow L}$ (cal/g-atom)	$\Delta F^{\alpha \rightarrow L}$			L	B	E (cal/g-atom)	A
Nb	2.0(2740−T)	2.8(1420−T)	2.85(1170−T)		Nb–Pd	−34,306	−35,172	−37,092	−37,122
Ta	2.0(3270−T)	2.8(1800−T)	2.85(1540−T)		Nb–Pt	−37,904	−38,926	−40,846	−40,876
Pd	2.8(820−T)	2.3(1470−T)	2.0(1820−T)		Ta–Pd	−37,466	−38,178	−40,098	−40,128
Pt	2.8(980−T)	2.3(1670−T)	2.0(2040−T)		Ta–Pt	−37,514	−38,394	−40,314	−40,344

FIG. 69. Regular-solution computed T_0 vs. x diagrams.

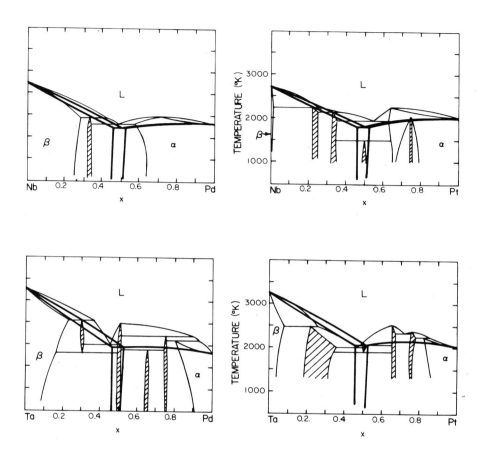

FIG. 70. Comparison of regular-solution computed and experimental phase diagrams (3), (3), (79), (90). Thick lines are computed and thin lines are observed.

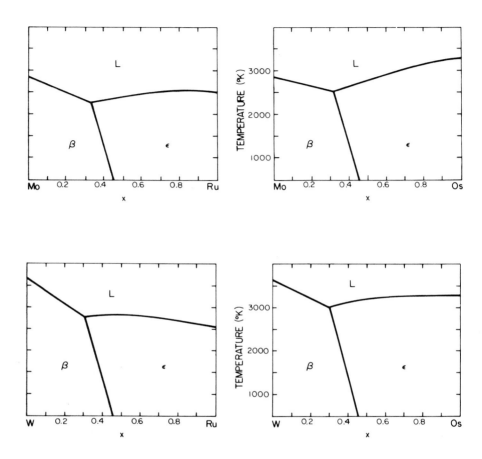

	$\Delta F^{\beta \rightarrow L}$	$\Delta F^{\epsilon \rightarrow L}$	$\Delta F^{\alpha \rightarrow L}$		L	B	E	A
		(cal/g-atom)				(cal/g-atom)		
Mo	2.0(2900−T)	2.0(1900−T)	2.15(1530−T)	Mo–Ru	−5662	−5721	−7401	−7521
W	2.0(3650−T)	2.0(2650−T)	2.15(2230−T)	Mo–Os	−5567	−5904	−7584	−7704
Ru	2.8(1420−T)	2.0(2550−T)	2.8(1780−T)	W–Ru	−5739	−5454	−7134	−7254
Os	2.8(1960−T)	2.0(3300−T)	2.8(2310−T)	W–Os	−5806	−5860	−7540	−7660

FIG. 71. Regular-solution computed T_0 vs. x diagrams.

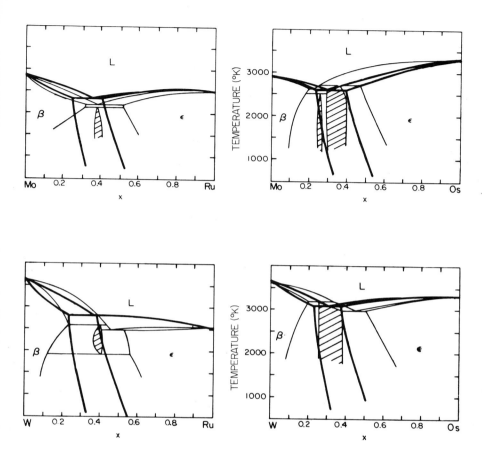

FIG. 72. Comparison of regular-solution computed and experimental phase diagrams (3), (59), (3), (60). Thick lines are computed and thin lines are observed.

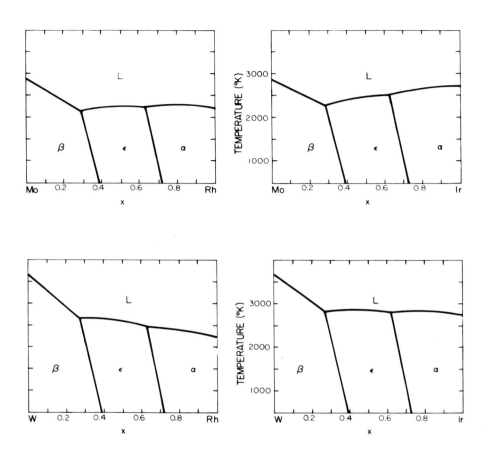

	$\Delta F^{\beta \to L}$	$\Delta F^{\epsilon \to L}$	$\Delta F^{\alpha \to L}$			L	B	E	A
		(cal/g-atom)					(cal/g-atom)		
Mo	2.0(2900−T)	2.0(1900−T)	2.15(1530−T)		Mo–Rh	−5915	−6267	−8287	−8267
W	2.0(3650−T)	2.0(2650−T)	2.15(2230−T)		Mo–Ir	−5806	−6357	−8377	−8357
Rh	3.05(930−T)	2.15(2010−T)	2.0(2240−T)		W–Rh	−4061	−4115	−6135	−6115
Ir	3.05(1260−T)	2.15(2490−T)	2.0(2750−T)		W–Ir	−5560	−5894	−7914	−7894

FIG. 73. Regular-solution computed T_0 vs. x diagrams.

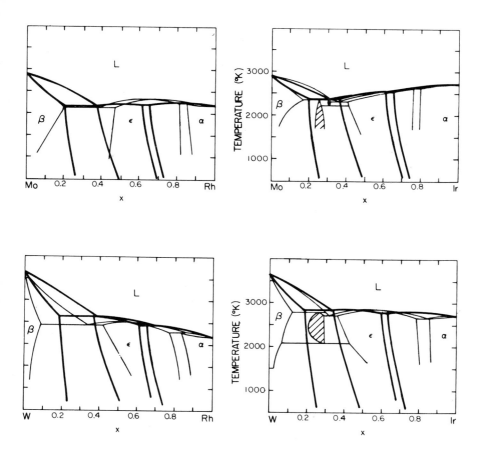

FIG. 74. Comparison of regular-solution computed and experimental phase diagrams (3), (53), (54), (54). Thick lines are computed and thin lines are observed.

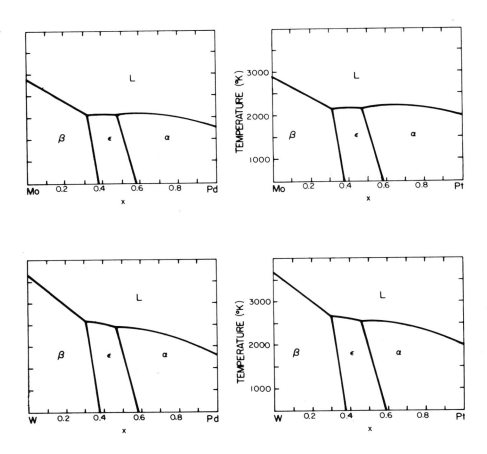

	$\Delta F^{\beta \to \text{L}}$	$\Delta F^{\epsilon \to \text{L}}$ (cal/g-atom)	$\Delta F^{\alpha \to \text{L}}$		L	B (cal/g-atom)	E	A
Mo	2.0(2900 − T)	2.0(1900 − T)	2.15(1530 − T)	Mo–Pd	−1391	−1682	−4822	−4832
W	2.0(3650 − T)	2.0(2650 − T)	2.15(2230 − T)	Mo–Pt	−5645	−6008	−9148	−9158
Pd	2.8(820 − T)	2.3(1470 − T)	2.0(1820 − T)	W–Pd	+4706	+4530	+1390	+1380
Pt	2.8(980 − T)	2.3(1670 − T)	2.0(2040 − T)	W–Pt	−3038	−3328	−6468	−6478

FIG. 75. Regular-solution computed T_0 vs. x diagrams.

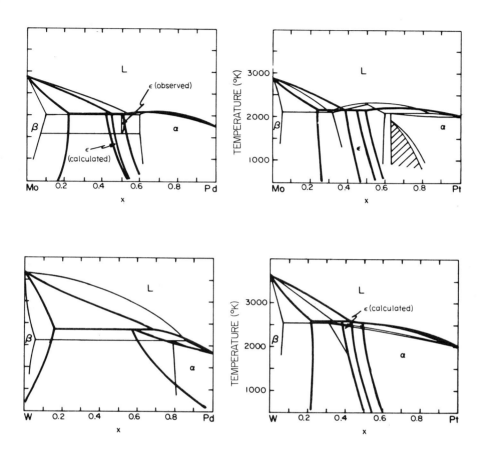

FIG. 76. Comparison of regular-solution computed and experimental phase diagrams (61), (3), (2), (62). Thick lines are computed and thin lines are observed.

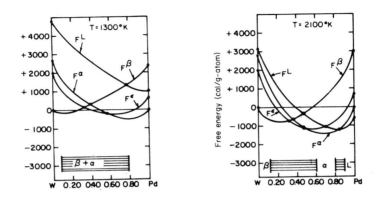

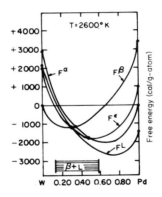

FIG. 77. Computed free energy–composition curves for liquid (L), bcc (β), hcp (ϵ), and fcc (α) phases at various temperatures in the tungsten–palladium system. Phase boundaries are indicated at selected temperatures. Figures 76 and 78 show the complete result.

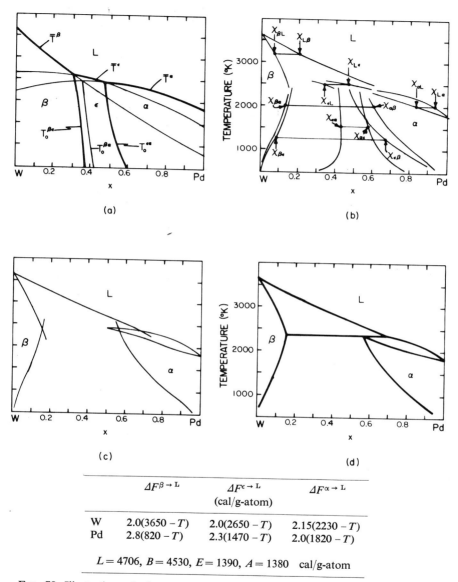

FIG. 78. Illustration of phase competition in the regular-solution computed phase diagram for the W–Pd system. Sequence (a)–(d) shows the series of steps employed in calculation of the phase diagram.

XII). As a consequence, the ϵ/L trajectory lies above the line connecting $\overline{T}_i^{\,\epsilon}$ and $\overline{T}_j^{\,\epsilon}$. Moreover, E is more negative than B as anticipated earlier from the comparison shown in Fig. 36. The (Mo/W) vs. (Rh/Ir) set (Fig. 73) shows maxima in the $\overline{T}^{\epsilon}\!-\!x$ curves for ϵ/L and $E < A < B < L$ as anticipated from Fig. 37. This behavior is repeated in the (Mo/W) vs. (Pd/Pt) grouping in Fig. 75, where E and A are nearly equal but both are less than L and B. The binary diagrams containing rhenium are shown in Figs. 79–84. As indicated earlier, all of the parameters required to compute technicium base diagrams are known, but few data are available for comparison. The $T_0\!-\!x$ curves for Zr–Re and Hf–Re in Fig. 79 display minima in the β/L and ϵ/L trajectories due to size effects and β stabilized at the expense of ϵ. The (Nb/Ta) vs. Re diagrams in Fig. 80 display nearly linear β/L trajectories owing to much smaller size differences than in the (Zr/Hf) vs. Re cases. Moreover, $B < E$ as anticipated from the earlier discussion of Fig. 35. The (Mo/W) vs. Re binaries given in Fig. 81 illustrate very small differences between the interaction parameters. Equally small differences are exhibited by (Ru/Os), (Rh/Ir), and (Pd/Pt) vs. Re in Figs. 82, 83, and 84. The $T_0\!-\!x$ curves for the (Ru/Os) vs. (Rh/Ir) set in Fig. 85 do not show $A < E$ as anticipated earlier. *This is a second deficiency of the analysis.* The $T_0\!-\!x$ curves for the (Ru/Os) vs. (Pd/Pt) set in Fig. 87 shows that L, E, and A are all nearly equal and are much less than B. The final $T_0\!-\!x$ set for (Rh/Ir) vs. (Pd/Pt) in Fig. 89 is trivial.

Thus, examination of the *differences* between L, B, E, and A for all of the foregoing sets (20 in all) discloses only two discrepancies out of a total of 120. These are the relative values of B and E in the (Zr/Hf) vs. (Pd/Pt) case (Fig. 61) and the relative values of E and A in the (Ru/Os) vs. (Rh/Ir) case (Fig. 85). In all other instances the differences $(L - B)$, $(L - E)$, $(L - A)$, $(B - E)$, $(E - A)$, and $(B - A)$ yield proper $T_0\!-\!x$ trajectories.

2. Computation of Regular-Solution Phase Diagrams

The $T_0\!-\!x$ computation is the first step in computing the phase diagram, since it discloses the temperature and composition range where the subject phases are stable. It also designates where one or more phases will be unstable. Thus, Fig. 51 indicates that the α phase need not be considered in computing the (Zr/Hf) vs. (Nb/Ta) phase diagrams. Consequently, the next step in the analysis is to calculate the two-phase boundaries associated with each $T_0\!-\!x$ curve by employing Eqs. (82) and (83). This procedure reflects the individual values of L, B, E, and A and not just the differences in parameters as in the $T_0\!-\!x$ case. Hence, the influence of e_0 and the internal pressure term e_p appears for the first time.

The (Zr/Hf) vs. (Nb/Ta) set shows positive values of L, B, E, and A due

(text continues on page 129)

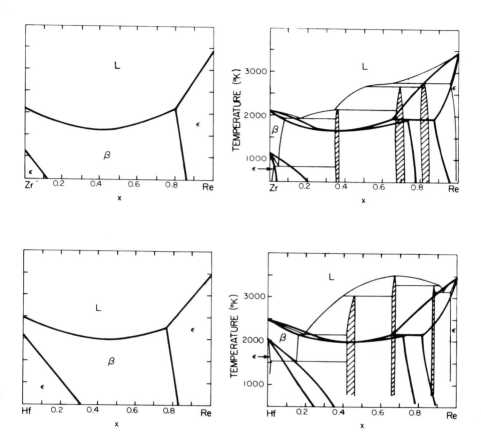

	$\Delta F^{\beta \to L}$	$\Delta F^{\epsilon \to L}$	$\Delta F^{\alpha \to L}$		L	B	E	A
		(cal/g-atom)				(cal/g-atom)		
Zr	2.0(2125−T)	2.9(1820−T)	2.9(1544−T)	Zr–Re	−17,298	−10,732	−5402	−5302
Hf	2.0(2495−T)	2.9(2351−T)	2.9(2076−T)	Hf–Re	−19,264	−13,931	−8601	−8501
Re	2.4(2710−T)	2.0(3450−T)	2.3(2890−T)					

FIG. 79. Regular-solution T_0 vs. x diagrams and comparison of regular-solution computed and experimental phase diagrams (3), (3). Thick lines are computed and thin lines are observed.

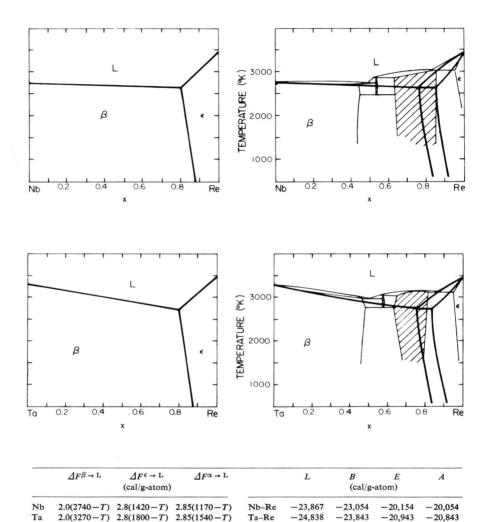

	$\Delta F^{\beta \rightarrow L}$	$\Delta F^{\epsilon \rightarrow L}$	$\Delta F^{\alpha \rightarrow L}$			L	B	E	A
		(cal/g-atom)					(cal/g-atom)		
Nb	$2.0(2740-T)$	$2.8(1420-T)$	$2.85(1170-T)$		Nb–Re	$-23,867$	$-23,054$	$-20,154$	$-20,054$
Ta	$2.0(3270-T)$	$2.8(1800-T)$	$2.85(1540-T)$		Ta–Re	$-24,838$	$-23,843$	$-20,943$	$-20,843$
Re	$2.4(2710-T)$	$2.0(3450-T)$	$2.3(2890-T)$						

FIG. 80. Regular-solution T_0 vs. x diagrams and comparison of regular-solution computed and experimental phase diagrams (3), (3). Thick lines are computed and thin lines are observed.

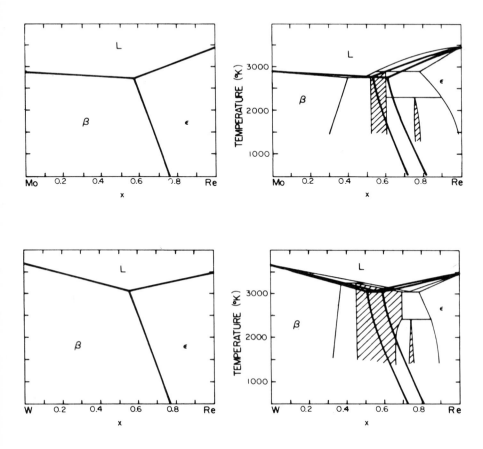

	$\Delta F^{\beta \rightarrow L}$	$\Delta F^{\epsilon \rightarrow L}$	$\Delta F^{\alpha \rightarrow L}$		L	B	E	A
		(cal/g-atom)				(cal/g-atom)		
Mo	2.0(2900−T)	2.0(1900−T)	2.15(1530−T)	Mo–Re	−10,767	−10,617	−10,617	−10,617
W	2.0(3650−T)	2.0(2650−T)	2.15(2230−T)	W–Re	−12,000	−11,704	−11,704	−11,704
Re	2.4(2710−T)	2.0(3450−T)	2.3(2890−T)					

FIG. 81. Regular-solution T_0 vs. x diagrams and comparison of regular-solution computed and experimental phase diagrams (58), (3). Thick lines are computed and thin lines are observed.

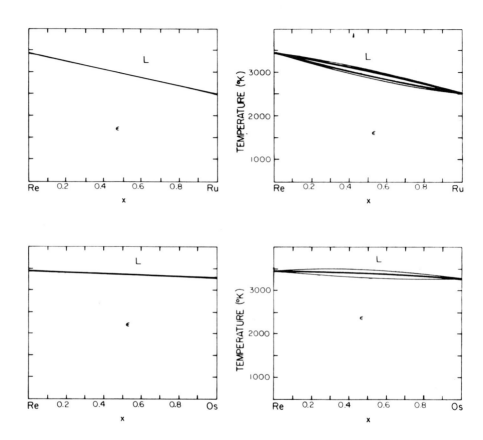

	$\Delta F^{\beta \rightarrow L}$	$\Delta F^{\epsilon \rightarrow L}$ (cal/g-atom)	$\Delta F^{\alpha \rightarrow L}$			L	B	E (cal/g-atom)	A
Re	$2.4(2710-T)$	$2.0(3450-T)$	$2.3(2890-T)$		Re–Ru	250	513	513	513
Ru	$2.8(1420-T)$	$2.0(2550-T)$	$2.8(1780-T)$		Re–Os	187	296	296	296
Os	$2.8(1960-T)$	$2.0(3300-T)$	$2.8(2310-T)$						

FIG. 82. Regular-solution T_0 vs. x diagrams and comparison of regular-solution computed and experimental phase diagrams (91), (91). Thick lines are computed and thin lines are observed.

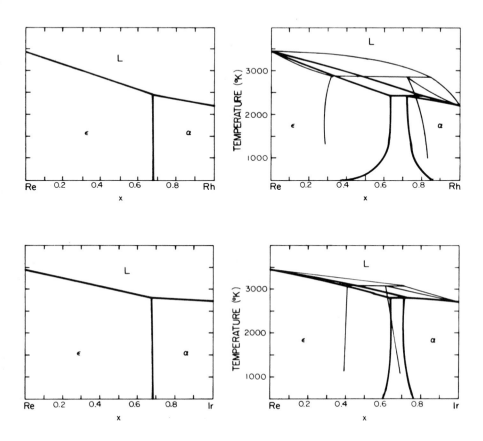

	$\Delta F^{\beta \to L}$	$\Delta F^{\epsilon \to L}$	$\Delta F^{\alpha \to L}$		L	B	E	A
		(cal/g-atom)				(cal/g-atom)		
Re	$2.4(2710-T)$	$2.0(3450-T)$	$2.3(2890-T)$	Re–Rh	1859	1923	1583	1723
Rh	$3.05(930-T)$	$2.15(2010-T)$	$2.0(2240-T)$	Re–Ir	424	372	32	172
Ir	$3.05(1260-T)$	$2.15(2490-T)$	$2.0(2750-T)$					

FIG. 83. Regular-solution T_0 vs. x diagrams and comparison of regular-solution computed and experimental phase diagrams (63), (63). Thick lines are computed and thin lines are observed.

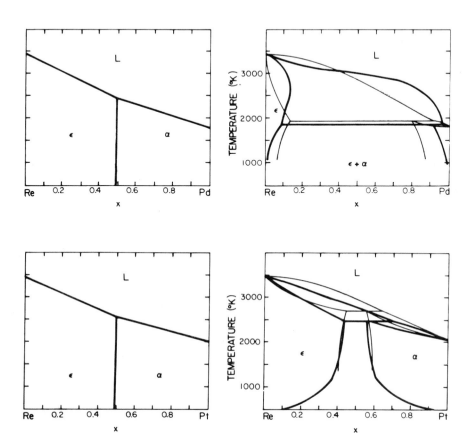

	$\Delta F^{\beta \to L}$	$\Delta F^{\epsilon \to L}$ (cal/g-atom)	$\Delta F^{\alpha \to L}$		L	B	E (cal/g-atom)	A
Re	2.4(2710 − T)	2.0(3450 − T)	2.3(2890 − T)	Re–Pd	10,300	10,800	9680	9650
Pd	2.8(820 − T)	2.3(1470 − T)	2.0(1820 − T)	Re–Pt	2,851	3,379	2259	2229
Pt	2.8(980 − T)	2.3(1670 − T)	2.0(2040 − T)					

FIG. 84. Regular-solution T_0 vs. x diagrams and comparison of regular-solution computed and experimental phase diagrams (63), (2). Thick lines are computed and thin lines are observed.

to the fact that e_0 is zero (Table XII) and the solubility parameters for the partners differ. Figure 52 compares experimental and computed diagrams, which show that the maximum gap temperature is overestimated slightly in Zr–Nb and Zr–Ta and underestimated slightly in Hf–Ta. In the Hf–Nb case, the bcc miscibility gap is not visible in either the experimental or computed phase diagrams. However, on the basis of the fact that the computed $\epsilon + \beta$ field is narrower than observed, we may conclude that the Hf–Nb value of T_c is underestimated. The interaction of the miscibility gap with the two-phase equilibria is shown in Fig. 55. Since the computed value of $T_c{}^\beta$ [which from Eq. (76) is defined as $B/2R$] is just below the single-phase β field in the computed diagram (Fig. 55) and no monotectoid appears in the experimental Hf–Nb system, we can conclude that B for the Hf–Nb system is only slightly underestimated in this case.

The free energy–composition curves for the β, ϵ, α, and L phases of the Hf–Ta systems presented earlier in Chapter III to illustrate the significance of the lattice stability terms were computed on the basis of Eqs. (65), (66), (77), (88), and (89), and Tables VIII and XIII. These curves shown in Figs. 24 and 25 are entirely consistent with Figs. 51 and 52. The curves for each phase shown in Figs. 24 and 25 are plotted by setting the free energy of the stable form of Hf and Ta equal to zero at each temperature.

Reference to Table XII shows that only the strain energy and internal pressure terms are finite in the (Zr/Hf) vs. (Nb/Ta) systems. The only other systems that permit a *direct* test to be made of the internal pressure calculation are the (Rh/Ir) vs. (Pd/Pt) quartet in Fig. 90. Reference to Table XII for these cases shows that only the internal-pressure contribution is significant in this quartet. Figure 90 indicates relatively good agreement for $T_c{}^\alpha$ computed and observed for Rh–Pd and Ir–Pd, but the Ir–Pt gap is computed near 300°K and observed near 1300°K.

Consequently, out of the six cases where clear comparisons can be made, i.e., Zr–Nb, Zr–Ta, Hf–Ta, Rh–Pd, Ir–Pd, and Ir–Pt, relatively good agreement is obtained in all but Ir–Pt. Of the remainder, two calculations are slightly high (Zr–Nb and Zr–Ta) and three are too low (Hf–Ta, Rh–Pd, and Ir–Pd). In the remaining two instances, no gap is observed in Rh–Pt ($T_c{}^\alpha$ computed is less than 100°K) and no gap is computed or observed in Hf–Nb. The latter system shows good agreement between the calculated and observed β/ϵ equilibria. Comparison of the computed and observed β/L equilibria in the (Zr/Hf) vs. (Nb/Ta) set in Fig. 52 is also quite good. This reflects the strain energy term alone (see Table XII). On this basis, we conclude that the present method for estimating e_p by means of Eq. (126) yields reasonable results.

The (Zr/Hf) vs. (Mo/W) set in Figs. 54 and 56 are dominated by positive interaction parameters due to the internal pressure e_p and "strain energy"

(text continues on page 136)

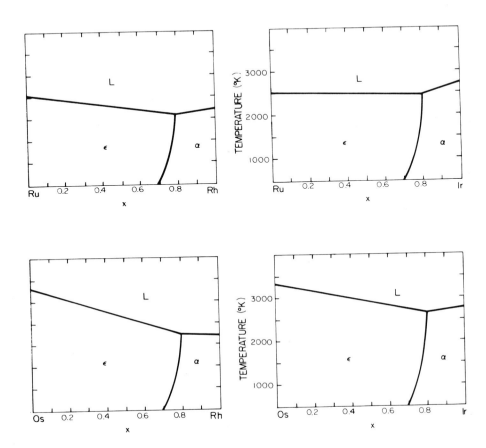

	$\Delta F^{\beta \to L}$	$\Delta F^{\epsilon \to L}$	$\Delta F^{\alpha \to L}$		L	B	E	A
		(cal/g-atom)				(cal/g-atom)		
Ru	2.8(1420 − T)	2.0(2550 − T)	2.8(1780 − T)	Ru–Rh	713	719	719	719
Os	2.8(1960 − T)	2.0(3300 − T)	2.8(2310 − T)	Ru–Ir	20	95	95	95
Rh	3.05(930 − T)	2.15(2010 − T)	2.0(2240 − T)	Os–Rh	848	856	856	856
Ir	3.05(1260 − T)	2.15(2490 − T)	2.0(2750 − T)	Os–Ir	46	55	55	55

FIG. 85. Regular-solution computation of T_0 vs. x diagram.

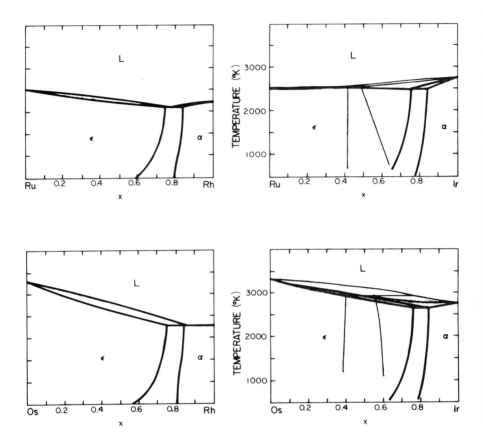

FIG. 86. Comparison of regular-solution computed and experimental phase diagrams (65), (66). Thick lines are computed and thin lines are observed.

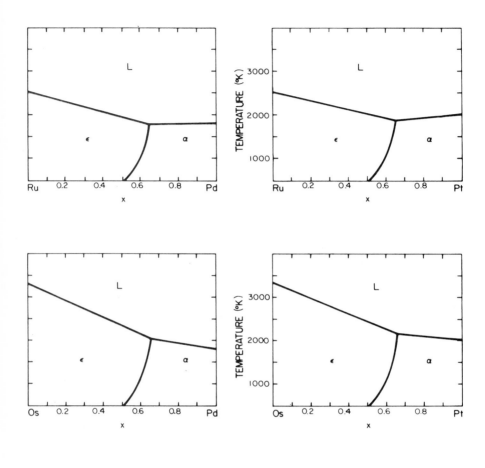

	$\Delta F^{\beta \to L}$	$\Delta F^{\epsilon \to L}$	$\Delta F^{\alpha \to L}$			L	B	E	A
		(cal/g-atom)					(cal/g-atom)		
Ru	$2.8(1420-T)$	$2.0(2550-T)$	$2.8(1780-T)$		Ru–Pd	7009	7800	7020	6850
Os	$2.8(1960-T)$	$2.0(3300-T)$	$2.8(2310-T)$		Ru–Pt	1330	2327	1547	1377
Pd	$2.8(820-T)$	$2.3(1470-T)$	$2.0(1820-T)$		Os–Pd	7494	8173	7393	7223
Pt	$2.8(980-T)$	$2.3(1670-T)$	$2.0(2040-T)$		Os–Pt	1520	2335	1555	1385

FIG. 87. Regular-solution computation of T_0 vs. x diagram.

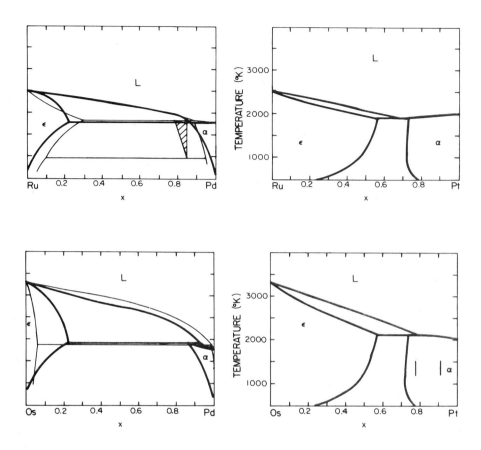

FIG. 88. Comparison of regular-solution computed and experimental phase diagrams (3), (2), (64). Thick lines are computed and thin lines are observed.

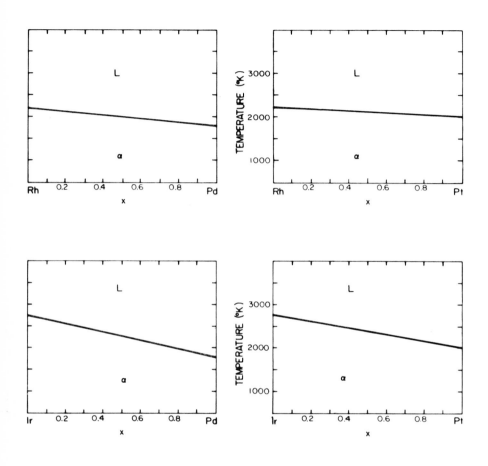

	$\Delta F^{\beta \to L}$	$\Delta F^{\epsilon \to L}$ (cal/g-atom)	$\Delta F^{\alpha \to L}$		L	B	E (cal/g-atom)	A
Rh	$3.05(930-T)$	$2.15(2010-T)$	$2.0(2240-T)$	Rh–Pd	3219	3330	3330	3330
Ir	$3.05(1260-T)$	$2.15(2490-T)$	$2.0(2750-T)$	Rh–Pt	84	355	355	355
Pd	$2.8(820-T)$	$2.3(1470-T)$	$2.0(1820-T)$	Ir–Pd	6406	6440	6440	6440
Pt	$2.8(980-T)$	$2.3(1670-T)$	$2.0(2040-T)$	Ir–Pt	1039	1179	1179	1179

FIG. 89. Regular-solution computation of T_0 vs. x diagram.

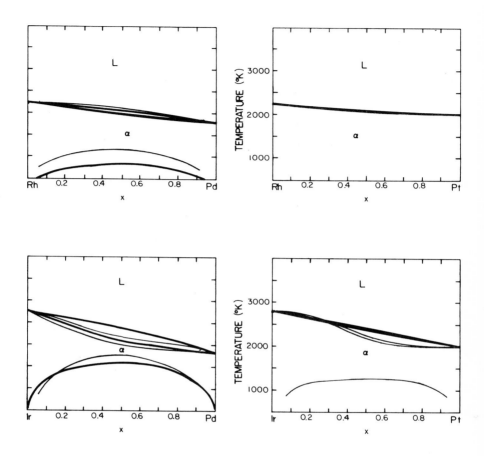

FIG. 90. Comparison of regular-solution computed and experimental phase diagrams
(3), (2), (2), (3). Thick lines are computed and thin lines are observed.

or "size effect" terms e_1 (see Table XII). All of these systems except Hf–Mo exhibit exposed miscibility gaps, as in Fig. 56. In fact, the values of $T_c{}^\beta$ for Zr–W and Hf–W lie in the temperature range where the liquid phase is stable. The experimental phase diagram shows the intrusion of Laves phases at $x_* = \frac{2}{3}$, thus limiting the extent to which experimental and theoretical diagrams can be compared in Fig. 54. The intrusion of Laves phase will be treated in Chapter VI in some detail; however, the present comparison of β/L and β/ϵ equilibria for the (Zr/Hf) vs. (Mo/W) systems is very good. As indicated earlier, this agreement further substantiates the present computation of e_1 on the basis of Eq. (142).

The (Zr/Hf) vs. (Ru/Os) and (Rh/Ir) quartets in Figs. 58 and 60 are dominated by the large negative values of e_0 as disclosed in Tables XI–XIII. This term leads to very large negative interaction parameters as required for compatibility with $ZrIr_3$, $HfIr_3$, etc. These compounds will be discussed in Chapter VI. It should be noted that the extremely narrow two-phase fields are a direct consequence of the large negative interaction parameters. This behavior was illustrated earlier in Figs. 43–45. Thus, as the interaction parameters become more negative, the phase boundaries converge on the T_0–x curve. The computed depression of the solid/liquid equilibria due to size differences and the predicted stabilization of the β phase in (Zr/Hf) due to additions of Ru, Os, Rh, and Ir are in agreement with the experimental Zr–Ru and Hf–Ir phase diagrams. In line with the large negative interaction parameters, all of these systems are dominated by compounds.

The (Zr/Hf) vs. (Pd/Pt) set shown in Fig. 62 is also characterized by large negative interaction parameters. The experimental phase diagrams are dominated by compound formation, which will be discussed in Chapter VI. The computed phase diagrams for Zr–Pd and Zr–Pt do not agree with the observed results on the β/ϵ equilibria, as indicated earlier, or on the width of the β/L field. Thus, although the calculations correctly predict the depression of the β/L range (i.e., $\overline{T}_{Pt}^\beta = 980°K$ and $L < B$) the observed width of the β/L field is much larger than calculated. This behavior can only be attributed to a deficiency in the regular-solution model for this system or to experimental error, for it is not possible to reconcile the large negative interaction parameters required by $ZrPt_3$ with the observed difference $x_L - x_\beta$.

The (Nb/Ta) vs. (Mo/W) set in Fig. 64 is trivial. The e_p terms lead to small positive interaction parameters which result in low-temperature miscibility gaps. In view of the refractory nature of these systems, observation of such gaps is improbable. The remaining (Nb/Ta) binary systems shown in Figs. 66, 68, and 70 are all characterized by large negative interaction parameters and are compound dominated. The $NbIr_3$ type of compound, which is the most stable, relative to the liquid, β, ϵ, and α phases in these systems, will be discussed in Chapter VI. For the present, it is of interest to note that notwithstanding the intrusion of the compound phases, the regular-solution computed diagrams predict solid/liquid equilibria quite well. In the (Nb/Ta) vs. (Pd/Pt) set, elevation of the melting point of the α phase is predicted, although the extent of this elevation is underestimated.

The 12 binary systems between (Mo/W) and (Ru/Os), (Rh/Ir), and (Pd/Pt) in Figs. 72, 74, and 76 offer striking substantiation of the calculations. The (Mo/W) vs. (Ru/Os) set is characterized by negative interaction parameters reflecting compound formation in these systems. Thus, the compounds that occur near $x = \frac{1}{3}$ in these binary systems and melt near 2500°K in equilibria with liquid phases having $L = -5600$ cal/g-atom would have free energies of formation of

$$\Delta F_f \approx -x(1-x)5600 + RT[x \ln x + (1-x) \ln(1-x)] \quad \text{cal/g-atom} \quad (151)$$

With $x = \frac{1}{3}$ at $T = 2500°K$, Eq. (151) yields

$$\Delta F_f \approx -1243 - 3157 = -4400 \quad \text{cal/g-atom}$$
$$= -13,200 \quad \text{cal/mole} \quad (152)$$

The comparison of computed (Mo/W) vs. (Ru/Os) in Fig. 72 agrees very well with observation if allowance is made for compound intrusion. As indicated earlier, the present regular-solution diagrams based on $E < B$ represent an improvement over the ideal-solution computations shown earlier in Fig. 36. The (Mo/W) vs. (Rh/Ir) set in Fig. 74 agrees very nicely with the observed diagrams, the only exception being the position of the ε/α phase boundaries, which are controlled by very small energy differences between the hcp and fcc phases. As indicated previously, the expansion of the ε field in the present regular-solution computations represents a distinct improvement over the ideal-solution computations shown in Fig. 37.

Figure 27 illustrates the free energy–composition curves of the β, ε, α, and L phases for W–Rh at 2400°K. These curves were computed on the basis of Eqs. (65), (66), (77), (88), and (89), and Tables VIII and XIII with $F_W^\beta = 0$ and $F_{Rh}^L = 0$ at 2400°K.

The (Mo/W) vs. (Pd/Pt) set in Fig. 76 is the most interesting group of the entire sequence. Calculation of the experimental diagrams is extremely demanding, since Mo–Pt contains a well-defined ε phase with ordering tendencies, Mo–Pd exhibits a very small hcp zone, while W–Pd shows evidence for positive interaction terms. The computed interaction parameters for these systems, which are given in Fig. 75 and Table XIII, yield precisely the required results. W–Pd shows positive parameters with A and E substantially smaller than B and L. This result is exactly that required by analysis of Fig. 38 as discussed earlier. In agreement with experiment, the computed Mo–Pt shows the most stable ε field, while the Mo–Pd ε field is narrow. The hcp phase is computed to be stable in the W–Pt system, even though it is not reported experimentally. Finally, the W–Pd system contains no ϵ phase and displays positive deviations as expected. Reference to Table XII shows that the computed sequence of interaction parameters arises because of the delicate balance between e_0 and e_p, the latter dominating the W–Pd system. The effect of positive interaction parameters in this system results in the unusual occurrence of a stable ϵ field on the T_0–x diagram for the W–Pd but *no stable* hcp field in the computed regular solution phase diagram.

Figure 77 shows the free energy–composition curves for the β, ϵ, α, and L phases of the W–Pd system as a function of temperature. These curves, which are computed on the basis of Eqs. (65), (66), (77), (88), (89), and Tables VIII and XIII with $F_W^\beta = F_{Pd}^\alpha = 0$ at each temperature, show this behavior graphically. Figure 78 illustrates the reason for this occurrence along different lines. Panel (a) shows the T_0–x curves for W–Pd, which generate the W–Pd diagram in Fig. 75. Panel (b) shows each of the six sets of phase boundaries corresponding to the six T_0–x curves. Thus, the phase boundaries for the β/L equilibria are labeled as $x_{\beta L}$ and $x_{L\beta}$ to denote the composition of the bcc phase in equilibrium with liquid and the liquid composition in equilibrium with the bcc phase. Because of the spreading of two-phase fields (resulting from positive interaction parameters) the ϵ/L boundaries are enclosed within the $\beta + L$ fields, indicating that ϵ/L equilibria do not lead to the minimum free energy. Moreover, the β/ϵ *and* ϵ/α fields are enclosed within the β/α fields. Thus, panel (c) discloses only the stable β/L, β/α, and α/L fields prior to insertion of the peritectic isotherm. The final result is given in panel (d) of Fig. 78.

The Zr–Re and Hf–Re systems are dominated by compound phases as shown in Fig. 79. The Laves phases at $x_* = \frac{2}{3}$ will be discussed in Chapter VI. The interaction parameters are negative, indicating large negative free energies of formation for these phases. Agreement of the β/L, ϵ/L, and ϵ/β equilibrium is good, except for the fact that the calculated solubility of Re in the ϵ phase of Hf–Re is much larger than observed. The Re vs. (Nb/Ta) and (Mo/W) sets illustrated in Figs. 80 and 81 are characterized by large negative interaction parameters and compound formation. As indicated earlier, the $B < E$ situation in the Re vs. (Nb/Ta) cases is an improvement over the ideal-solution computation shown in Fig. 35. Consequently, the agreement between computed and observed behavior is quite good, in spite of the compound intrusion.

The Re vs. (Ru/Os) cases in Fig. 82 are trivial, since internal pressure and strain energies are small in these systems. Comparison of the computed and observed Re–Rh and Re–Ir phase diagrams shown in Fig. 83 indicates that the interaction parameters are not large enough. Although the Re–Rh parameters are larger than the Re–Ir values of L, B, E, and A, both sets are too small. Reference to Tables XII and XIII show that the size differences and strain-energy terms are negligibly small and that only the internal-pressure term e_p makes a significant contribution. However, the present values of e_p would have to be increased by a factor of 6 in order to generate $\epsilon + \alpha$ fields comparable with those observed in Re vs. (Rh/Ir).

On the other hand, the Re–Pd and Re–Pt systems in Fig. 84 show excellent agreement with experiment. Here again, the internal-pressure term is the source of the difference. The (Ru/Os) vs. (Rh/Ir) quartet in Fig. 86 are nearly ideal. However, as pointed out earlier, A should be less than, rather than equal to E. The (Ru/Os) vs. (Pd/Pt) group clearly shows the effects of the internal-pressure term, with excellent agreement for the Ru–Pd and Ru–Os cases. Reference to the limited Os–Pt diagram suggests that E should be less than, rather than greater than A in these cases. Unfortunately, the energy differences between the ϵ and α phases are very small in this range of group numbers.

CHAPTER VI

Consideration of Compound Phases

In order to illustrate the intrusion of compound phases in the binary phase diagrams of the transition metals, it is convenient to consider the Laves phases λ and the AuCu$_3$-type phase that is observed for ZrIr$_3$, HfIr$_3$, NbIr$_3$, etc., which will be denoted by ξ. The simplest discussion of these compounds follows when they are treated as being stoichiometric at $x_* = \frac{2}{3}$ (λ) and $x_* = \frac{3}{4}$ (ξ) without considering departures from stoichiometry (92–94). Rudman (95) has considered the formation of Laves phases in the systems of interest and noted their occurrence in (Zr/Hf) vs. (Mo/W), Re, and (Ru/Os). In addition, ZrIr$_2$ is a λ phase. These phases have the MgZn$_2$(C-14) and MgCu$_2$(C-15) structures in which the j atoms are arranged in hexagonal layers and two out of four j atoms are removed to be replaced by one Zr or Hf atom, yielding ij_2 or $x_* = \frac{2}{3}$ as the stoichiometry (95). Therefore, in the ideal case (95) the volume of i should be twice that of j and the λ phase forms when the size differences between partners is substantial and λ phase formation can lead to a reduction in strain energy. Rudman has noted that the ratio $2V_j/V_i$ can actually vary between 0.6 and 1.4. Reference to the atomic volumes in Table XI suggests that only Zr/Hf systems would qualify for Laves phase (λ) formation under these circumstances.

The description of the compound phases employed here has been given in general terms by Eqs. (91)–(94) in Chapter III. Thus, as a first approximation and with reference to Figs. 54 and 79, we have chosen to represent the free energy of the Laves phase (λ) in these systems as follows.

$$F^\lambda = \tfrac{1}{3}F_i^\epsilon + \tfrac{2}{3}F_j^\epsilon + \tfrac{2}{9}(L - 18{,}000) \quad \text{cal/g-atom} \tag{153}$$

where $i = 4$, $j = 6$, 7, 8, 9, or 10. The $j = 5$ case (Nb/Ta) is ruled out because of the volume requirement. Therefore, Eq. (153) represents the free energy

140

of the λ phase as a function of temperature only (at $x_* = \frac{2}{3}$), as if it had no positional entropy, and is characterized by an interaction parameter which is 18,000 cal/g-atom less than that for the liquid in the i–j system. Moreover, Eq. (153) defines the free energy of the λ phase relative to that of the hcp form of the pure elements.

In order to compute the phase boundary composition of the liquid phase in a system i–j that is in equilibrium with the λ phase, we equate the partial molar free energies as follows:

$$\bar{F}_i^{\,L}\big|_{x_{L\lambda}} = \bar{F}_i^{\,\lambda}\big|_{x_{\lambda L}} = \bar{F}_i^{\,\lambda}\big|_{x_* = 2/3} \tag{154}$$

and

$$\bar{F}_j^{\,L}\big|_{x_{L\lambda}} = \bar{F}_j^{\,\lambda}\big|_{x_{\lambda L}} = \bar{F}_j^{\,\lambda}\big|_{x_* = 2/3} \tag{155}$$

where $\bar{F}_i^{\,L}$ and $\bar{F}_j^{\,L}$ have already been defined by Eqs. (69) and (70). Multiplying Eq. (154) by $\frac{1}{3}$ and Eq. (155) by $\frac{2}{3}$ and adding yields the following:

$$\tfrac{1}{3}F_i^{\,L} + \tfrac{2}{3}F_j^{\,L} + (RT/3)\ln(1 - x_{L\lambda})\, x_{L\lambda}^2 + (L/3)(2 - 4x_{L\lambda} + 3x_{L\lambda}^2)$$
$$= \tfrac{1}{3}F_i^{\,\epsilon} + \tfrac{2}{3}F_j^{\,\epsilon} + \tfrac{2}{9}(L - 18{,}000) \tag{156}$$

Equation (156) provides solutions for the composition of the liquid $x_{L\lambda}$ in equilibrium with the λ phase as a function of temperature. These solutions are double valued, corresponding to values of $x_{L\lambda}$ that are less than $\frac{2}{3}$ and $x_{L\lambda}$ that are greater than $\frac{2}{3}$. The latter are termed $\bar{x}_{L\lambda}$ to distinguish the solutions for $x_{L\lambda} > \frac{2}{3}$ (96).

Similar sets of equations are generated for the β/λ, ϵ/λ, and α/λ equilibria.

$$\tfrac{1}{3}F_i^{\,\beta} + \tfrac{2}{3}F_j^{\,\beta} + (RT/3)\ln(1 - x_{\beta\lambda})\, x_{\beta\lambda}^2 + (B/3)(2 - 4x_{\beta\lambda} + 3x_{\beta\lambda}^2)$$
$$= \tfrac{1}{3}F_i^{\,\epsilon} + \tfrac{2}{3}F_j^{\,\epsilon} + \tfrac{2}{9}(L - 18{,}000) \tag{157}$$

$$\tfrac{1}{3}F_i^{\,\epsilon} + \tfrac{2}{3}F_j^{\,\epsilon} + (RT/3)\ln(1 - x_{\epsilon\lambda})\, x_{\epsilon\lambda}^2 + (E/3)(2 - 4x_{\epsilon\lambda} + 3x_{\epsilon\lambda}^2)$$
$$= \tfrac{1}{3}F_i^{\,\epsilon} + \tfrac{2}{3}F_j^{\,\epsilon} + \tfrac{2}{9}(L - 18{,}000) \tag{158}$$

and

$$\tfrac{1}{3}F_i^{\,\alpha} + \tfrac{2}{3}F_j^{\,\alpha} + (RT/3)\ln(1 - x_{\alpha\lambda})\, x_{\alpha\lambda}^2 + (A/3)(2 - 4x_{\alpha\lambda} + 3x_{\alpha\lambda}^2)$$
$$= \tfrac{1}{3}F_i^{\,\epsilon} + \tfrac{2}{3}F_j^{\,\epsilon} + \tfrac{2}{9}(L - 18{,}000) \tag{159}$$

where B, E, and A refer to the previously defined interaction parameters for the β, ϵ, and α phases. Equations (156)–(159) are expanded versions of Eq. (94).

The free energy of the ξ phase at $x = \frac{3}{4}$ is defined in an analogous fashion to Eq. (153). However, since ξ does not form principally to reduce the volume strain, we have restricted its consideration to systems which exhibit very large negative interaction parameters (i.e., where L is less than $-15{,}000$ cal/g-atom). Thus, on the basis of Table XIII, we describe the ξ phase in (Zr/Hf) vs. Re, (Ru/Os), (Rh/Ir), and (Pd/Pt) and in (Nb/Ta) vs. Re, (Ru/Os), (Rh/Ir), and (Pd/Pt) as follows:

$$F^{\xi} = \tfrac{1}{4}F_i^{\,\alpha} + \tfrac{3}{4}F_j^{\,\alpha} + \tfrac{3}{16}(L - 18{,}000) \quad \text{cal/g-atom} \tag{160}$$

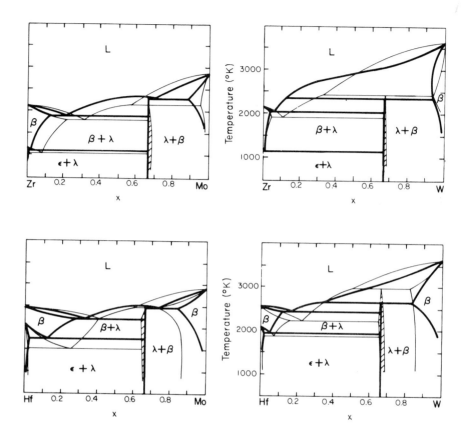

FIG. 91. Comparison of observed and computed phase diagrams with compound intrusion (2), (2), (3), (3). Thick lines are computed and thin lines are observed.

As with the λ phase, the ξ phase is described as having no positional entropy, with an interaction parameter 18,000 cal/g-atom less than that of the liquid phase in the i–j system. The free energy of the ξ phase is defined relative to the α phase. The conditions for L/ξ, β/ξ, ϵ/ξ, and α/ξ equilibria are defined as before by Eqs. (161)–(164).

$$\tfrac{1}{4}F_i^{L} + \tfrac{3}{4}F_j^{L} + (RT/4)\ln(1 - x_{L\xi})\,x_{L\xi}^3 + (L/4)(3 - 6x_{L\xi} + 4x_{L\xi}^2)$$
$$= \tfrac{1}{4}F_i^{\alpha} + \tfrac{3}{4}F_j^{\alpha} + \tfrac{3}{16}(L - 18{,}000) \tag{161}$$

$$\tfrac{1}{4}F_i^{\beta} + \tfrac{3}{4}F_j^{\beta} + (RT/4)\ln(1 - x_{\beta\xi})\,x_{\beta\xi}^3 + (B/4)(3 - 6x_{\beta\xi} + 4x_{\beta\xi}^2)$$
$$= \tfrac{1}{4}F_i^{\alpha} + \tfrac{3}{4}F_j^{\alpha} + \tfrac{3}{16}(L - 18{,}000) \tag{162}$$

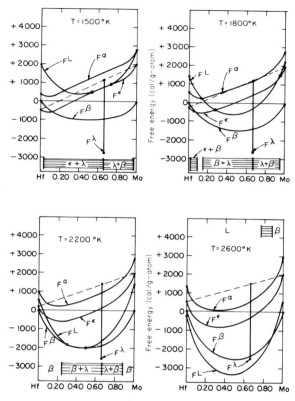

FIG. 92. Computed free energy–composition curves for liquid (L), bcc (β), hcp (ϵ), fcc (α), and Laves (λ) phases at various temperatures in the hafnium–molybdenum system. Phase boundaries are indicated at selected temperatures. Complete results are given in Figs. 91 and 93.

$$\tfrac{1}{4}F_i^\epsilon + \tfrac{3}{4}F_j^\epsilon + (RT/4)\ln(1 - x_{\epsilon\xi})x_{\epsilon\xi}^3 + (E/4)(3 - 6x_{\epsilon\xi} + 4x_{\epsilon\xi}^2)$$
$$= \tfrac{1}{4}F_i^\alpha + \tfrac{3}{4}F_j^\alpha + \tfrac{3}{16}(L - 18{,}000) \qquad (163)$$

and

$$\tfrac{1}{4}F_i^\alpha + \tfrac{3}{4}F_j^\alpha + (RT/4)\ln(1 - x_{\alpha\xi})\, x_{\alpha\xi}^3 + (A/4)(3 - 6x_{\alpha\xi} + 4x_{\alpha\xi}^2)$$
$$= \tfrac{1}{4}F_i^\alpha + \tfrac{3}{4}F_j^\alpha + \tfrac{3}{16}(L - 18{,}000) \qquad (164)$$

Equations (156)–(159) were employed to compute the L/λ, β/λ, ϵ/λ, and α/λ equilibria in the (Zr/Hf) vs. (Mo/W) quartet. Solutions of these equations, which represent "line compound–regular-solution equilibria," were obtained by means of a computer program designated as LCRSE. Details of the program are contained in Appendix 1. Equations (156)–(164) were used to compute L/λ, β/λ, ϵ/λ, α/λ, L/ξ, β/ξ, ϵ/ξ, α/ξ, and λ/ξ equilibria in the 14 (Zr/Hf) vs. Re, (Ru/Os), (Rh/Ir), and (Pd/Pt) systems. Finally, L/ξ, β/ξ, ϵ/ξ, and α/ξ solutions were obtained in the 14 (Nb/Ta) vs. Re, (Ru/Os), (Rh/Ir), and (Pd/Pt) sets. The results are summarized in Figs. 91–105. As before, the computed diagrams are drawn with thick lines, while the observed diagrams are drawn with thin lines. Experimentally observed compound phases are shown cross hatched. The reference sources for the experimental diagrams are contained in the figure legends, starting in the upper left corner and going clockwise.

Figure 91 contains the (Zr/Hf) vs. (Mo/W) quartet after insertion of the λ phase. Comparison with Fig. 54, which shows the same set before λ intrusion, reveals the changes. The computed monotectoid in the Zr–Mo system has been eliminated by the λ phase, which is computed to decompose via a eutectic rather than a peritectic reaction, as is observed. The computed decomposition temperature exceeds the observed temperature by about 200°K or 10%. Agreement between the computed and observed $L/\beta + \lambda$ and $\beta/\epsilon + \lambda$ eutectic (-oid) in this system is much better. Moreover, the $x_{L\lambda}$ and $x_{L\lambda}$ curves are displaced too far toward the zirconium side of the diagram. This indicates that the present set of L, B, E, and A values are all slightly higher than they should be. However, the β/L equilibria on both sides of the diagram agree well with experiment as noted earlier. The remaining members of this set exhibit similar features, except that in the Zr–W and Hf–W cases peritectic melting of the λ phase is correctly predicted. The Hf–W $(L + \beta)/\lambda$ peritectic is computed to lie 300°K below the observed peritectic. In the Zr–W and Hf–Mo, the computed $(L + \beta)/\lambda$ decomposition temperatures lie much closer to the observed values.

The free energy–composition curves for the β, ϵ, α, and L phases in addition to the free energy of the Laves phase are shown in Fig. 92 as function of temperature in order to illustrate the basis for the phase diagram calculation graphically. At each temperature, the free energy of the bcc forms of Hf and Mo are set equal to zero.

Figure 93 illustrates the sequence of computational steps taken to accommodate compound intrusion on the basis of Eqs. (156)–(159). Reference to Fig. 54 shows that the α phase is less stable than β, ϵ, and L over the entire temperature–composition range of interest. Consequently, α/λ equilibria are not relevant in the Hf–Mo system. Panel (a) of Fig. 93 reproduces the computed equilibria between the ϵ, β, and L phases. Panel (b) shows the loci of $x_{\epsilon\lambda}$, $x_{\beta\lambda}$, and $\bar{x}_{\beta\lambda}$, and the $x_{L\lambda}$–$\bar{x}_{L\lambda}$ pair, which are generated by Eqs. (158), (157), and (156) as a function of temperature. The $\bar{x}_{\epsilon\lambda}$ curve is omitted because β is more stable than ϵ on the Mo side of the Hf–Mo diagram. The next step is to superimpose Figs. 93a and 93b to locate the isothermal transformation temperatures. This is illustrated in panel (c) of Fig. 93. Reference

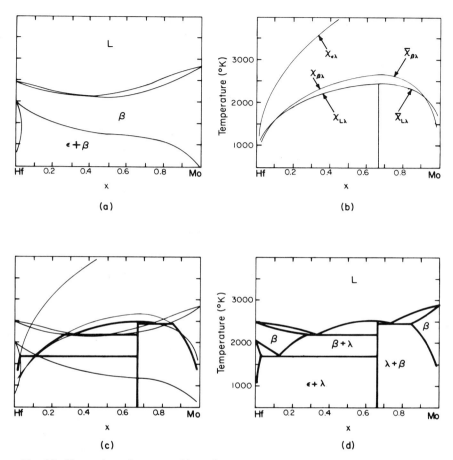

FIG. 93. Illustration of compound intrusion computation in the Hf–Mo system. Sequence (a)–(d) shows the series of steps employed in computing the compound intrusion.

to panel (c) shows that the computed melting temperature of the HfMo$_2$ λ phase is 2480°K. With decreasing temperature, the $x_{L\lambda}$–T curve (denoting the composition of the liquid phase that is in equilibrium with stoichiometric λ) shows decreasing values of $x_{L\lambda}$ with decreasing temperature. At 2200°K $x_{L\lambda} = x_{L\beta}$, or the composition of the liquid phase in equilibrium with the λ phase is also in equilibrium with the β phase. Thus β, liquid, and λ are all in equilibrium, defining the eutectic. Note also that the composition $x_{\beta L}$ coincides with $x_{\beta\lambda}$ at this temperature. Figure 93c shows all of the " new " phase diagram curves and lines generated by the intrusion of the λ phase, panel (b) on the ϵ, β, L phase diagram shown in Figure 93a. A similar procedure results from considering the L/λ equilibria for liquid compositions in excess of $\frac{2}{3}$, i.e., the $\bar{x}_{L_\lambda}$ vs. T curve in Figure 93b. Here, a eutectic occurs near 2470°K, where the $\bar{x}_{L_\lambda}$ curve intersects the $x_{L\beta}$ curve and the $x_{\beta L}$ curve coincides with the $\bar{x}_{\beta\lambda}$ curve. Finally, the eutectoid reaction at 1750°K is defined by the intersection of the $x_{\beta\lambda}$ curve with the $x_{\beta\epsilon}$ curve. Again at 1750°K, $x_{\epsilon\lambda}$ is equal to $x_{\epsilon\beta}$. Thus, the (Zr/Hf) vs. (Mo/W) set illustrates intrusion of the Laves phase, which can form to lower the large volume strain energy in this group of binary systems.

The Zr–Re and Hf–Re cases are given in Fig. 99. These systems are characterized by moderately large negative interaction parameters (where L is less than $-15,000$ cal/g-atom). Thus, in these systems the ξ phase must also be considered. However, when this is done, the ξ turns out to be unstable as will be shown below. Reference to the comparison between the computed and observed Zr–Re and Hf–Re diagrams in Fig. 99 indicates that in the former case the melting point of the λ phase is overestimated while in the latter case it is underestimated. The discrepancies are about 10% in the former case and 15% in the latter case. As indicated above, other compounds in Zr–Re and Hf–Re have not been considered; nevertheless, the computed versions of the Zr–Re and Hf–Re systems shown in Fig. 99 agree quite well with experiment.

Figure 100 shows free energy–composition curves for the solution phases and compounds in the Hf–Re system at various temperatures. These curves have been calculated on the basis of $F^\beta_{Hf} = F^\epsilon_{Re} = 0$ at each temperature.

Figure 101 presents the details of the Hf–Re phase-diagram computation in order to illustrate the λ/ξ competition. As before, panel (a) shows the ϵ, β, and liquid phase equilibria before intrusion of λ or ξ is considered. Panel (b) shows the equilibrium between ϵ, β, L, and λ, as well as ϵ, L, and ξ. The α phase is not stable in this system, nor is the β phase relevant for compositions greater than 90 at.% Re [see panel (a)]. Comparison of the $\bar{x}_{L\lambda}$ and $\bar{x}_{L\xi}$ vs. T curves shows that in the temperature range where the liquid is stable, $\bar{x}_{L\xi} < \bar{x}_{L\lambda}$. Moreover, in the temperature range where ϵ is the stable phase, $\bar{x}_{\epsilon\xi} < \bar{x}_{\epsilon\lambda}$. Thus, in the Hf–Re system the ξ phase is computed to be unstable. A similar situation prevails in Zr–Re. Panel (c) of Fig. 101 shows the superimposition of (a) and (b) and the resultant new lines due to λ intrusion, while panel (d) shows the result. The (Zr/Hf) vs. (Ru/Os) quartet is similar to the (Zr/Hf) vs. Re pair in that the ξ phase is computed to be unstable. Figure 94 displays the results for the (Zr/Hf) vs. (Ru/Os) cases. The Zr–Ru case is the only one where comparisons between observed and computed results can be made at present. The present computation overestimates the melting point of the λ-ZrRu$_2$ by about 10%. In addition, it should be noted that the effects of the CsCl-type ZrRu compound are not included in the present calculations.

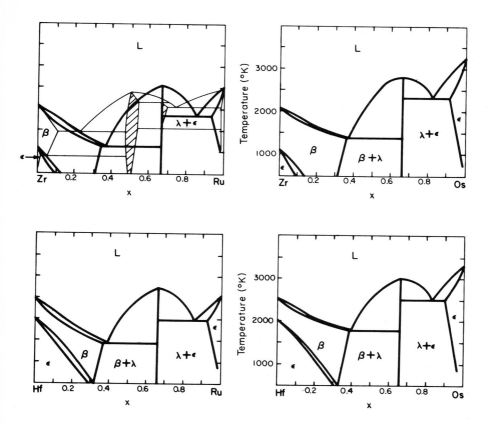

FIG. 94. Computed phase diagrams with compound intrusion (84). Thick lines are computed and thin lines are observed.

The (Zr/Hf) vs. (Rh/Ir) quartet is shown in Fig. 95. In these cases, the ξ phase is stable. Figure 96 shows the free energy–composition curves for the β, ϵ, α, and liquid phases of the Zr–Ir system at 2000°K. These curves were computed on the basis of $F_{Zr}^{L} = F_{Ir}^{\alpha} = 0$. Details of the Zr–Ir case are included in Fig. 97 in order to demonstrate the stability of the ξ phase (AuCu$_3$). As before, panel (a) gives the β, ϵ, α, and L equilibria, while panel (b) shows the relevant equilibria between the foregoing and the λ and ξ phases. Reference to panel (b) indicates that $\bar{x}_{L\xi} > \bar{x}_{L\lambda}$ and $\bar{x}_{\alpha\xi} > \bar{x}_{\alpha\lambda}$ over the entire temperature range. Thus, in contrast to the six (Zr/Hf) vs. Re and (Ru/Os) cases described earlier, where ξ is unstable relative to λ, the ξ phase (AuCu$_3$) is stable in the (Zr/Hf) vs. (Rh/Ir) quartet. The same condition prevails in the

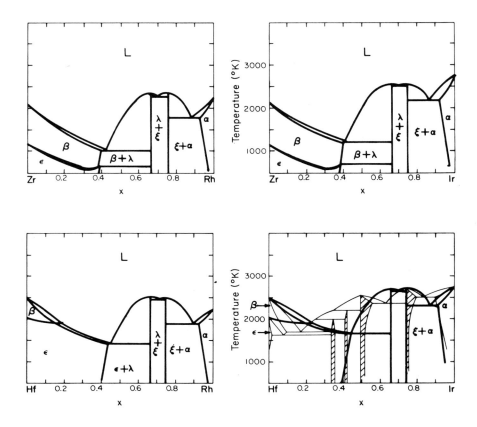

FIG. 95. Computed phase diagrams with compound intrusion (*85*). Thick lines are computed and thin lines are observed.

(Zr/Hf) vs. (Pd/Pt) cases shown in Fig. 98. The Hf–Ir case in Fig. 95 and the Zr–Pd and Zr–Pt cases in Fig. 98 alone provide a basis for comparing the experimental and computed results. Experimental counterparts of the remaining computed diagrams in Figs. 95 and 98 are not available at present. Although the (Zr/Hf) vs. (Pd/Pt) set exhibit the proper volume relations for Laves-phase formation (*95*), the Zr–Pd and Zr–Pt systems do not display Laves phases. The ZrPd$_2$ phase that is observed (*3*) has a tetragonal MoSi$_2$-type structure. Moreover, the ZrPd$_3$ and ZrPt$_3$ compounds are hcp and are isotypic with TiNi$_3$ rather than with the fcc (AuCu$_3$) ξ arrangement (*3*). The ZrPd and ZrPt structures are quite complex (*3*). The latter, in equilibria with the ZrPt$_3$ phase, is sufficiently stable to eliminate the ZrPt$_2$ phase from contention as a stable phase in the Zr–Pt system.

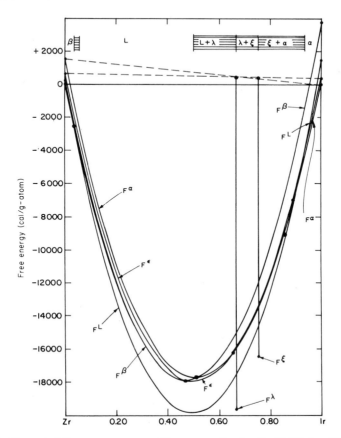

FIG. 96. Computed free energy–composition curves for the liquid (L), bcc (β), hcp (ϵ), fcc (α), Laves (λ), and AuCu$_3$ (ξ) phases in the Zr–Ir system at 2000°K. Phase boundaries are indicated. Complete results are given in Figs. 95 and 97.

In summary, Figs. 91–101 contain recalculations of eighteen (Zr/Hf) vs. (Mo/W), Re, (Ru/Ir), and (Pd/Pt) binary phase diagrams in which the λ (at $x_* = \frac{2}{3}$) and ξ (at $x_* = \frac{3}{4}$) compound phases were introduced. The stability of each phase was defined by a single parameter as shown in Eqs. (153) and (160). In ten of the 18 cases, the ξ phase (at $x_* = \frac{3}{4}$) was computed to be unstable relative to the λ phase. In these systems, (Zr/Hf) vs. (Mo/W), Re, and (Ru/Os), the Laves phase is observed in seven cases. In the remaining instances no experimental data are available. The seven binary diagrams that exhibit the Laves phase (Figs. 91, 94, and 99) display good comparison between computed and observed results.

In the eight remaining (Zr/Hf) binaries shown in Figs. 95–98, both λ and ξ

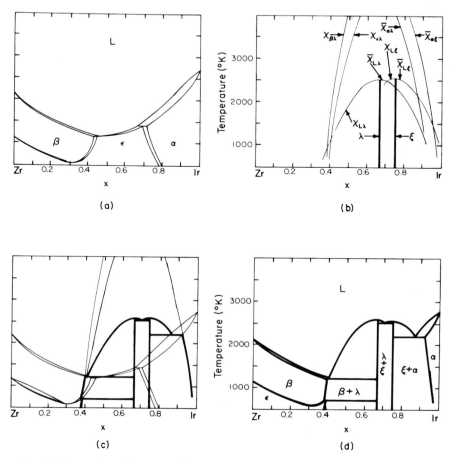

FIG. 97. Illustration of compound intrusion computations in the Zr–Ir system. Sequence (a)–(d) shows the series of steps employed in computing the compound intrusion.

forms are computed to be stable. The (Zr/Hf) vs. (Rh/Ir) quartet forms a stable $AuCu_3$ phase at $x_* = \frac{3}{4}$, and $ZrIr_2$ is a stable λ phase (95). The observed Hf–Ir binary (Fig. 95) does not contain a stable $HfIr_2$ λ phase. In this case, however, the computed melting temperature of $HfIr_3$ agrees closely with observation. The set of four (Zr/Hf) vs. (Pd/Pt) diagrams shown in Fig. 98 are computed to exhibit stable λ ($x_* = \frac{2}{3}$) and ξ ($x_* = \frac{3}{4}$) phases. In this group, no experimental phase diagram data are available for Hf–Pd and Hf–Pt, although the both systems form hcp $HfPd_3$ and $HfPt_3$ compounds isotypic with $TiNi_3$ (3). The partial phase diagram available for Zr–Pd shows a $ZrPd_2$ phase at $x_* = \frac{2}{3}$ and a $ZrPd_3$ phase ($x_* = \frac{3}{4}$). However, these are of the $MoSi_2$

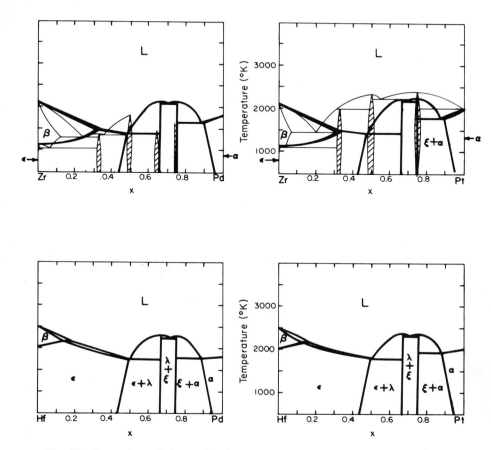

FIG. 98. Comparison of observed and computed phase diagrams with compound intrusion (3), (3). Thick lines are computed and thin lines are observed.

and TiNi₃ type rather than the λ and ξ (AuCu₃) forms. In the Zr–Pt case, no compound is observed at $x_* = \frac{2}{3}$ and the $x_* = \frac{3}{4}$ phase is of the TiNi₃ (hcp) form rather than the CuAu₃ ξ form. However, reference to Fig. 98, which contains the (Zr/Hf) vs. (Pd/Pt) quartet, shows that the computed melting point of the AuCu₃ type of ZrPt₃ lies fairly close to the observed melting point of the TiNi₃-type ZrPt₃. As a consequence, one would expect that the free energies of the TiNi₃ and AuCu₃-type phases do not differ markedly in this system.

The results obtained by insertion of the ξ phase (at $x_* = \frac{3}{4}$) in the (Nb/Ta) vs. Re, (Ru/Os), (Rh/Ir), and (Pd/Pt) systems are shown in Figs. 99–105. In the (Nb/Ta) vs. (Rh/Ir) set, the AuCu₃-type ξ phase is the stable form and

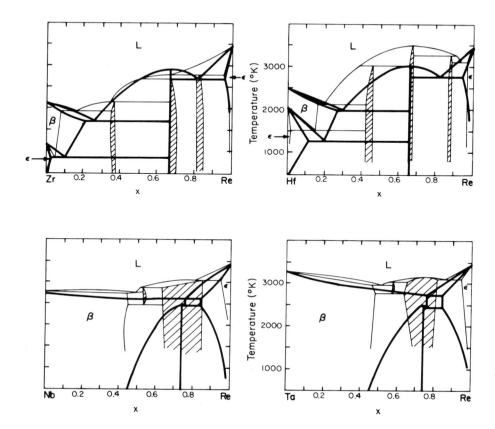

FIG. 99. Comparison of observed and computed phase diagrams with compound intrusion (3), (3), (3), (3). Thick lines are computed and thin lines are observed.

actually dominates the system. Reference to Fig. 103 shows excellent agreement between the computed and observed melting points for $NbRh_3$, $NbIr_3$, $TaRh_3$, and $TaIr_3$. Although the computed diagrams in Fig. 103 do not agree precisely with the observed phase diagrams because of the intrusion of other compounds (eight more compound phases exist in the Nb–Rh system), the delineation between the liquid and solid ranges is computed in excellent agreement with observations. Figure 104 illustrates the details of the Ta–Ir calculation. As before, panel (a) shows the β, ϵ, α, and liquid equilibria. Since the β phase is not stable in the region near $x = \frac{3}{4}$, only the equilibria between α, ϵ, L, and ξ need be computed. Equations (162)–(164) yield panel (b) of Fig. 104. Superposition of (b) and (a) leads to panel (c), where the new lines are shown. Finally, Fig. 104d shows the computed phase diagram, which exhibits a small fcc (α) field near $x = 0.6$ that has been isolated by ξ intrusion.

The (Nb/Ta) vs. (Pd/Pt) cases shown in Fig. 105 are similar to the foregoing (Nb/Ta) vs. (Rh/Ir) set except that the AuCu$_3$ ξ phase is not the stable form in these systems. The $x_* = \frac{3}{4}$ compounds that form here are denoted as α TaPt$_3$ and β TaPt$_3$ (79), which are complex close-packed structures containing 8 and 48 atoms per unit cell, respectively. The α-TaPt$_3$ structure is orthorhombic, while the β is monoclinic with an angle that is 32.4 minutes greater than 90° and a unit cell six times the height of α-TaPt$_3$. On the basis of Fig. 105, we may conclude that the free energy of the α and β forms of the $x_* = \frac{3}{4}$ compound phases in the (Nb/Ta) vs. (Pd/Pt) systems have free energies similar to the ξ phase. However, these systems do not exhibit as many additional phases as the (Nb/Ta) vs. (Rh/Ir) quartet that produce additional distortions. Nevertheless, as in the (Nb/Ta) vs. (Rh/Ir) cases, Fig. 105 illustrates that the computed liquid and solid ranges agree quite well with observation, once ξ phase intrusion is included.

The (Nb/Ta) vs. Re and (Ru/Os) systems shown in Figs. 99 and 102 do not contain ξ phases at $x_* = \frac{3}{4}$. The NbRu$_3$ phase is hcp, while the broad phase fields in the Nb–Re and Ta–Re systems that extend over a wide compositional range in the vicinity of NbRe$_3$ and TaRe$_3$ are α-Mn structures. In the latter cases (Fig. 99), the calculated ξ phase is less stable than the α-Mn phases, since it does not impinge on the β and ϵ phase as strongly as the α-Mn phases do. Nonetheless, the computed $x_{\beta\xi}$ and $\bar{x}_{\epsilon\xi}$ curves show clearly the extent to which compound intrusion restricts the β and ϵ fields (see Fig. 80 for comparison). Insertion of the free energy for the α-Mn phase in Nb–Re and Ta–Re, which is more stable than the ξ phase, should predict phase boundaries for the β and ϵ phases in good agreement with observed results.

The (Nb/Ta) vs. (Ru/Os) systems in Fig. 102 do not contain ξ phases. The $x_* = \frac{3}{4}$ phase in Nb–Ru (hcp) is comparable in stability with the computed ξ phase, however; thus the $x_{\beta\xi}$ and $\bar{x}_{\epsilon\xi}$ curves derived from Eqs. (162) and (163) agree quite well with those observed. In the Ta–Ru and Ta–Os cases, intermediate phases form for $0.3 < x < 0.6$. The Ta–Ru system exhibits two ordered bcc phases, while the Ta–Os system forms a sigma phase for x near 0.3 and an α-Mn near $x = 0.6$. In these cases insertion of an ξ phase near $x_* = \frac{3}{4}$ on the computed phase diagram does not yield improvement over the diagram derived solely on the basis of β, ϵ, and L equilibria.

Thus, insertion of a ξ (AuCu$_3$) phase at $x_* = \frac{3}{4}$ in 14 (Nb/Ta) binaries has been performed using the simple representation of this phase [Eq. (160)]. In four of these cases, (Nb/Ta) vs. (Rh/Ir), given in Figure 68, the ξ phase is stable and the computed diagram is in good agreement with the experimental diagrams. In seven of the remaining ten cases, i.e., (Nb/Ta) vs. (Pd/Pt) (Fig. 105), (Nb/Ta) vs. Re (Fig. 99), and Nb–Ru (Fig. 102), compound phases are

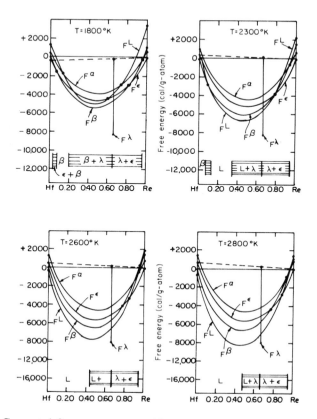

FIG. 100. Computed free energy–composition curves for liquid (L), bcc (β), hcp (ϵ), cc (α), and Laves (λ) phases at various temperatures in the hafnium–rhenium system. Phase boundaries are indicated at selected temperatures. Complete results are shown in Figs. 99 and 101.

observed at $x_* = \frac{3}{4}$ which do not exhibit the AuCu$_3$ structure but are of comparable stability. In these cases, the computed diagram reflecting ξ intrusion agrees more closely with experiment than the corresponding cases where ξ intrusion is ignored. In one case, Nb–Os (Fig. 102), no experimental data are available for comparison. The final two systems, Ta–Ru and Ta–Os (Fig. 102), do not contain compounds at $x = \frac{3}{4}$, and insertion of the ξ phase leads to poor results.

In summary then, 32 of the 72 binary diagrams have been recomputed in order to show the effects of compound intrusion. Simplified, one-parameter descriptions of the free energies of Laves phases (λ) at $x_* = \frac{2}{3}$ and AuCu$_3$ phases (ξ) at $x_* = \frac{3}{4}$ were employed. Comparisons of computed and observed

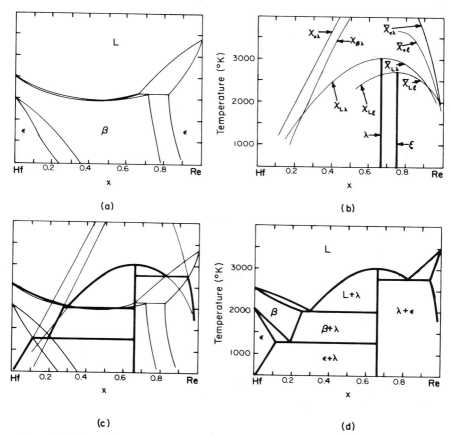

FIG. 101. Illustration of compound intrusion computations in the Hf–Re system. Sequence (a)–(d) shows the series of steps employed in computing the compound intrusion.

phase diagrams is limited to 23 cases where experimental data are available. In 12 of the 23 cases—(Zr/Hf) vs. (Mo/W) in Fig. 91, Zr–Ru in Fig. 94, Hf–Ir in Fig. 95, (Zr/Hf) vs. Re in Fig. 99, and (Nb/Ta) vs. (Rh/Ir) in Fig. 103—the recomputed diagrams are in reasonable agreement with observations. In nine of the remaining cases—Zr vs. (Pd/Pt) in Fig. 98, (Nb/Ta) vs. Re in Fig. 99, Nb–Os in Fig. 102, and (Nb/Ta) vs. (Pd/Pt) in Fig. 105—compound phases are present at $x_* = \frac{3}{4}$ that have different structures than the ξ phase but are of comparable stability. In these nine cases the recomputed diagrams are improvements over those that neglected compound intrusion. The two remaining diagrams—Ta–Ru and Ta–Os in Fig. 102—do not exhibit $x_* = \frac{3}{4}$ compounds. In these cases, the recomputed diagrams are poorer than those originally calculated.

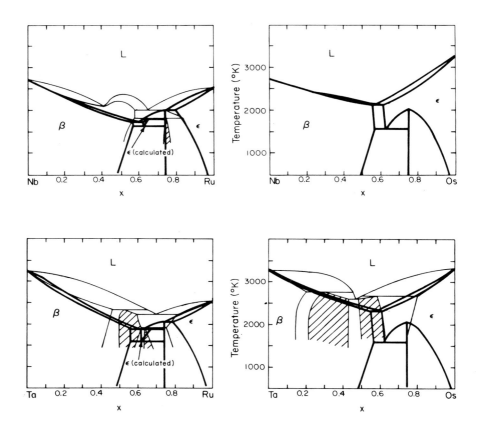

FIG. 102. Comparison of observed and computed phase diagrams with compound intrusion (57), (3), (3). Thick lines are computed and thin lines are observed.

1. CALCULATION OF INTERACTION PARAMETERS BASED ON ELECTRONEGATIVITY

In the discussion of interaction parameters for the liquid phase presented in Section 1 of Chapter IV, reference was made to the computation of e_0 on the basis of electronegativities [Eq. (131)]. It was noted that such a calculation does not yield uniformly satisfactory results for the systems under consideration. It is now appropriate to return to this discussion in order to demonstrate the inconsistencies. Table XIVa contains the values of X_i for groups 4–10 based on the recent complete tabulation of Furakawa (72). Other tabulations are available (70, 71, 73) which differ slightly from the former set. However, these differences do not affect the present argument.

Table XIVa contains a compilation of interaction parameters for the systems involved based on Eq. (131) and the X_i values in Table XIVa. These values are denoted as W to distinguish them from the present values of e_0 based on Eq. (139). Comparison of the W terms and the e_0 terms in Table XIVa leads to the following conclusions.

In the (Mo/W) vs. (Ru/Os), (Rh/Ir), and (Pd/Pt) cases, good agreement is obtained. Similar agreement is observed in the Re vs. (Ru/Os), (Rh/Ir), and (Pd/Pt); in (Rh/Ir) vs. (Pd/Pt); in (Zr/Hf) vs Re; and in (Nb/Ta) vs. (Mo/W).

TABLE XIVa

COMPARISON OF REGULAR-SOLUTION PARAMETER RESULTS WITH
ELECTRONEGATIVITY CALCULATION [a]

Metal	X [b]	System	W (cal/g-atom)	e_0 [c] (cal/g-atom)
Zr/Hf	1.5	Zr/Hf–Nb/Ta	−10,377	0
		–Mo/W	−18,448	−4,000
		–Re	−28,825	−30,000
		–Ru/Os	−41,508	−62,000
		–Rh/Ir	−41,508	−80,000
		–Pd/Pt	−41,508	−74,000
Nb/Ta	1.8	Nb/Ta–Mo/W	−1,153	0
		–Re	−4,612	−26,000
		–Ru/Os	−10,377	−44,000
		–Rh/Ir	−10,377	−40,000
		–Pd/Pt	−10,377	−40,000
Mo/W	1.9	Mo/W–Re	−1,153	−12,000
		–Ru/Os	−4,612	−6,000
		–Rh/Ir	−4,612	−6,000
		–Pd/Pt	−4,612	−6,000
Re	2.0	Re–Ru/Os	−1,153	0
		–Rh/Ir	−1,153	0
		–Pd/Pt	−1,153	0
Ru/Os	2.1	Ru/Os–Rh/Ir	0	0
		–Pd/Pt	0	0
Rh/Ir	2.1	Rh/Ir–Pd/Pt	0	0
Pd/Pt	2.1			

[a] $W = -23,060\bar{n}(X_i - X_j)^2$ cal/g-atom [Eq. (131)], $\bar{n} = 5$.
[b] Furakawa (72).   [c] See Table XII.

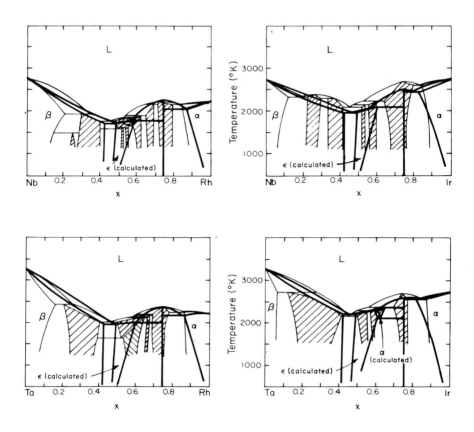

FIG. 103. Comparison of observed and computed phase diagrams with compound intrusion (86), (87), (88), (89). Thick lines are computed and thin lines are observed.

Thus, in half of the binary systems investigated, little difference would be noted if Eq. (131) were used to compute e_0 in place of Eq. (45).

In the remaining instances, however, serious differences exist. The first of these occur in the eight (Zr/Hf) vs. (Nb/Ta) and (Mo/W) binaries, where the electronegativity computation [Eq. (131)] leads to values of W (or e_0) that are much too negative. Table XIVb computes L' and B' which result from the electronegativity computation of W (or e_0), the internal pressure term e_p, the size effect term e_1, and the energy e_2. A second computation of L'' and B'' is also displayed which eliminates the 40% reduction of e_p discussed prior to Eq. (126). The results in Table XIVb show that the values of B' and/or B'' for the (Zr/Hf) vs. (Nb/Ta) set are all negative (except for B'' in

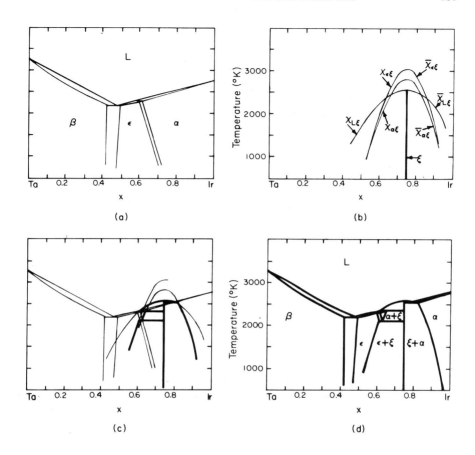

FIG. 104. Illustration of compound intrusion computation in the Ta–Ir system.

Zr–Ta). Such values are inconsistent with Figs. 51, 52, and 55, since miscibility gaps in these systems demand positive interaction parameters for the bcc phase in these systems. The Zr–Ta value of B'' is too small, since the Zr–Ta miscibility gap demands an interaction parameter of at least 8000 cal/g-atom. A similar discrepancy exists in the (Zr/Hf) vs. (Mo/W) set as shown in Table XIVb. Again, the large values of W (or e_0) generated by Eq. (131) leads to negative values of B', negative values of B'' for (Zr/Hf) vs. Mo, and positive values of (Zr/Hf) vs. W which are too small to be consistent with the phase diagrams (see Figs. 53, 54, and 56).

The remaining 12 (Zr/Hf) binary cases in Table XIVb yield values of W based on electronegativity computations which are not as negative as the e_0

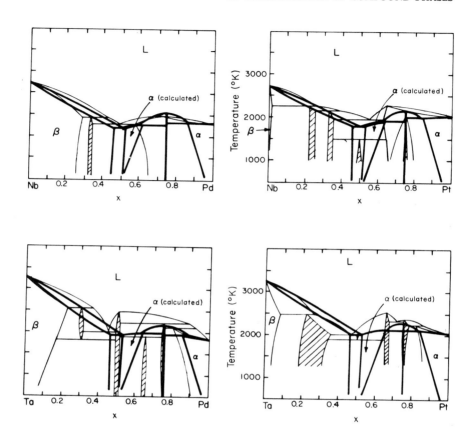

FIG. 105. Comparison of observed and computed phase diagrams with compound intrusion (3), (3), (79), (90). Sequence (a)–(d) shows the series of steps employed in computing the compound intrusion. Thick lines are computed and thin lines are observed.

values. If the W values were used to compute interaction parameters for these systems in place of the e_0 values and the resultant L, B, E, and A parameters employed for phase diagrams, the results would be virtually indistinguishable from Figs. 58, 61, 62, 94, 95, and 97. This is due to the fact that very large negative values of L, B, E, and A shrink the width of the two-phase fields. Thus, a set of interaction parameters in the $-70,000$-cal/g-atom range will produce slightly narrower two-phase fields than a set of parameters in the $-40,000$-cal/g-atom range. Of course, the differences in $L - B$, $L - E$, and $L - A$ must be the same in both instances as it is in the present case. The principal reason for rejecting the values of W generated by Eq. (131)

lies in the fact that $W = -41,508$ cal/g-atom for the (Zr/Hf) vs. (Rh/Ir) and (Pd/Pt) cases is not consistent with the formation of HfIr$_3$, etc., from HfC and iridium [see the discussion following Eq. (127)].

Similarly, the values of W are not negative enough in the (Nb/Ta) vs. (Rh/Ir) and (Pd/Pt) cases because of the reactions of NbC with Ir and Pt discussed earlier [see Eqs. (127)–(130)].

In the four remaining binary systems—(Nb/Ta) and (Mo/W) vs. Re— the electronegativity calculation of W (or e_0) is not negative enough, since reference to Figs. 80 and 81 indicates compound formation.

TABLE XIVb

CALCULATION OF REGULAR-SOLUTION PARAMETERS [a]
USING ELECTRONEGATIVITY VALUES

System	L'	B'	L''	B''
Zr–Nb	−6,080	−3,443	−3,214	−577
Zr–Ta	−4,078	−1,457	+123	+2,744
Hf–Nb	−7,158	−5,240	−5,011	−3,093
Hf–Ta	−5,429	−3,538	−2,129	−238
Zr–Mo	−12,936	−7,897	−9,260	−4,221
Zr–W	−5,339	−23	+3,404	+8,720
Hf–Mo	−14,148	−10,129	−11,280	−7,261
Hf–W	−7,376	−3,190	+9	+4,195

$$L' = W + e_p \qquad (165)\ ^b$$
$$B' = L' + e_1 + e_2 \qquad (166)$$
$$L'' = W + 1.67 e_p \qquad (167)$$
$$B'' = L'' + e_1 + e_2 \qquad (168)$$

[a] In calories per gram atom.
[b] See Table XII for values of e_p, e_1, and e_2.

CHAPTER VII

Estimation of the Heat of Formation of Compound Phases

The foregoing thermodynamic description of transition metal binary diagrams permits estimates to be made of the heat of formation of compound phases. For the Laves phases, Eq. (153) implies that the heat of formation at 298°K, is

$$\Delta H_f{}^\lambda[298°\text{K}] = H^\lambda[298°\text{K}] - \tfrac{1}{3}H_i{}^\circ - \tfrac{2}{3}H_j{}^\circ \quad \text{cal/g-atom} \tag{169}$$

where $H_i{}^\circ$ and $H_j{}^\circ$ are the enthalpies of the stable form of i and j at 298°K. Since i is always Zr or Hf in the present set of cases,

$$H_i{}^\circ[298°\text{K}] = H_i{}^\epsilon[298°\text{K}]$$

and

$$\begin{aligned}
\Delta H_f{}^\lambda[298°\text{K}] &= \tfrac{2}{3}(H_j{}^\epsilon - H_j{}^\circ) + \tfrac{2}{9}(L - 18{,}000) \quad \text{cal/g-atom} \\
&= \tfrac{2}{3}\,\Delta H_j{}^{\circ \to \epsilon} + \tfrac{2}{9}(L - 18{,}000) \quad \text{cal/g-atom}
\end{aligned} \tag{170}$$

Since L is given in Table XIII and $\Delta H_j{}^{\circ \to \epsilon}$ is contained in Table VIII for all of the relevant metals, Eq. (170) defines the heat of formation of Laves phases in explicit terms.

Similarly, the heat of formation of the AuCu$_3$ ξ phase is implied by Eq. (160) as

$$\Delta H_f{}^\xi[298°\text{K}] = \tfrac{1}{4}\,\Delta H_i{}^{\circ \to \alpha} + \tfrac{3}{4}\,\Delta H_j{}^{\circ \to \alpha} + \tfrac{3}{16}(L - 18{,}000) \quad \text{cal/g-atom} \tag{171}$$

The heats of formation for compound phases computed from Eqs. (170) and (171) are shown in Table XV. The computations were performed for phases that are known to exhibit the λ and ξ structures or have similar free energies to the ξ structure on the basis of the discussion of Chapter VI. Table XV also shows the figure number of the relevant binary phase diagram exhibiting the compound in question.

TABLE XV
ESTIMATED HEATS OF FORMATION OF COMPOUNDS

Compound	Estimation method	Figure	$\Delta H_f[298°\text{K}]$ (kcal/g-atom)
ZrMo$_2$ (λ)	Eq. (170)	91	-2.3
ZrW$_2$ (λ)	Eq. (170)	91	-0.6
HfMo$_2$ (λ)	Eq. (170)	91	-2.6
HfW$_2$ (λ)	Eq. (170)	91	-1.1
ZrRe$_2$ (λ)	Eq. (170)	99	-7.8
HfRe$_2$ (λ)	Eq. (170)	99	-8.2
ZrRu$_2$ (λ)	Eq. (170)	94	-15.8
ZrOs$_2$ (λ)	Eq. (170)	94	-15.7
HfRu$_2$ (λ)	Eq. (170)	94	-16.2
HfOs$_2$ (λ)	Eq. (170)	94	-16.1
ZrRh$_2$ (λ)	Eq. (170)	95	-20.8
ZrIr$_2$ (λ)	Eq. (170)	95	-19.9
HfRh$_2$ (λ)	Eq. (170)	95	-21.0
HfIr$_2$ (λ)	Eq. (170)	95	-20.2
ZrRh$_3$ (ξ)	Eq. (171)	95	-17.4
ZrIr$_3$ (ξ)	Eq. (171)	95	-16.6
HfRh$_3$ (ξ)	Eq. (171)	95	-17.6
HfIr$_3$ (ξ)	Eq. (171)	95	-16.9
Zr$_2$Pd (MoSi$_2$)	Mp (176)	98	-16.9
ZrPd (?)	Mp (176)	98	-21.3
ZrPd$_2$ (λ)	Eq. (170)	98	-20.2
ZrPd$_3$ (ξ)	Eq. (171)	98	-17.0
Zr$_2$Pt (?)	Mp (176)	98	-18.9
ZrPt (?)	Mp (176)	98	-22.1
ZrPt$_3$ (?)	Mp (176)	98	-17.0
HfPd$_2$ (λ)	Eq. (170)	98	-20.2
HfPt$_2$ (λ)	Eq. (171)	98	-19.8
HfPd$_3$ (ξ)	Eq. (171)	98	-17.0
HfPt$_3$ (ξ)	Eq. (171)	98	-16.6
NbRe$_3$ (α-Mn)	Mp (176)	99	-7.3
TaRe$_3$ (α-Mn)	Mp (176)	99	-7.4
NbRu$_3$ (hcp)	Eq. (171)	102	-10.8
TaRu (?)	Mp (176)	102	-13.6
Ta$_2$Os (σ)	Mp (176)	102	-12.6
TaOs (α-Mn)	Mp (176)	102	-13.5

TABLE XV (*continued*)

Compound	Estimation method	Figure	$\Delta H_f [298°K]$ (kcal/g-atom)
NbRh$_3$ (ξ)	Eq. (171)	103	−10.0
NbIr$_3$ (ξ)	Eq. (171)	103	−9.9
TaRh$_3$ (ξ)	Eq. (171)	103	−9.9
TaIr$_3$ (ξ)	Eq. (171)	103	−9.9
Nb$_2$Rh (σ)	Mp (176)	103	−9.6
Nb$_2$Ir (σ)	Mp (176)	103	−10.4
Ta$_2$Rh (σ)	Mp (176)	103	−10.0
Ta$_2$Ir (σ)	Mp (176)	103	−10.7
NbPd$_3$	Eq. (171)	105	−9.3
NbPt$_3$	Eq. (171)	105	−9.9
TaPt$_3$	Eq. (171)	105	−9.9
TaPd$_3$	Eq. (171)	105	−9.9
Nb$_2$Pd (σ)	Mp (176)	105	−9.6
Ta$_2$Pd (σ)	Mp (176)	105	−11.6
TaPd (tetrag)	Mp (176)	105	−12.8
Ta$_2$Pt (σ)	Mp (176)	105	−10.8
TaPt$_2$ (ortho)	Mp (176)	105	−11.6
Mo$_2$Re$_3$ (σ)	Mp (176)	81	−5.5
MoRe$_3$ (α-Mn)	Mp (176)	81	−2.9
WRe (σ)	Mp (176)	81	−6.7
WRe$_3$ (α-Mn)	Mp (176)	81	−3.8
Mo$_5$Ru$_3$ (σ)	Mp (176)	72	−3.1
W$_5$Ru$_3$ (σ)	Mp (176)	72	−3.4
Mo$_5$Os$_3$ (σ)	Mp (176)	72	−4.5
W$_5$Os$_3$ (σ)	Mp (176)	72	−5.3
Mo$_3$Ir (β-W)	Mp (176)	74	−3.2
W$_3$Ir (σ)	Mp (176)	74	−3.2

In addition to the foregoing, the heats of formation of other compounds can be estimated as follows. Consider a compound phase Ψ having a stoichiometric composition x_* in the i–j system. As a first approximation, we estimate the enthalpy of this phase $H^\Psi[T]$ at any temperature by

$$H^\Psi[T] = (1 - x_*)H_i° + x_* H_j° + \Delta H_f^\Psi[298°K] \quad \text{cal/g-atom} \quad (172)$$

The entropy of this compound is approximated by

$$S^\Psi[T] = (1 - x_*)S_i° + x_* S_j° \quad (173)$$

where $H_i°$, $S_i°$, $H_j°$, and $S_j°$, are the enthalpies and entropies of the forms of the pure metals which are stable at 298°K. Under these conditions, the free energy of the phase becomes

$$F^{\Psi}[T] = (1 - x_*) F_i^{\circ}[T] + x_* F_j^{\circ}[T] + \Delta H_t^{\Psi}[298°K] \quad \text{cal/g-atom} \quad (174)$$

where $F_i^{\circ}[T]$ and $F_j^{\circ}[T]$ are the free energies of the forms of the pure metals that are stable at 298°K (i.e., the β, ϵ, or α forms). If the melting temperature of the compound is known (or can be estimated in the case of peritectic decomposition) as $\overline{T}^{\Psi}$, then

$$F^{\Psi}[\overline{T}^{\Psi}] = F^{L}[x_*, \overline{T}^{\Psi}] \quad (175)$$

Equations (65) and (175) yield

$$\Delta H_t^{\Psi}[298°K] = (1 - x_*) \Delta F_i^{\circ \to L}[\overline{T}^{\Psi}] + x_* \Delta F_j^{\circ \to L}[\overline{T}^{\Psi}] + Lx_*(1 - x_*)$$
$$+ R\overline{T}^{\Psi}[x_* \ln x_* + (1 - x_*) \ln(1 - x_*)] \quad (176)$$

where L is the appropriate interaction parameter for the i–j system (Table XIII) and $\Delta F_i^{\circ \to L}$ and $\Delta F_j^{\circ \to L}$ are the free energy differences between the form of i and j stable at 298°K and the liquid (Table VIII).

Equation (176) has been used to estimate the heats of formation of the remaining compounds listed in Table XV. Reference to the estimated heats of formation indicates relatively small values for the $ZrMo_2$ set of Laves phases. However, larger negative heats of formation are noted as one proceeds to the $ZrRe_2$, $ZrRu_2$, $ZrRh_2$, and $ZrPt_2$ sets. The maximum negative values in the $ZrRh_2$ set reflects the fact that these systems, (Zr/Hf) vs. (Rh/Ir), have the most negative values of L. Thus, all of the compounds formed between partners in Groups 4–8, 4–9, and 4–10 exhibit large negative heats of formation.

Similarly, the compounds in (Nb/Ta) binaries exhibit the least negative heats of formation in the $NbRe_3$ and $TaRe_3$ cases. However, when one proceeds to Group 5–8, 5–9, and 5–10 interactions, larger negative values are encountered. However, these are not quite as large as in the corresponding 4–8, 4–9, and 4–10 systems.

Finally, the compounds in the (Mo/W) binaries have the smallest (negative) heats of formation.

CHAPTER VIII

Computation of Component Vapor Pressures in Binary Alloy Systems

When a compound or alloy is exposed to very high temperatures below its melting point, vaporization of the components can occur. If the vaporization of individual components occurs at differing rates, the composition of the alloy or compound phase can change. Under certain conditions this can lead to melting or changes in phase constitution. The thermodynamic description presented here permits computation of the vapor pressure of component elements in all of the binary systems as a function of temperature and composition. The development detailed earlier in connection with Eqs. (65)–(102) provides the framework for this calculation.

In order to illustrate the computations, vapor pressure vs. composition calculations have been performed for selected systems at high temperatures. The results have been obtained by employing published values of p_i° (7, 74), which are contained in Table XVI. The results are shown in Figs. 106–127.

In Figures 106–127, the curves p_i^ϕ and p_j^ϕ (where ϕ denotes the L, β, ϵ, or α phase) were calculated for all of the phases present at the temperature noted on the top of each diagram. Two-phase regions where the vapor pressure of i and j are constant are indicated by horizontal pressure–composition curves in line with Eqs. (80) and (81). This composition is shown in Fig. 113 for the system Ta–Ru at 1800°K. The first step is to calculate the pressure–composition curves for p_{Ru}^β, p_{Ru}^ϵ, p_{Ta}^β, and p_{Ta}^ϵ, since only the β and ϵ phases are stable at this temperature. The β phase terminates at $x_{\beta\epsilon} = 0.560$, while the limit of the ϵ phase is $x_{\epsilon\beta} = 0.620$. Thus, horizontal segments of both Ta and Ru vapor pressure curves occur across the $\beta + \epsilon$ field from $x = 0.560$ to $x = 0.620$ as the stable phase changes from β to ϵ. The intrusion of the TaRu$_3$ (ξ) compound in the ϵ field causes additional steps in the ϵ portion of the pressure curves, as indicated in Fig. 113. The vertical lines

166

TABLE XVI

VAPOR PRESSURES OF THE ELEMENTS [a,b]

T (°K)	Zr	Hf [c]	Nb	Ta	Mo	W	Re
1000	24.4	25.6	29.8	33.4	26.8	36.9	32.9
1500	13.9	14.7	17.2	19.9	15.3	22.0	20.0
2000	8.7	9.2	11.0	13.1	9.7	14.5	12.8
2500	5.6	6.0	7.2	9.0	6.3	10.0	8.8
3000	3.6	4.0	4.8	6.2	4.1	7.0	6.2
3500	2.3	2.2	3.1	4.6	2.6	4.9	4.1
4000	1.1	1.0	1.8	2.8	1.5	3.3	3.0

T (°K)	Ru	Os [d]	Rh	Ir	Pd	Pt
1000	25.7	26.7	21.2	26.9	11.9	21.6
1500	13.7	15.2	11.0	15.3	5.8	11.9
2000	8.8	9.4	6.8	9.5	2.9	7.0
2500	4.5	5.9	4.0	6.0	1.2	4.2
3000	3.3	3.6	2.2	3.9	0.0	2.4
3500	1.8	1.8	0.9	2.3	−0.9	1.1
4000	0.7	0.4	0.0	1.1	−1.7	0.1

[a] Minus log pressure in atmospheres.
[b] Hultgren, Orr, Anderson, and Kelley (7).
[c] Stull and Sinke (74); corrected for an enthalpy of vaporization of −150 kcal/g-atom.
[d] Stull and Sinke (74).

at $x_*^\xi = 0.750$ are approximate. The TaRu$_3$ compound phase has, in reality, a finite compositional range of stability across which the vapor pressure curves change continuously. However, the present approximation treats TaRu$_3$ as a line compound having an infinitely narrow range of stability at all temperatures below its melting point; the resulting vapor pressure curves are vertical steps. The final pressure curves for Ta–Ru are shown in the preceding diagram, Fig. 112. The remaining diagrams, Figs. 106–127, were constructed in the same fashion.

The vapor pressures for Hf–Ta at 2300°K (Fig. 106), Nb–Mo at 2500°K (Fig. 111), Re–Ru at 2500°K (Fig. 125), and Ir–Pt at 2000°K (Fig. 127) are all smooth curves because only one phase is stable for these systems over the entire range of composition at the temperatures selected.

The systems containing two stable phases (and therefore a single discontinuity in the pressure curves) are seen in the case of Mo–Ru at 2200°K (Fig. 116), W–Os at 2500°K (Fig. 117), Ta–Re at 2500°K (Fig. 121), W–Re at 3000°K (Fig. 122), Re–Ir at 2500°K (Fig. 123), Re–Pd at 1800°K (Fig. 124), and Ru–Pt at 1800°K (Fig. 126). The large two-phase field ($\epsilon + \alpha$) occurring in the Re–Pd system is to be noted. Systems containing three

(text continues on page 179)

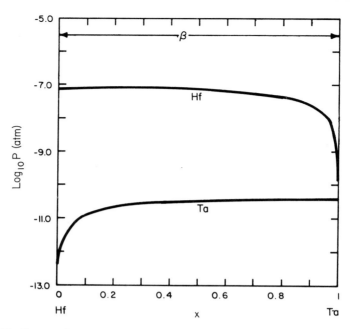

FIG. 106. Computed vapor pressure–composition relations in the Hf–Ta system at 2300 °K.

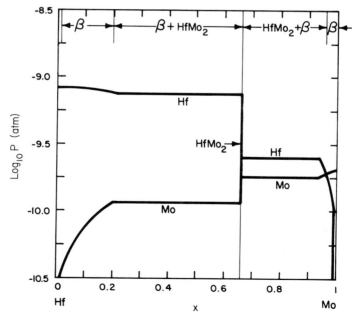

FIG. 107. Computed vapor pressure–composition relations in the Hf–Mo system at 2000 °K.

168

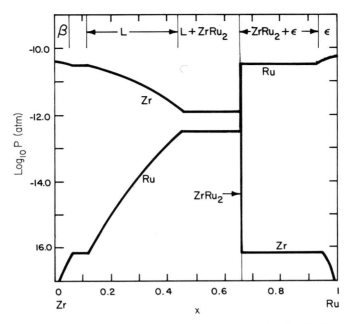

FIG. 108. Computed vapor pressure–composition relations in the Zr–Ru system at 1800°K.

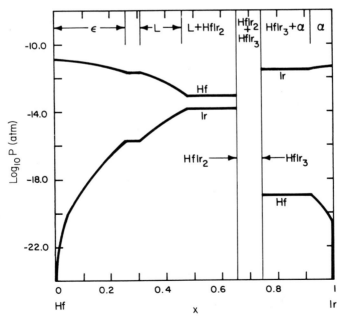

FIG. 109. Computed vapor pressure–composition relations in the Hf–Ir system at 1800°K.

169

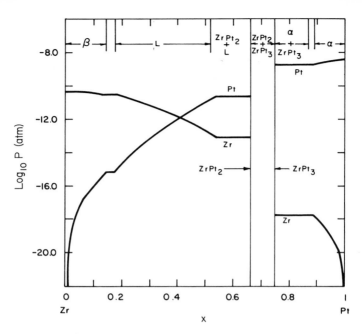

Fig. 110. Computed vapor pressure–composition relations in the Zr–Pt system at 1800°K.

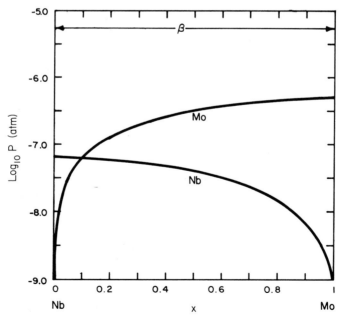

Fig. 111. Computed vapor pressure–composition relations in the Nb–Mo system at 2500°K.

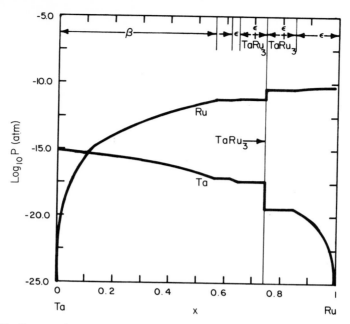

FIG. 112. Computed vapor pressure–composition relations in the Ta–Ru system at 1800°K.

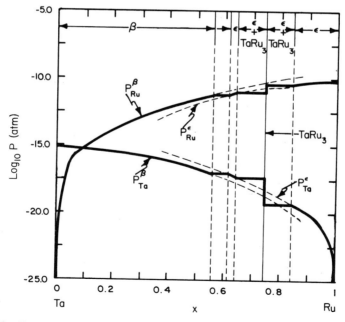

FIG. 113. Computed vapor pressure–composition relations in the Ta–Ru system at 1800°K.

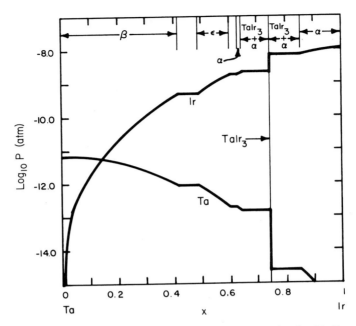

FIG. 114. Computed vapor pressure–composition relations in the Ta–Ir system at 2200°K.

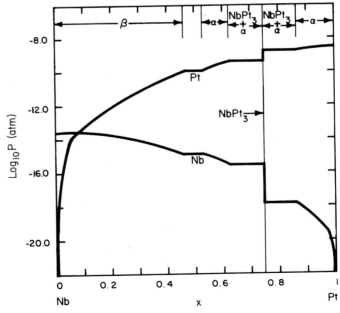

FIG. 115. Computed vapor pressure–composition relations in the Nb–Pt system at 1800°K.

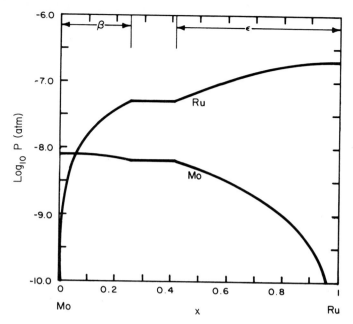

FIG. 116. Computed vapor pressure–composition relations in the Mo–Ru system at 2200 °K.

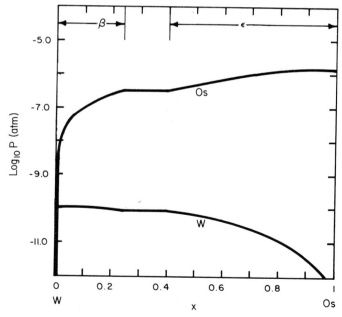

FIG. 117. Computed vapor pressure–composition relations in the W–Os system at 2500 °K.

173

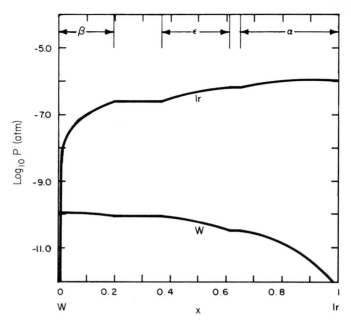

Fig. 118. Computed vapor pressure–composition relations in the W–Ir system at 2500°K.

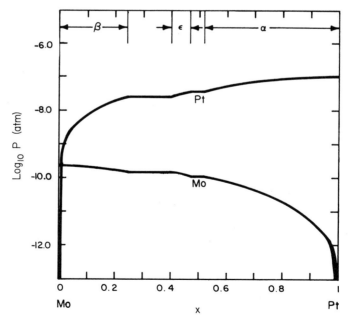

Fig 119. Computed vapor pressure–composition relations in the Mo–Pt system at 2000°K.

174

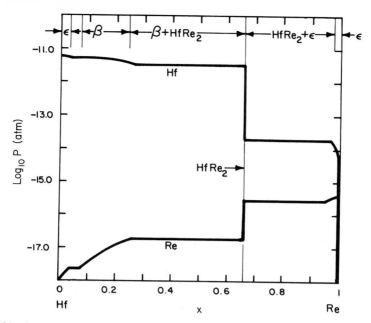

FIG. 120. Computed vapor pressure–composition relations in the Hf–Re system a 1800°K.

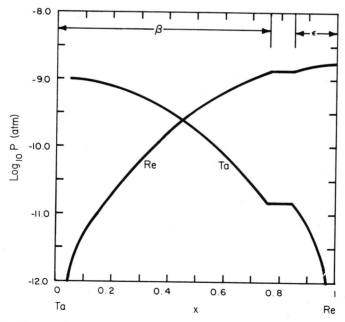

FIG. 121. Computed vapor pressure–composition relations in the Ta–Re system at 2500°K.

175

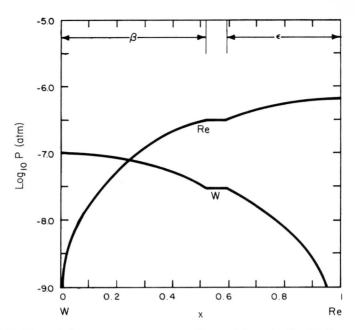

FIG. 122. Computed vapor pressure–composition relations in the W–Re system at 3000 °K.

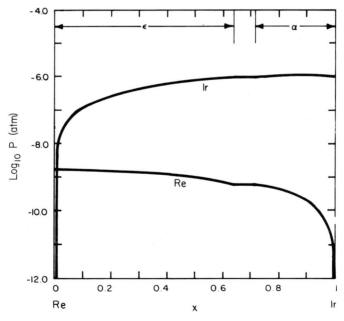

FIG. 123. Computed vapor pressure–composition relations in the Re–Ir system at 2500 °K.

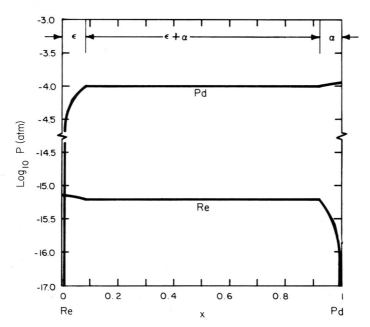

Fig. 124. Computed vapor pressure–composition relations in the Re–Pd system at 1800°K.

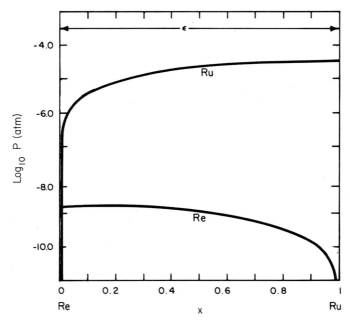

Fig. 125. Computed vapor pressure–composition relations in the Re–Ru system at 2500°K.

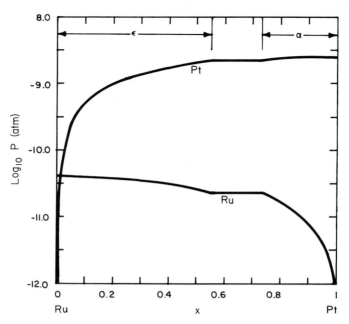

FIG. 126. Computed vapor pressure–composition relations in the Ru–Pt system at 1800°K.

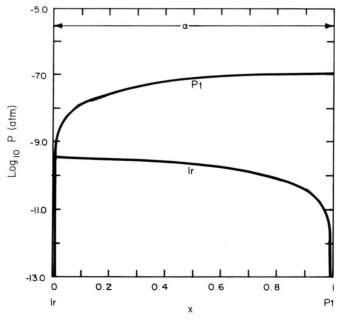

FIG. 127. Computed vapor pressure–composition relations in the Ir–Pt system at 2000°K.

178

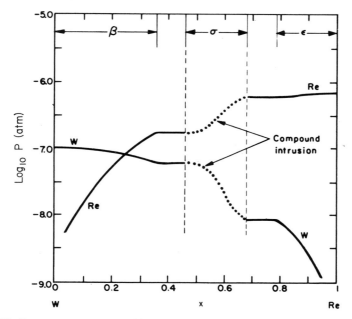

FIG. 128. Vapor pressure–composition relations in the W–Re system at 3000 °K due to sigma phase intrusion.

stable phases and therefore two discontinuities in the pressure curves are W–Ir at 2500°K (Fig. 118) and Mo–Pt at 2000°K (Fig. 119).

A single compound phase (λ) is encountered in the case of Hf–Mo at 2000°K (Fig. 107), Zr–Ru at 1800°K (Fig. 108), Ta–Ru at 1800°K (Fig. 112), Ta–Ir at 2200°K (Fig. 114), Nb–Pt at 1800°K (Fig. 115), and Hf–Re at 1800°K (Fig. 120). In these cases the discontinuous step in the pressure curves of i and j occur as previously described. Hf–Ir at 1800°K (Fig. 109) and Zr–Pt at 1800°K (Fig. 110) have both the λ and ξ compound phases present. Note that the pressure curves are not connected between the stoichiometric compositions $x_* = 0.667$ and $x_* = 0.750$. The reason for this is that the present line compound approximation does not specify the free energy–composition curve for the λ or ξ phases. Thus, the composition limits $x_{\lambda\xi}$ and $x_{\xi\lambda}$ cannot be calculated. As a result, it is impossible to compute the level of the pressure curves across the $\lambda + \xi$ field.

In all of Figs. 106–127 the pressure curve of element i decreases and the pressure curve of element j increases with increasing composition either smoothly or in steps, in accordance with the relative changes in activity of the elements. In fact p_i must go to zero at $x = 1.0$ and p_j must go to zero at $x = 0.0$. Since the ordinate of the pressure curves is log p, the curves remain finite until very close to the limits $x = 0$ and 1, at which points they fall rapidly to minus infinity.

The question of congruent vaporization can now be considered. Congruent vaporization occurs within a single-phase region when the rates of vaporization of i and j are in the same ratio as their respective atomic concentrations. The Langmuir vaporization equation specifies the individual rates of vaporization* of i and j as follows:

$$W_i{}^\phi = [2\pi RTM_i]^{-1/2} p_i{}^\phi [x, T]$$
$$= [2\pi RTM_i]^{-1/2} p_i{}^\circ (1 - x) \exp[(\Phi/RT)x^2 + \Delta F_i{}^{\circ \to \phi}/RT] \frac{\text{g-atom}}{\text{cm}^2\text{-sec}} \quad (177)$$

$$W_j{}^\phi = [2\pi RTM_j]^{-1/2} p_j{}^\phi [x, T]$$
$$= [2\pi RTM_j]^{-1/2} p_j{}^\circ x \exp[(\Phi/RT)(1 - x)^2 + \Delta F_j{}^{\circ \to \phi}/RT] \frac{\text{g-atom}}{\text{cm}^2\text{-sec}} \quad (178)$$

where ϕ denotes the phase L, β, ϵ, or α, and Φ denotes the corresponding interaction parameter (i.e., L, B, E, or A). The condition for congruent vaporization is given by Eq. (179).

$$W_i{}^\phi / W_j{}^\phi = (1 - x_c{}^\phi)/x_c{}^\phi \quad (179)$$

where $x_c{}^\phi [T]$ is the composition for congruent vaporization associated with the phase ϕ at temperature T. Solving Eq. (179) after insertion of Eqs. (177) and (178) yields the following expression:

$$x_c{}^\phi = \frac{1}{2}\left\{1 - \frac{RT}{\Phi}\left[\frac{1}{2}\ln\left(\frac{M_j}{M_i}\right) + \ln\left(\frac{p_i{}^\circ}{p_j{}^\circ}\right)\right]\right\} + \frac{\Delta F_j{}^{\circ \to \phi} - \Delta F_i{}^{\circ \to \phi}}{2\Phi} \quad (180)$$

Since the atomic weights M_i and M_j are not very different for the elements in question, $x_c{}^\phi$ will not fall within the range $0 \leqslant x_c{}^\phi \leqslant 1$ unless Φ, RT, and the vapor pressures of the pure elements are comparable in magnitude. The composition $x_e{}^\phi$ at which $p_i{}^\phi$ and $p_j{}^\phi$ are equal is given by Eq. (181).

$$\ln\left(\frac{x_e{}^\phi}{1 - x_e{}^\phi}\right) + \frac{\Phi}{RT}(1 - 2x_e{}^\phi) = \ln\left(\frac{p_i{}^\circ}{p_j{}^\circ}\right) + \frac{\Delta F_i{}^{\circ \to \phi} - \Delta F_j{}^{\circ \to \phi}}{RT} \quad (181)$$

It then follows that

$$x_c{}^\phi = x_e{}^\phi + \frac{RT}{2\Phi}\left[\ln\left(\frac{1 - x_e{}^\phi}{x_e{}^\phi}\right) + \frac{1}{2}\ln\left(\frac{M_i}{M_j}\right)\right] \quad (182)$$

Equation (182) indicates that a necessary but not sufficient condition for a congruency condition to exist is that the pressure curves p_i and p_j cross at

---

* Equation (177) becomes $W = 44.4 \, (MT)^{-1/2} p$ when T is in degrees Kelvin and p is in atmospheres.

the temperature of interest. Among the systems in Figs. 106–127 one might anticipate a congruency to occur in such cases as Ta–Re at 2500°K within the β phase (Fig. 121) or Zr–Pt at 1800°K within the liquid phase (Fig. 110), but no congruencies among such cases as Hf–Ta at 2800°K (Fig. 106) or Re–Ru at 2500°K (Fig. 125) where the vapor pressures of the elements are different by orders of magnitude. It is emphasized that x_c is temperature dependent, so that failure to observe congruency at one temperature within a particular phase does not mean that it cannot be observed at another temperature.

Equation (180) was used to calculate x_c at $T = 2000°K$ for the L, β, ϵ, and α phases in all of the binary systems of interest. Congruency was deemed impossible if:

1. The computed value of x_c was negative or greater than unity.
2. The value of x_c fell outside the compositional range of stability of the phase ϕ at 2000°K.
3. The particular phase ϕ was unstable at 2000°K or entirely absent from the phase diagram.

Of the 72 systems examined, only 14 indicated a congruency condition at 2000°K. These are listed in Table XVII along with the particular phase in which the congruency will be found. When a compound phase occurs, the vapor pressure curves can cross, as for example, in Zr–Ru at 1800°K (Fig. 108). However, the present "line compound" approximation precludes detailed computations of x_c for such cases at this time. Modifications of the

TABLE XVII

COMPUTED COMPOSITIONS FOR
CONGRUENT VAPORIZATION AT 2000°K

System	x_c	Phase
Zr–Rh	0.385	L
Zr–Pd	0.145	L
Zr–Pt	0.402	L
Hf–Rh	0.354	L
Hf–Pd	0.095	ϵ
Hf–Pt	0.368	L
Nb–Ru	0.228	β
Nb–Os	0.311	β
Nb–Ir	0.301	β
Nb–Pt	0.010	β
Ta–Re	0.418	β
Ta–Os	0.071	β
Ta–Ir	0.032	β
Mo–Ir	0.556	ϵ

line compound description along the lines employed earlier (*90–92*) could eliminate this difficulty.

It should be pointed out that the foregoing discussion is based solely on the computed phase relations between L, β, ϵ, α, λ, and ξ. No modifications have been introduced to allow for differences between the calculated phase diagrams and those observed experimentally. Moreover, no allowance has been made for finite compound phase fields. In order to illustrate the procedure for making such adjustments, it is instructive to reconsider the vapor pressure–composition curves computed for the W–Re system at 3000°K, which are shown in Fig. 122.

Reference to Fig. 81 shows that a broad compound phase (σ) intrudes over the range $0.45 \leqslant x \leqslant 0.68$ at this temperature restricting the β field to $0 \leqslant x \leqslant 0.36$ and the ϵ field to $0.78 \leqslant x \leqslant 1$. Figure 122, based on β/ϵ equilibria, does not reflect this intrusion. In order to modify Fig. 122 to reflect σ phase intrusion, it is necessary to introduce the β/σ and σ/ϵ phase boundaries as shown in Fig. 128. The latter illustrates the method for considering the σ phase. The dotted portions of the vapor pressure–composition curves for tungsten and rhenium are merely guesses at the compositional dependence across the σ field.

CHAPTER IX

Calculation of Regular-Solution Phase Diagrams for Titanium-Base Binary Systems

The regular-solution description developed in Chapters III–VIII and applied to refractory transition metals has been extended to titanium-base binary systems (96). Twenty phase diagrams between titanium and other transition metals have been computed on the basis of the lattice stability parameters given in Figs. 31 and 32 and using the methods for estimating the free energies of solution and compound phases described in Chapters IV and VI.

The formulation afforded by Eqs. (65)–(94) in Chapter III was employed to describe the thermodynamic properties of the Ti–Zr, Ti–Hf, Ti–V, Ti–Nb, Ti–Ta, Ti–Cr, Ti–Mo, Ti–W, Ti–Re, Ti–Ru, Ti–Os, Ti–Co, Ti–Rh, Ti–Ir, Ti–Ni, Ti–Pd, Ti–Pt, Ti–Cu, Ti–Al, and V–Al systems. The lattice stability parameters specifying the relative stability of the β, ϵ, α, and liquid forms of all the pure metals involved are shown in Figs. 31 and 32. The only exception is aluminium, which was not considered earlier. Table XVIII contains the free energy differences required to describe the lattice stability of the principal forms of the metals of interest (16) that are not contained in Table VIII. The lattice stability parameters for aluminum were obtained by combining an analysis of the Al–Mg system with the assumption that the entropy differences between the β and ϵ forms of Mg and Be [see Eq. (29)] are identical as shown in Table XVIII.

At this point it is of interest to digress and compare the values for the enthalpy differences between the bcc, hcp, and fcc forms of Zn, Al, and Mg derived from pseudopotential calculations (Table VI) and those derived empirically (Tables VIII and XVIII). In the case of Mg excellent agreement is observed, since the pseudopotential method yields values of $\Delta H^{\beta \to \epsilon} = -1345$ cal/g-atom and $\Delta H^{\alpha \to \epsilon} = -715$ cal/g-atom, while the present calcu-

β = bcc
α = fcc

TABLE XVIII

SUMMARY OF FREE ENERGY DIFFERENCES FOR THE PURE METALS [a]

Metal	Free energy difference (cal/g-atom)	Temperature (°K)	Metal	Free energy difference (cal/g-atom)	Temperature (°K)
Ti	$\Delta F^{\beta \to L} = 3880 - 2.0T$	$T^\beta = 1940$	Cr	$\Delta F^{\beta \to L} = 4350 - 2.0T$	$T^\beta = 2175$
	$\Delta F^{\epsilon \to L} = 4920 - 2.9T$	$T^\epsilon = 1697$		$\Delta F^{\epsilon \to L} = 2350 - 2.0T$	$T^\epsilon = 1175$
	$\Delta F^{\alpha \to L} = 4120 - 2.9T$	$T^\alpha = 1421$		$\Delta F^{\alpha \to L} = 1850 - 2.15T$	$T^\alpha = 860$
	$\Delta F^{\beta \to \epsilon} = -1040 + 0.9T$	$T_0^{\beta\epsilon} = 1155$		$\Delta F^{\beta \to \epsilon} = 2000$	
	$\Delta F^{\alpha \to \epsilon} = -800$			$\Delta F^{\alpha \to \epsilon} = -500 - 0.15T$	
	$\Delta F^{\alpha \to \beta} = +240 - 0.90T$	$T_0^{\alpha\beta} = 265$		$\Delta F^{\alpha \to \beta} = -2500 - 0.15T$	
V	$\Delta F^{\beta \to L} = 4360 - 2.0T$	$T^\beta = 2180$	Co	$\Delta F^{\beta \to L} = 2710 - 3.37T$	$T^\beta = 805$
	$\Delta F^{\epsilon \to L} = 2860 - 2.80T$	$T^\epsilon = 1020$		$\Delta F^{\epsilon \to L} = 4210 - 2.47T$	$T^\epsilon = 1705$
	$\Delta F^{\alpha \to L} = 2210 - 2.85T$	$T^\alpha = 780$		$\Delta F^{\alpha \to L} = 4100 - 2.32T$	$T^\alpha = 1768$
	$\Delta F^{\beta \to \epsilon} = +1500 + 0.80T$			$\Delta F^{\beta \to \epsilon} = -1500 - 0.90T$	
	$\Delta F^{\alpha \to \epsilon} = -650 - 0.05T$			$\Delta F^{\alpha \to \epsilon} = -110 + 0.15T$	$T_0^{\alpha\epsilon} = 730$
	$\Delta F^{\alpha \to \beta} = -2150 - 0.85T$			$\Delta F^{\alpha \to \beta} = +1390 + 1.05T$	

Ni

$\Delta F^{\beta \to L} = 2860 - 3.24T$
$\Delta F^{\epsilon \to L} = 3960 - 2.74T$
$\Delta F^{\alpha \to L} = 4210 - 2.44T$
$\Delta F^{\beta \to \epsilon} = -1100 - 0.50T$
$\Delta F^{\alpha \to \epsilon} = +250 + 0.30T$
$\Delta F^{\alpha \to \beta} = +1350 + 0.80T$

$\bar{T}^{\beta} = 885$
$\bar{T}^{\epsilon} = 1445$
$\bar{T}^{\alpha} = 1725$

Cu

$\Delta F^{\beta \to L} = 2270 - 2.10T$
$\Delta F^{\epsilon \to L} = 2970 - 2.60T$
$\Delta F^{\alpha \to L} = 3120 - 2.30T$
$\Delta F^{\beta \to \epsilon} = -700 + 0.50T$
$\Delta F^{\alpha \to \epsilon} = 150 + 0.30T$
$\Delta F^{\alpha \to \beta} = 850 - 0.20T$

$\bar{T}^{\beta} = 1081$
$\bar{T}^{\epsilon} = 1142$
$\bar{T}^{\alpha} = 1357$

Al

$\Delta F^{\beta \to L} = 150 - 1.60T$
$\Delta F^{\epsilon \to L} = 1250 - 2.32T$
$\Delta F^{\alpha \to L} = 2560 - 2.75T$
$\Delta F^{\beta \to \epsilon} = -1100 + 0.72T$
$\Delta F^{\alpha \to \epsilon} = 1310 - 0.43T$
$\Delta F^{\alpha \to \beta} = 2410 - 1.15T$

$\bar{T}^{\beta} = 94$
$\bar{T}^{\epsilon} = 539$
$\bar{T}^{\alpha} = 931$
$T_0^{\beta\epsilon} = 1528$
$T_0^{\alpha\epsilon} = 3047$
$T_0^{\alpha\beta} = 2080$

Be

$\Delta F^{\beta \to L} = 2490 - 1.60T$
$\Delta F^{\epsilon \to L} = 3590 - 2.32T$
$\Delta F^{\alpha \to L} = 3125 - 2.75T$
$\Delta F^{\beta \to \epsilon} = -1100 + 0.72T$
$\Delta F^{\alpha \to \epsilon} = -465 - 0.43T$
$\Delta F^{\alpha \to \beta} = 635 - 1.15T$

$\bar{T}^{\beta} = 1556$
$\bar{T}^{\epsilon} = 1547$
$\bar{T}^{\alpha} = 1136$
$T_0^{\beta\epsilon} = 1528$
$T_0^{\alpha\beta} = 552$

Mg

$\Delta F^{\beta \to L} = 1040 - 1.60T$
$\Delta F^{\epsilon \to L} = 2140 - 2.32T$
$\Delta F^{\alpha \to L} = 1675 - 2.75T$
$\Delta F^{\beta \to \epsilon} = -1100 + 0.72T$
$\Delta F^{\alpha \to \epsilon} = -465 - 0.43T$
$\Delta F^{\alpha \to \beta} = 635 - 1.15T$

$\bar{T}^{\beta} = 650$
$\bar{T}^{\epsilon} = 922$
$\bar{T}^{\alpha} = 610$
$T_0^{\beta\epsilon} = 1528$
$T_0^{\alpha\beta} = 552$

lations yield -1100 cal/g-atom and -465 cal/g-atom, respectively. For Al the agreement is poorer, since the pseudopotential method yields $\Delta H^{\beta \rightarrow \epsilon} = -3200$ cal/g-atom and $\Delta H^{\alpha \rightarrow \epsilon} = +1170$ cal/g-atom, while the present results indicate -1100 cal/g-atom and $+1310$ cal/g-atom, respectively. The worst agreement is observed for zinc, where the pseudopotential method incorrectly predicts that the fcc form is more stable than the hcp form. In this case, the pseudopotential method yields $\Delta H^{\beta \rightarrow \epsilon} = -1700$ cal/g-atom and $\Delta H^{\alpha \rightarrow \epsilon} = +2325$ cal/g-atom, while the values in Table VII are -690 cal/g-atom and -440 cal/g-atom, respectively.

Returning to the titanium-base systems under discussion, specification of the lattice stability parameters contained in Tables VIII and XVIII is followed by estimation of the numerical values of the interaction parameters L, B, E, and A for the liquid, β, ϵ, and α phases for these cases along the lines described in Chapter IV. The results are contained in Tables XIX–XXI.

TABLE XIX

CALCULATION OF e_p AND e_1 PARAMETERS [a]

Element	H (cal/g-atom)	V (cm³/g-atom)	System	e_p (cal/g-atom)	e_1 (cal/g-atom)
Ti	$-113,000$	10.72	Ti–Zr	3	2,328
Zr	$-146,000$	14.04	Ti–Hf	55	1,732
Hf	$-150,000$	13.50	Ti–V	1,971	1,788
V	$-123,000$	8.37	Ti–Nb	3,621	4
Nb	$-173,000$	10.84	Ti–Ta	5,179	11
Ta	$-187,000$	10.91	Ti–Cr	694	3,912
Cr	$-94,000$	7.23	Ti–Mo	4,260	581
Mo	$-157,000$	9.40	Ti–W	11,020	497
W	$-202,000$	9.58	Ti–Re	10,471	1,350
Re	$-186,000$	8.86	Ti–Ru	6,910	2,399
Ru	$-155,000$	8.19	Ti–Os	7,428	1,966
Os	$-162,000$	8.43	Ti–Co	2,262	5,760
Co	$-102,000$	6.69	Ti–Rh	3,245	1,973
Rh	$-133,000$	8.31	Ti–Ir	6,706	1,713
Ir	$-160,000$	8.56	Ti–Ni	2,636	6,222
Ni	$-103,000$	6.57	Ti–Pd	11	920
Pd	$-91,000$	8.86	Ti–Pt	2,176	829
Pt	$-135,000$	9.10	Ti–Cu	89	3,975
Cu	$-81,000$	7.11	Ti–Al	1,384	115
Al	$-77,000$	10.00	V–Al	6,218	787
			V–Cr	243	579

[a] Where $e_p = 0.3(V_{Ti} + V_j)[(-H_{Ti}/V_{Ti})^{1/2} - (-H_j/V_j)^{1/2}]^2$
$e_1 = -0.5(H_{Ti} + H_j)(V_{Ti} - V_j)^2(V_{Ti} + V_j)^{-2}$.

As indicated in Table XIX, the calculation of e_p and e_1 was performed on the basis of Eqs. (126) and (142) described in Chapter IV. In this regard the computations of e_p and e_1 for titanium base binary systems are identical with those performed for refractory metal binary systems shown in Tables XI and XII.

Reference to Eqs. (140)–(150) and Tables XI–XIII of Chapter IV show that e_0, e_2, e_3, and e_4 must be evaluated to compute the L, B, E, and A parameters shown in Table XX. The relationships between the e's and the inter-

TABLE XX

CALCULATION OF L, B, E, AND A PARAMETERS [a]

System	e_0	e_p	e_1	e_2	e_3	e_4
Ti–Zr	$-1,300$	3	2,328	0	$+1,060$	0
Ti–Hf	$-1,300$	55	1,732	0	$+1,060$	0
Ti–V	$-1,100$	1,971	1,788	0	0	0
Ti–Nb	-500	3,621	4	0	0	0
Ti–Ta	$-2,400$	5,179	11	0	0	0
Ti–Cr	$+800$	694	3,912	-900	$+2,430$	0
Ti–Mo	$-2,700$	4,260	581	-900	$+2,430$	0
Ti–W	$-2,700$	11,020	497	-900	$+2,430$	0
Ti–Re	$-30,000$	10,471	1,350	$-1,900$	$+8,600$	$+100$
Ti–Ru	$-22,000$	6,910	2,399	$-3,700$	$+8,550$	$+80$
Ti–Os	$-22,000$	7,428	1,966	$-3,700$	$+8,550$	$+80$
Ti–Co	$-40,000$	2,262	5,760	-600	$+9,230$	$-8,900$
Ti–Rh	$-40,000$	3,245	1,973	-600	$+9,230$	$-8,900$
Ti–Ir	$-40,000$	6,706	1,713	-600	$+9,230$	$-8,900$
Ti–Ni	$-34,000$	2,636	6,222	$+2,100$	$+5,130$	$-10,050$
Ti–Pd	$-34,000$	11	920	$+2,100$	$+5,130$	$-10,050$
Ti–Pt	$-34,000$	2,176	829	$+2,100$	$+5,130$	$-10,050$
Ti–Cu	$-4,400$	89	3,975	$+2,000$	$+2,000$	$-3,550$
Ti–Al	$-10,500$	1,384	115	$-4,000$	$-1,000$	$+4,800$
V–Al	$-13,400$	6,218	787	$-6,200$	$-1,000$	$+4,800$
V–Cr	$-4,000$	243	579	0	0	0

[a] In calories per gram atom: $L = e_0 + e_p$, $B = L + e_1 + e_2$, $E = B + e_3$, $A = E + e_4$.

action parameters employed in Chapter IV [Eqs. (123), (140), (147), and (149)] were used in deriving the values contained in Table XXI. However, the estimation procedures for e_0, e_2, e_3, and e_4 described in Tables XI and XII are insufficient to specify these parameters for the case of the titanium base alloy systems. Consequently, the current values (shown in Table XX) were derived by using the earlier results as a guide and employing the observed phase diagrams to provide supplemental information.

TABLE XXI

SUMMARY OF L, B, E, AND A PARAMETERS [a]

System	L	B	E	A
Ti–Zr	− 1,297	+ 1,031	+ 2,091	+ 2,091
Ti–Hf	− 1,245	+ 487	+ 1,547	+ 1,547
Ti–V	+ 871	+ 2,659	+ 2,659	+ 2,659
Ti–Nb	+ 3,121	+ 3,125	+ 3,125	+ 3,125
Ti–Ta	+ 2,779	+ 2,790	+ 2,790	+ 2,790
Ti–Cr	+ 1,494	+ 4,506	+ 6,936	+ 6,936
Ti–Mo	+ 1,560	+ 1,241	+ 3,671	+ 3,671
Ti–W	+ 8,320	+ 7,917	+ 10,347	+ 10,347
Ti–Re	− 19,529	− 20,079	− 11,479	− 11,379
Ti–Ru	− 15,090	− 16,391	− 7,841	− 7,761
Ti–Os	− 14,572	− 16,306	− 7,756	− 7,676
Ti–Co	− 37,738	− 32,578	− 23,348	− 32,248
Ti–Rh	− 36,755	− 35,382	− 26,152	− 35,052
Ti–Ir	− 33,294	− 32,181	− 22,951	− 31,851
Ti–Ni	− 31,364	− 23,042	− 17,912	− 27,962
Ti–Pd	− 33,989	− 30,969	− 25,839	− 35,889
Ti–Pt	− 31,824	− 28,895	− 23,765	− 33,815
Ti–Cu	− 4,311	+ 1,664	+ 3,664	+ 114
Ti–Al	− 9,116	− 13,001	− 14,001	− 9,201
V–Al	− 7,182	− 12,595	− 13,595	− 8,795
V–Cr	− 3,759	− 3,178	− 3,178	− 3,178

[a] In calories per gram atom.

1. CONSIDERATION OF COMPOUND PHASES IN TITANIUM-BASE BINARY SYSTEMS

In addition to examining the competition between liquid, bcc, hcp, and fcc phases in each system, some of the dominant compound phases are considered. In line with the description offered earlier, these phases are treated as line compounds in order to simplify the description. Thus, according to Eq. (91) the free energy of the compound phase Ψ characterized by a composition x_* is defined as

$$F^{\Psi} = (1 - x_*)F_i^{\theta} + x_* F_j^{\theta} + x_*(1 - x_*)(L - C) \quad \text{cal/g-atom} \qquad (183)$$

where, as in Eq. (91), L is the interaction parameter for the liquid phase in the system i–j, and $\theta = \beta$, ϵ, or α. Thus, as shown in Table XXII, the TiRu and TiOs phases ($\Psi = \mu$) exhibit the CsCl structure, and $\theta = \beta$; i.e., the base is bcc. Thus on the basis of Eq. (183) and Tables XXI and XXII, the free energy of the TiRu phase is defined as

$$F^{\mu} = \tfrac{1}{2}F_{Ti}^{\beta} + \tfrac{1}{2}F_{Ru}^{\beta} - \tfrac{1}{4}(33,090) \quad \text{cal/g-atom} \qquad (184)$$

in the present development.

The identification of the base phase θ in Table XXII was performed by computing the relative stability of the β, ϵ, and α phases at composition x_* and identifying θ with the most stable form. The value of C in Table XXII was fixed by using the known melting temperature of the compound phase.

Thus, the present calculation of titanium base binary phase diagrams differs from the calculation of refractory metal phase diagrams in that the experimental diagrams were employed to a larger extent in fixing the interaction parameters and compound constants than was the case in Chapters IV–VI. The description of lattice stability, interaction, and compound parameters contained in Tables VIII, XVIII, XXI, and XXII were employed to

TABLE XXII

Description of Compound Phases

Compound	Crystal structure	Designation	Base phase	C (cal/g-atom)	Computed heat of formation (kcal/g-atom)
TiCr$_2$	Laves	λ	ϵ	13,000	-1.2
Ti$_5$Re$_{24}$	bcc (α-Mn)	π	β	26,000	-6.0
TiRu	CsCl	μ	β	18,000	-7.2
TiOs	CsCl	μ	β	18,000	-7.1
TiCo	CsCl	μ	β	14,000	-11.7
TiCo$_2$	Laves	λ	ϵ	7,000	-9.9
TiCo$_3$	AuCu$_3$	ξ	α	6,000	-7.9
TiRh$_3$	AuCu$_3$	ξ	α	14,000	-9.3
TiIr$_3$	AuCu$_3$	ξ	α	14,000	-8.7
TiNi	CsCl	μ	β	11,000	-9.1
TiNi$_3$	Hexagonal	ρ	ϵ	12,000	-7.9
TiPd$_3$	TiNi$_3$	ρ	ϵ	18,000	-9.6
TiPt$_3$	AuCu$_3$	ξ	α	18,000	-9.1
TiCu	CuAu	λ	α	6,000	-2.2
TiAl	CuAl	λ	α	15,000	-5.6
TiAl$_3$	TiAl$_3$	δ	ϵ	18,000	-4.1
V$_5$Al$_8$	bcc (γ-brass)	ν	β	17,000	-4.2
VAl$_3$	TiAl$_3$	δ	ϵ	22,000	-4.1

compute the titanium base binary phase diagrams on the basis of Eqs. (65)–(94). Solutions were obtained by employing computer programs TRSE and LCRSE contained in Appendix 1 along the lines previously described in Chapters V and VI.

2. COMPARISON OF COMPUTED AND OBSERVED PHASE DIAGRAMS

Figures 129–133 compare 20 computed phase diagrams with those observed according to the compilations of Hansen and Anderko (2) and Elliot (3). The observed Ti–Ru phase diagram is due to Raub and Roeschel (84). In each case, the observed phase diagrams are shown as thin lines, while the computed diagrams are shown by thick lines. The *observed* fields of compound stability are shown as hatched areas on the phase diagrams. By contrast the *computed* compounds are restricted to lines (at fixed composition) as indicated earlier. This restriction could be readily removed (92). However, additional parameters would be required to describe F^{Ψ}. Although there is no experimental version of the Ti–Os phase diagram for comparison with the computed version, the comparisons of the computed and experimental Ti–Zr, Ti–Hf, and Ti–Ru cases are quite satisfactory. The stability of the TiRu phase is slightly underestimated. A closer agreement between the observed and computed melting temperature could be effected by increasing C by 5% for TiRu. Agreement between the computed and observed phase diagrams for Ti–V, Ti–Nb, and Ti–Ta shown in Fig. 130 is excellent. The computed Ti–Al diagram includes only the TiAl and $TiAl_3$ compound phases and does not include the Ti_2Al compound phases. Nevertheless, the computed diagram reproduces most of the salient features of the observed diagram.

The Ti–W case shown in Fig. 131 illustrates a situation where the simple regular-solution model is simply inadequate to describe the observed diagram accurately. This is due to the fact that Eqs. (66), (77), (88), and (89) are restrictive. A larger interaction parameter for the β phase than the present $B = +7917$ would result in a higher value of T_c in better agreement with the observed diagram. However, the low-temperature solubility would be restricted in disagreement with observations. A smaller value of B would increase the low-temperature solubility but depart further from the $\beta/L-\beta$-gap peritectic that is observed. The only solution would be a temperature-dependent interaction parameter (increasing with increasing temperature) or compositional departures from regularity. Subsequent to the publication of these calculations, Rudy and Windisch published a revised version of the Ti–W phase diagram (97), which is shown in Figs. 134 and 135. Their new results are in closer agreement with the computed version of the Ti–W phase diagram than with the diagram shown by Hansen (2). The remaining dia-

(text continues on page 196)

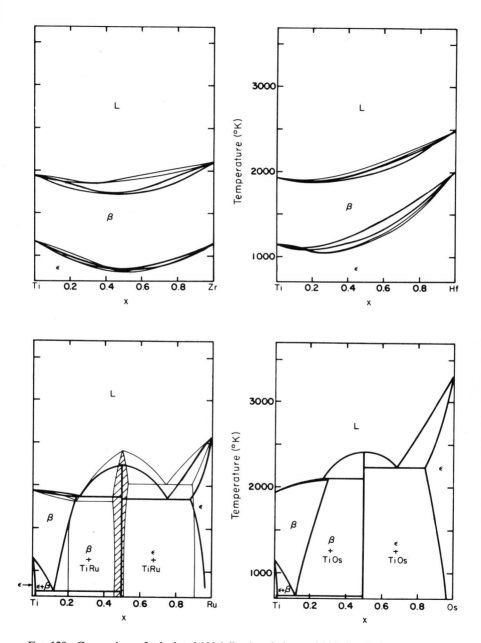

FIG. 129. Comparison of calculated (thick lines) and observed (thin lines) phase diagrams.

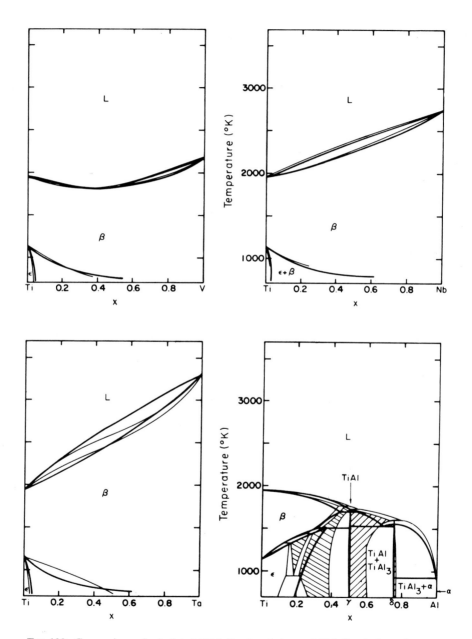

FIG. 130. Comparison of calculated (thick lines) and observed (thin lines) phase diagrams.

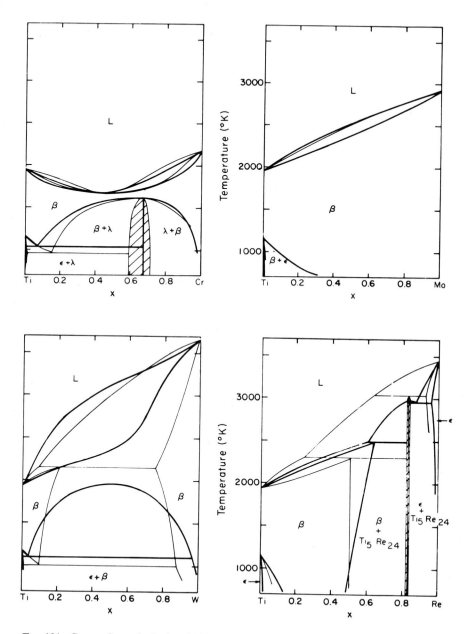

FIG. 131. Comparison of calculated (thick lines) and observed (thin lines) phase diagrams.

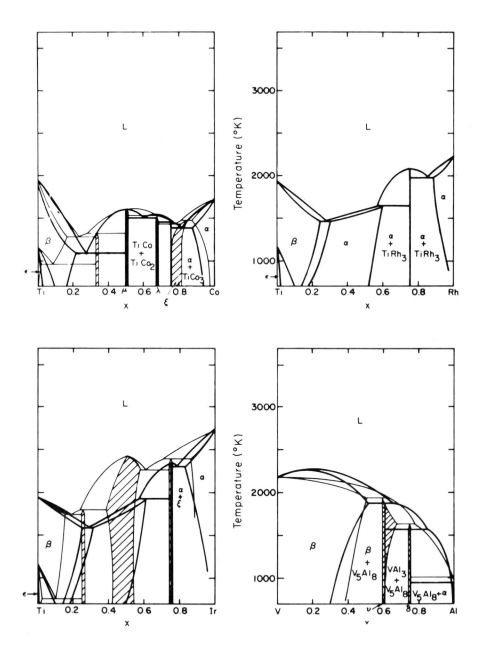

FIG. 132. Comparison of calculated (thick lines) and observed (thin lines) phase diagrams.

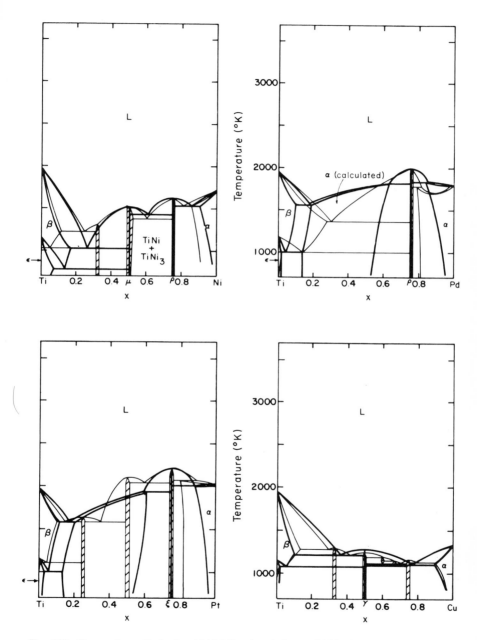

FIG. 133. Comparison of calculated (thick lines) and observed (thin lines) phase diagrams.

grams displayed in Fig. 131 are in good agreement with observation. In the additional phase diagrams shown in Figs. 132 and 133, Ti–Co, V–Al, Ti–Ni, and Ti–Cu are all seen to agree with observations. The computed Ti–Ir and Ti–Pt cases are incomplete only because the equiatomic compounds have not been inserted; TiIr and TiPt are reported to be complex phases. Finally, there is some controversy over the nature of the observed phase diagram for Ti–Pd.

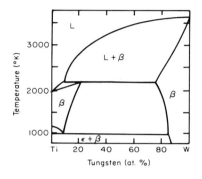

FIG. 134. Observed Ti–W phase diagram (2).

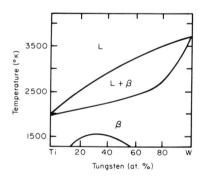

FIG. 135. Revision of the observed Ti–W phase diagram [after Rudy and Windisch (97)]. Compare Figs. 134 and 135 with the lower left panel of Fig. 131.

In general, reference to Figs. 129–133 shows a rather good correspondence between computed and observed phase diagrams.

3. CALCULATION OF THERMODYNAMIC PROPERTIES

The foregoing description permits computation of the activity of each component as a function of temperature and composition in all of the above mentioned systems. Application of Eqs. (95)–(102) defines the activity.

Although few data are available for direct comparison, the limited data in Ti–Cu and Ti–Al (98), Ti–Cr (99), and V–Al (100) are in satisfactory agreement with the calculated activity relations. The heat of formation of compound phases follows directly from Eqs. (169)–(171) or (183) as

$$\Delta H_t = (1 - x_*) \, \Delta H_i^{\circ \to \theta} + x_* \, \Delta H_j^{\circ \to \theta} + x_*(1 - x_*)(L - C) \quad \text{cal/g-atom} \quad (185)$$

where H_i° and H_j° refer to the stable form of i and j at 298°K (i.e., β, ϵ, or α). Equation (185) has been employed to compute the heat of formation of the compounds shown in Table XXII. Although few experimental data are available for comparison (86), relatively good agreement is found in the $TiCr_2$, TiNi, and $TiNi_3$ cases. However, the computed values for TiAl and $TiAl_3$ are smaller in magnitude than those observed by 3–4 kcal/g-atom.

The chief advantage of the current approach is that it can be employed to compute phase equilibria and stability in ternary and higher-order systems as indicated in Chapters XI–XIII. In addition, metastable equilibria and reactions between metastable phases can be readily considered. Application of the currently available programs for computation of ternary equilibria along with the simple parametric description of the thermodynamic properties offers a rapid and relatively inexpensive method for considering the effects of simultaneously varying several alloying ingredients. This approach may thus prove to be an expeditious means of systematizing development of advanced alloy and composite material systems.

CHAPTER X

Consideration of Alternative Descriptions of Phase Stability

Before proceeding with the application of the present formulation of binary phase equilibria in transition metal systems to ternary and multi-component systems, it is instructive to review the current description and compare it with other contemporary approaches to the general problem. The essence of the present development was presented at the beginning of Chapter III following the discussion of Figs. 23–27. This discussion empha-sized the aspect of phase competition as providing a universal basis for considering the stability of metallic phases. As indicated in Chapter III, implementation of this framework in developing an operational means for assessing phase stability required evolution of procedures for determining the lattice stability of the principal forms of the pure metals (bcc, hcp, fcc), as well as a description of the free energy of mixing of stable and unstable phases over complete ranges of temperature and composition.

Since most of the required data are currently unavailable, procedures for estimating the required information were developed and applied in Chapters III–IX. This description is by no means unique. Indeed, other routes toward solution of the general problem, or specific aspects thereof, have been taken by various authors. Although this chapter will not present an exhaustive review of these alternative approaches, several other viewpoints will be reviewed in order to provide some insight into the relation of these alternative routes to the current development.

The present discussion will include experimental measurements of thermo-dynamic properties, correlation of thermodynamic properties with physical properties, expanded analytical descriptions of the free energy of mixing, and relation of phase stability to electron/atom ratios.

1. EXPERIMENTAL MEASUREMENTS OF THERMODYNAMIC PROPERTIES

The most extensive survey of experimental studies of the thermodynamics of metallic systems is currently being undertaken at the University of California under the direction of R. Hultgren (7). The interaction between phase equilibria and thermodynamic properties is considered (7) where it relates to the problem of choosing between competing sets of data for a given system and occasional free energy–composition curves of the type shown in Figs. 24, 25, and 27 are displayed. However, these curves are limited to the regions where a phase is stable. Consideration of metastable and unstable regions are absent. As a consequence (55) utilization of the results of this survey (7) for the prediction of phase equilibria is limited.

Kubaschewski and co-workers (8, 22) and Lumsden (18) have taken a broader approach in correlating thermodynamic measurements and phase equilibria. Kubaschewski and Chart (22) have employed measurements of the enthalpy and free energy of formation to construct free energy–composition curves for a significant number of systems. The noteworthy aspect of this work is that the free energy–composition curves cover stable, metastable, and unstable regions. Indeed, this work has demonstrated several instances where phase boundaries calculated from measurements of the thermodynamic properties are more reliable than boundaries obtained from phase equilibria studies. Kubaschewski has also investigated the deviations of the integral excess entropy and enthalpy from the regular-solution behavior implied by Eqs. (66), (77), (88), and (89). These equations approximate the integral excess free energy as depending parabolically upon composition [i.e., proportional to $x(1-x)$] and independent of temperature. Kubaschewski's result for a wide variety of solutions exhibiting complete mutual solubility indicate that the parabolic representation is an oversimplification. If the maximum values of the excess entropy of mixing divided by the gas constant, $S_E(\text{max})/R$, are plotted against the quantity $H_E(\text{max})/0.5R(T_i^b + T_j^b)$ a straight line is obtained. This relation is based on the boiling points of i and j, T_i^b, and T_j^b, and the maximum value of the heat of mixing $H_E(\text{max})$. In the regular-solution model, $S_E/R = 0$, and $H_E(\text{max})/0.5R$ is equal to 0.25 times the interaction parameter on the basis of Eqs. (66), (77), (88), and (89). Although most of the points of Kubaschewski's plot lie between $0.50 \geqslant S_E(\text{max})/R \geqslant -0.50$ and $-0.50 \leqslant H_E(\text{max})/0.5R(T_i^b + T_j^b) \leqslant 0.50$, Eq. (186) is presented as being representative of the data:

$$S_E(\text{max}) = 1.28 H_E(\text{max})/(T_i^b + T_j^b) \tag{186}$$

In the current framework, this result would modify Eq. (66) as follows:

$$F_E^L = x(1-x)L[1 - 1.28T/(T_i^b + T_j^b)] \tag{187}$$

with similar changes introduced in Eqs. (77), (88), and (89). Equation (187) retains the parabolic representation utilized earlier, which is at variance with Kubaschewski's conclusion that parabolic behavior represents an over-simplification, but includes the finite excess entropy representation described by Eq. (186) as a refinement. This modification could be readily included in the present formulation. Moreover, different values of $T_i^{\,b}$ could be computed for the L, β, ϵ, and α form of each metal with no difficulties on the basis of Tables VIII and XVIII.

Kubaschewski (8) has reported no success in predicting the deviation of $H_E(max)$ and $S_E(max)$ from $x = 0.50$. Lumsden (18) has analyzed a number of metallic systems and presented several clear descriptions of the interaction between thermodynamic properties and phase equilibria. As indicated in Chapter I, Lumsden's calculation of the lattice stability of the bcc, hcp, and fcc forms of thallium (18) was employed successfully (19, 20) to predict the fcc form at high pressure. This calculation (18) was performed for the Tl–Pb and Tl–In systems along the lines illustrated in Fig. 29 for the Fe–Ru system.

2. Correlation of Thermodynamic and Physical Properties

In order to circumvent some of the restrictions related to experimental characterization of the thermodynamic properties, several attempts have been made to correlate these properties with physical phenomena (13, 101–110). The formulation of this method is analogous to the development of Eqs. (39) and (40) for the free energy of a pure metal. In general, the free energy of the ϵ phase of an alloy in the i–j system can be represented by Eq. (188).

$$F^\epsilon[x, T] = H_0^\epsilon[x] + F^\epsilon[\Theta^\epsilon[x]/T] + F^{\epsilon c}[\Theta^\epsilon[x], T] + F^{\epsilon\mu}[x, T]$$
$$- \tfrac{1}{2}\, \gamma^\epsilon[x]\, T^2 - T\, S_p[x, T] \tag{188}$$

Most of the terms in Eq. (188) have the same physical meaning as the counterparts in Eq. (40) except that they are composition dependent. The main difference is the introduction of positional entropy, $S_p[x, T]$, which could include ordering effects. In principle, the specific heat effects associated with the loss of ordering with increases in temperature could be incorporated into $F^{\epsilon\mu}[x, T]$. Combination of Eq. (188) with the definition of the excess free energy given in Eq. (65) yields Eq. (189).

$$F_E^\epsilon[x, T] = \{H_0^\epsilon[x] - (1 - x)H_{i0}^\epsilon - xH_{j0}^\epsilon\}$$
$$+ \{F^{\epsilon\mu}[x, T] - (1 - x)\, F_i^{\epsilon\mu}\,[T] - x\, F_j^{\epsilon\mu}[T]\}$$
$$+ \{F^\epsilon[\Theta^\epsilon[x]/T] - (1 - x)\, F_i^\epsilon\,[\Theta_i^\epsilon/T] - x\, F_j^\epsilon[\Theta_j^\epsilon/T]\}$$
$$+ \{F^{\epsilon c}[\Theta^\epsilon[x], T] - (1 - x)\, F_i^{\epsilon c}[\Theta_i^\epsilon, T] - x\, F_j^{\epsilon c}[\Theta_j^\epsilon, T]\}$$

$$-\tfrac{1}{2}T^2\{\gamma^\epsilon[x] - (1-x)\gamma_i^\epsilon - x\gamma_j^\epsilon\}$$
$$-T\{S_p[x, T] + R[x \ln x + (1-x) \ln(1-x)]\} \tag{189}$$

In the regular-solution approximation, Eq. (189) implies that

$$H_0^\epsilon[x] - (1-x)H_{i0}^\epsilon - xH_{j0}^\epsilon = Ex(1-x) \tag{190}$$

and that all of the other terms on the right of Eq. (189) vanish. In general, the representation afforded by Eq. (190) can be considered to be the first term in a truncated series expansion. Thus a more generalized description of Eq. (190) would be

$$H_0^\epsilon[x] - (1-x)H_{i0}^\epsilon - xH_{j0}^\epsilon = x(1-x)(E + E_1x + E_2x^2 + E_3x^3 + \cdots) \tag{191}$$

where E_1, E_2, E_3, etc. are constants. Before considering the effect of replacing Eq. (190) by (191) on the description presented in Chapters III–IX, it is instructive to consider the procedures that have been developed to estimate the remaining terms in Eq. (189). In the course of these considerations, it must be remembered that the description is *required* for *all* of the phases under consideration. Thus, analogues of Eq. (189) must be written and defined for the β, α, and L phases in order to advance the problem to a higher level of accuracy than considered in Section 1 of Chapter III.

Tauer and Weiss (37) have developed an explicit procedure for estimating the magnetic contributions which is based on separating the magnetic, vibrational, and electronic components of the specific heat as indicated by Eq. (39) and Fig. 13. In cases where adequate measurements are unavailable or cannot be made, a method of approximation has been developed which is based on a knowledge of the Curie temperature, T_C, and $2\bar{S}$, the average magnetic moment per atom of the metal or alloy under consideration. This development is based on approximating the magnetic specific heat as

and
$$C_p^\mu = MT^3 \quad \text{for} \quad T < T_C \tag{192}$$
$$C_p^\mu = N/T^2 \quad \text{for} \quad T > T_C \tag{193}$$

In addition, several other empirical relations are employed, including the following relation between the Curie temperature and average moment (102–105):

$$T_C = 113Z \ln(2\bar{S} + 1) \tag{194}$$

where $Z = 8$ for a bcc ferromagnetic phase and $Z = 12$ for a fcc ferromagnetic phase. The constants M and N in Eq. (192) and (193) are defined by two additional empirical relations:

and
$$RT_C\bar{S}^{1/3} = \int_0^\infty C_p^\mu \, dT \tag{195}$$

$$R \ln(2\bar{S} + 1) = \int_0^\infty C_p^\mu T^{-1} \, dT \tag{196}$$

The constraints imposed by Eqs. (195) and (196) define M and N in (192) and (193) as follows:

$$C_p{}^\mu = 4.8RT_C^{-3}[\ln(2\bar{S}+1) - 0.5\bar{S}^{1/3}]T^3 \qquad \text{for} \quad T < T_C \qquad (197)$$

and

$$C_p{}^\mu = 1.6RT_C^2[\bar{S}^{1/3} - 0.75\ln(2\bar{S}+1)]T^{-2} \qquad \text{for} \quad T > T_C \qquad (198)$$

Equations (197) and (198) can be employed to compute the magnetic free energy, since

$$F^\mu[T] = H^\mu[0°K] + \int_0^T C_p{}^\mu \, dT - T\int_0^T C_p{}^\mu T^{-1} \, dT \qquad (199)$$

Since the total energy required to uncouple all of the spins is approximately $RT_C\bar{S}^{1/3}$ as indicated by Eq. (195), then

$$H^\mu[0°K] = -RT_C\bar{S}^{1/3} \qquad (200)$$

Equations (197)–(200) yield

$$F^\mu[T] = -RT_C\bar{S}^{1/3} - RT_C(T/T_C)^4[0.4\ln(2\bar{S}+1) - 0.2\bar{S}^{1/3}]$$
$$\text{for} \quad T \leqslant T_C \qquad (201)$$

and

$$F^\mu[T] = -RT\ln(2\bar{S}+1) - RT_C^2[0.8\bar{S}^{1/3} - 0.6\ln(2\bar{S}+1)]T^{-1}$$
$$\text{for} \quad T \geqslant T_C \qquad (202)$$

Equations (201) and (202) were employed to estimate the magnetic free energies of the various forms of manganese in constructing Fig. 20. A simplification of Eqs. (201) and (202) based on the results for iron shown in Table V is given by Eqs. (203) and (204) as

$$F^\mu \cong -RT_C \qquad \text{for} \quad 0 \leqslant T \leqslant T_C/2 \qquad (203)$$

and

$$F^\mu \cong -RT(T_C/113Z) - 40R(T_C/T)^4 \qquad \text{for} \quad T > T_C/2 \qquad (204)$$

Table XXIII compares the results of Eqs. (203) and (204) obtained for the magnetic free energy of α iron ($T_C = 1043°K$) with the experimental results derived from Table V. Agreement to within 5% is obtained at most temperatures. Equations (203) and (204) suggest that the magnetic free energy is dependent upon the Curie temperature. Thus, if $T_C^\epsilon[x]$ represents the composition dependence of the Curie temperature of i–j alloys, while T_{Ci}^ϵ and T_{Cj}^ϵ are the Curie temperatures of the ϵ forms of metal i and j, then the excess magnetic free energy at high temperatures implied by Eq. (204) is

$$F_E^{\epsilon\mu}[x, T] = -RT\{T_C^\epsilon[x] - (1-x)T_{Ci}^\epsilon - xT_{Cj}^\epsilon\}/113Z^\epsilon$$
$$- 40RT^{-4}\{(T_C^\epsilon[x])^4 - (1-x)(T_{Ci}^\epsilon)^4 - x(T_{Cj}^\epsilon)^4\} \qquad (205)$$

Thus, Eq. (205) represents a reasonable approximation to the second term on the right of Eq. (189). In most of the refractory transition metal systems discussed in Chapters IV–IX, magnetic effects are of no consequence since all of the Curie temperatures vanish. The Ti–Co and Ti–Ni cases shown in Figs. 132 and 133 are exceptions where finite contributions would arise from the magnetic free energy. It goes without saying that analogues of Eq. (205) would be required for each phase, i.e., β, ϵ, α. Weiss and Tauer (105) have provided estimation procedures for computing T_C for unstable forms of the transition metals and for alloys.

TABLE XXIII

COMPARISON OF EXPERIMENTAL MAGNETIC FREE ENERGY OF bcc IRON
WITH ESTIMATED VALUES

T (°K)	Magnetic free energy (cal/mole)		T (°K)	Magnetic free energy (cal/mole)	
	Experimental	Estimated [a]		Experimental	Estimated [a]
0	− 2072	− 2072	1000	− 2397	− 2385
100	− 2072	− 2072	1100	− 2544	− 2585
200	− 2073	− 2072	1200	− 2719	− 2795
300	− 2078	− 2072	1300	− 2909	− 3012
400	− 2081	− 2072	1400	− 3107	− 3234
500	− 2095	− 2072	1500	− 3312	− 3457
600	− 2119	− 2105	1600	− 3523	− 3681
700	− 2154	− 1996	1700	− 3739	− 3906
800	− 2206	− 2061	1800	− 3958	− 4135
900	− 2288	− 2219			

[a] Equations (203) and (204).

The effects of magnetic contributions are most profound in systems based on iron and alloys between transition metals of the first long period. In such cases, it is likely that such effects would dominate in controlling phase stability. Weiss and Tauer (13, 109, 110) have developed an analytical designation for the third and fourth terms in Eq. (189) that is valid for temperatures greater than 2Θ. Their result (exclusive of second-order terms) yields

$$F_E^{\epsilon}[\Theta/T] = -5.11T\{(0.221 + T^{-1}\Theta^{\epsilon}[x])^{-1}$$
$$- (1-x)(0.221 + T^{-1}\Theta_i^{\epsilon})^{-1} - x(0.221 + T^{-1}\Theta_j^{\epsilon})^{-1}\} \quad (206)$$

Consequently, when

$$S_p[x, T] = -R[x \ln x + (1-x) \ln(1-x)] \quad (207)$$

and specific heat effects due to "disordering" are absent, the excess free energy given by Eq. (189) can be approximated at high temperatures by Eq. (208) as

$$F_E^\epsilon[x, T] = x(1-x)(E + E_1 x + E_2 x^2 \ldots) - 0.5T^2\{\gamma^\epsilon[x] - (1-x)\gamma_i^\epsilon - x\gamma_j^\epsilon\}$$
$$- RT\{T_C^\epsilon[x] - (1-x)T_{Ci}^\epsilon - xT_{Cj}^\epsilon\}/113Z^\epsilon$$
$$- 5.11T\{(0.221 + T^{-1}\Theta^\epsilon[x])^{-1} - (1-x)(0.221 + T^{-1}\Theta_i^\epsilon)^{-1}$$
$$- x(0.221 + T^{-1}\Theta_j^\epsilon)^{-1}\} \tag{208}$$

Equation (208) describes the excess free energy in terms of the enthalpy of mixing at 0°K, the electronic specific heat coefficients, Curie temperature, and Debye temperature. As indicated by Eq. (208), each of these quantities must be defined as a function of composition (i.e., $T_C^\epsilon[x]$, $\gamma^\epsilon[x]$, $\Theta^\epsilon[x]$) for $0 \leqslant x \leqslant 1$. Moreover, as indicated earlier, Eq. (208) must be supplemented by analogous expressions *for each of the phases under consideration.*

3. EXPANDED ANALYTICAL MODELS FOR THE EXCESS FREE ENERGY OF MIXING

A number of purely mathematical descriptions of $F_E^\epsilon[x, T]$ have been developed in order to cover situations where explicit values of $T_C^\epsilon[x]$, $\gamma^\epsilon[x]$ and $\Theta^\epsilon[x]$ cannot be obtained (*111–116*). These models describe $F_E^\epsilon[x, T]$ by Eq. (209):

$$F_E^\epsilon[x, T] = x(1-x)[E + E_{01}T + E_{02}T^2 + \cdots$$
$$+ x(E_1 + E_{11}T + E_{12}T^2 + \cdots)$$
$$+ x^2(E_2 + E_{21}T + E_{22}T^2 + \cdots)] \tag{209}$$

Hardy (*111*) has truncated this series representation and defined Eq. (210) as the subregular model:

$$F_E^\epsilon[x, T] = x(1-x)[E + E_{01}T + x(E_1 + E_{11}T)] \tag{210}$$

The definitions of the partial molar free energies presented in Eqs. (65)–(68) when combined with Eq. (210) yield

$$\bar{F}_i^\epsilon = F_i^\epsilon + RT \ln(1-x) + x^2[(E - E_1) + (E_{01} - E_{11})T$$
$$+ 2x(E_1 + E_{11}T)] \tag{211}$$

and

$$\bar{F}_j^\epsilon = F_j^\epsilon + RT \ln x + (1-x)^2[E + E_{01}T + 2x(E_1 + E_{11}T)] \tag{212}$$

Wriedt (*112*) and Hillert (*113*) have applied Eqs. (210)–(212) to analysis of a number of binary alloy systems. In each case phase diagrams were employed to compute the values of E, E_{01}, E_1, and E_{11}. These parameters were checked by calculating the activity of both components for comparison with experimental data. When E_{01} and E_{11} are equal to zero, the conditions for miscibility gap formation specified following Eq. (75) define T_c and x_c as

$$\frac{RT_c}{x_c(1-x_c)} = 6E_1 x_c - 2(E_1 - E) \tag{213}$$

and

$$\frac{RT_c}{x_c^2(1-x_c)^2} = \frac{6E_1}{2x_c - 1} \tag{214}$$

In contrast to the earlier results [Eqs. (75) and (76)], x_c is not equal to 0.5 unless E_1 is zero, which reduces the model to the regular case. The gap boundaries $x_{\epsilon 1}$ and $x_{\epsilon 2}$ are defined by Eqs. (71), yielding Eqs. (215) and (216).

$$RT \ln\left(\frac{1-x_{\epsilon 1}}{1-x_{\epsilon 2}}\right) = x_{\epsilon 2}^2[(E-E_1) + 2E_1 x_{\epsilon 2}] - x_{\epsilon 1}^2[(E-E_1) + 2E_1 x_{\epsilon 1}] \tag{215}$$

and

$$RT \ln\left(\frac{x_{\epsilon 1}}{x_{\epsilon 2}}\right) = (1-x_{\epsilon 2})^2[E + 2E_1 x_{\epsilon 2}] - (1-x_{\epsilon 1})^2[E + 2E_1 x_{\epsilon 1}] \tag{216}$$

Finally the subregular-solution analogs to Eqs. (80)–(83) for equilibrium between a β and an L phase are given by Eqs. (217) and (218) as follows

$$\Delta F_i^{\beta \to L} + RT \ln\left(\frac{1-x_L}{1-x_\beta}\right) = x_\beta^2[(B-B_1) + 2B_1 x_\beta] - x_L^2[(L-L_1) + 2L_1 x_L] \tag{217}$$

$$\Delta F_j^{\beta \to L} + RT \ln\left(\frac{x_L}{x_\beta}\right) = (1-x_\beta)^2[B + 2B_1 x_\beta] - (1-x_L)^2[L + 2L_1 x_L] \tag{218}$$

In order to illustrate the application of Eqs. (213)–(219) to analysis of a real system it is of interest to consider the recent measurements of the activity of manganese and copper in the Cu–Mn system performed by Krenzer and Pool at 1075°K (117). The observed phase diagram is shown in Fig. 136. Surprisingly, the activity data showed rather "flat" activity–composition curves which resembled the curves shown for the activity of Hf and Ta in the Hf–Ta system illustrated in Fig. 106. In the latter system (see Fig. 52) a miscibility gap is observed, indicating the origin of the "flat" activity–composition curves. By contrast no gap is evident in the Cu–Mn system.

Fortunately an independent analysis of the properties of fcc (γ) Cu–Mn alloys can be made on the basis of the α/β and γ/α phase equilibria shown in Fig. 136, the numerical values of the lattice stability of γ, β, and α Mn of Tauer and Weiss (37) shown in Fig. 20, and Eqs. (213)–(218). In the case at hand, the free energy of the γ phase is given by Eq. (219) as

$$F^\gamma = (1-x)F_{Cu}^\gamma + xF_{Mn}^\gamma + x(1-x)(G_0 + G_1 x)$$
$$+ RT[x \ln x + (1-x) \ln(1-x)] \tag{219}$$

where F_{Cu}^γ and F_{Mn}^γ are the free energies of fcc Cu and Mn, respectively, and

G_0 and G_1 are constants as a first approximation. Under these circumstances, the partial molar free energy of Mn in the γ phase is given by Eq. (220):

$$\bar{F}^{\gamma}_{\text{Mn}} = F^{\gamma}_{\text{Mn}} + RT \ln x + (1 - x)^2 (G + 2G_1 x) \tag{220}$$

If x_{γ} is the composition of the $\gamma/(\gamma + \beta)$ phase boundary at or above

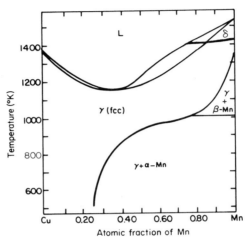

FIG. 136. The copper–manganese phase diagram (2).

$1000°\text{K}$, and the composition of the $\gamma/(\gamma + \alpha)$ boundary below $1000°\text{K}$ (see Fig. 136), then since the solubility of copper in Mn is nil,

$$\bar{F}^{\gamma}_{\text{Mn}}[x_{\gamma}] = F^{\beta}_{\text{Mn}} \qquad \text{for} \quad T \geqslant 1000°\text{K} \tag{221a}$$

and

$$\bar{F}^{\gamma}_{\text{Mn}}[x_{\gamma}] = F^{\alpha}_{\text{Mn}} \qquad \text{for} \quad T \leqslant 1000°\text{K} \tag{221b}$$

Hence,

$$(1 - x_{\gamma})^2 (G + 2G_1 x_{\gamma}) = \Delta F^{\gamma \to \beta}_{\text{Mn}} - RT \ln x_{\gamma} \qquad \text{for} \quad T \geqslant 1000°\text{K} \tag{222}$$

and

$$(1 - x_{\gamma})^2 (G + 2G_1 x_{\gamma}) = \Delta F^{\gamma \to \alpha}_{\text{Mn}} - RT \ln x_{\gamma} \qquad \text{for} \quad T \leqslant 1000°\text{K} \tag{223}$$

Employing the lattice stability values for $\Delta F^{\gamma \to \beta}_{\text{Mn}}$ and $\Delta F^{\gamma \to \alpha}_{\text{Mn}}$ reported by Tauer and Weiss permits evaluation of $G = 1200$ and $G_1 = 3150$, as shown in Table XXIV.

As a consequence, these values define Eqs. (219) and (220) explicitly as well as the partial molar free energy of copper given by Eq. (224):

$$\bar{F}^{\gamma}_{\text{Cu}} = F^{\gamma}_{\text{Cu}} + RT \ln(1 - x) + x^2(-1950 + 6300x) \tag{224}$$

On this basis, the activity of copper (relative to pure fcc copper) and the activity of manganese (relative to pure β Mn) in the γ phase can be computed for comparison with the experimental results (117). At 1075°K, $F_{Mn}^{\gamma \to \beta} = -150$ cal/g-atom. Thus,

$$RT \ln a_{Mn}^{\gamma}|_{Mn}^{\beta} = \bar{F}_{Mn}^{\gamma} - F_{Mn}^{\beta}$$
$$= \Delta F_{Mn}^{\beta \to \gamma} + RT \ln x + (1-x)^2(1200 + 6300x) \quad (225)$$

or

$$a_{Mn}^{\gamma}|_{Mn}^{\beta} = x \exp\{[150 + (1-x)^2(1200 + 6300x)](RT)^{-1}\} \quad (226)$$

and

$$RT \ln a_{Cu}^{\gamma}|_{Cu}^{\gamma} = \bar{F}_{Cu}^{\gamma} - F_{Cu}^{\gamma} = RT \ln(1-x) + x^2(-1950 + 6300x) \quad (227)$$

or

$$a_{Cu}^{\gamma}|_{Cu}^{\gamma} = (1-x) \exp[x^2(-1950 + 6300x)(RT)^{-1}] \quad (228)$$

TABLE XXIV

EVALUATION OF THE SUBREGULAR-SOLUTION PARAMETERS [a]

Temperature (°K)	x_γ	Co-existing phase ϕ	$\Delta F_{Mn}^{\gamma \to \phi}$	$-RT \ln x_\gamma$	Right side of Eqs. (222) and (223)	$(1-x_\gamma)^2$ $\times (1200 + 6300x_\gamma)$	Difference across Eqs. (222) and (223)
1100	0.89	β	−135	+255	+120	+82	+38
1000	0.76	β	−200	+545	+345	+345	0
950	0.54	α	−280	+1309	+1029	+1087	−58
900	0.40	α	−360	+1638	+1278	+1339	−61
800	0.26	α	−510	+2142	+1632	+1555	+77
700	0.20	α	−680	+2239	+1559	+1574	−15

[a] In calories per gram atom.

Table XXV compares the results of Eqs. (226) and (228) with the experimental results.

Table XXV shows that the computed values for the activity of Mn approaches unity at $x = 0.87$ as required by the phase diagram. In addition, at low values of x (i.e., $x = 0.13, 0.25$) where surface depletion effects would have the greatest effect on the measurements (36), the calculated values are higher than observed. This is in keeping with the fact that if the composition at the surface of these alloys were really lower than $x = 0.13$ and $x = 0.25$, Eq. (226) would lead to lower results (i.e., at $x = 0.09$ and $x = 0.20$), which would more closely compare with the measured values.

Finally, the explicit form of Eq. (219)

$$F^{\gamma} = (1-x)F_{Cu}^{\gamma} + xF_{Mn}^{\gamma} + x(1-x)(1200 + 3150x)$$
$$+ RT[x \ln x + (1-x) \ln(1-x)] \quad (229)$$

implies a miscibility gap in the γ phase. The critical temperature and composition of this gap [see Eqs. (67)–(76)] are defined by

$$\frac{\partial^2 F^\gamma}{\partial x^2} = \frac{\partial^3 F^\gamma}{\partial x^3} = 0 \qquad \text{at} \quad x = x_c \quad \text{and} \quad T = T_c$$

These conditions [see Eqs. (213) and (214)] yield

$$0 = (RT_c)/x_c(1 - x_c) + 3900 - 6(3150)x_c$$

and

$$0 = (RT_c)/x_c^2(1 - x_c)^2 - 6(3150)/(2x_c - 1)$$

or

$$x_c = 0.707 \qquad \text{and} \quad T_c = 987°K = 714°C \tag{230}$$

Examination of these results relative to Fig. 136 strongly suggest that the γ phase contains a miscibility gap that is just barely unstable with respect to

TABLE XXV

COMPARISON OF COMPUTED AND OBSERVED ACTIVITIES
OF COPPER AND MANGANESE
IN fcc COPPER–MANGANESE ALLOYS AT 1075°K

x	$a_{Cu}^\gamma \vert_{Cu}^\gamma$		$a_{Mn}^\gamma \vert_{Mn}^\beta$	
	Measured	Eq. (228)	Measured	Eq. (226)
0.13	0.866	0.86	0.179	0.28
0.25	0.755	0.74	0.375	0.55
0.33	0.675	0.67	0.501	0.70
0.44	0.589	0.60	0.670	0.84
0.53	0.514	0.56	0.719	0.90
0.73	0.369	0.52	0.844	0.95
0.83	0.313	0.49	0.914	0.97
0.91	0.255	0.40	0.942	1.00

the γ/α equilibrium [i.e., a miscibility gap would be observed if the α form of Mn did not become stable below 1000°K (727°C)]. This result also explains the relatively flat activity–composition data for manganese shown in Table XXV for $0.40 \leqslant x \leqslant 0.60$ at 1075°K.

Lumsden has developed an alternate description for the excess free energy that differs from Hardy's subregular model (111) shown in Eq. (210). In Lumsden's model,

$$F_E^\epsilon[x, T] = x(1 - x)\{(E_2 + E_{21}T)[(1 - x) + xr^2]^{-1} + (E_5 + E_{51}T)[(1 - x) + xr^5]^{-1}$$
$$- x(1 - x)(E_2 + E_{21}T)^2[Z^\epsilon RT(1 - x + xr^{8/3})^3]^{-1}\} \tag{231}$$

where r is the ratio of the atomic radius of component j to component i, Z^ϵ

s the coordination number, and E_2, E_{21}, E_5, and E_{51} are parameters to be determined from the phase diagram or from activity data. Lumsden (18) applied Eq. (231) to analysis of the Pb–Zn system. Sundquist (114) applied Eqs. (231) and (210) to a number of systems and concluded that Eq. (231) was more suitable than the subregular model. Sundquist's analysis of the Al–Zn system (114) on the basis of Eqs. (210) and (231) at 548°K yielded values of 275 cal/mole and 320 cal/mole, respectively, for the free energy difference between hcp and fcc zinc. For comparison, Eq. (106) in Table VII yields a value of 220 cal/mole at this temperature.

Hillert (116) has reviewed a variety of approaches for computing phase equilibria employing equations of the form exemplified by Eqs. (217) and (218). In Hillert's equations the B_{01}, B_{11}, L_{01}, and L_{01} terms are retained so that the temperature-dependent terms in the excess free energy of mixing are included. In dealing with α (bcc)/γ (fcc) equilibria in iron base alloys within this framework, Hillert considers situations where the analog of $\Delta F_i^{\beta \to L}$ is known (i.e., $\Delta F_{Fe}^{\alpha \to \gamma}$) and the phase boundaries occur at small values of composition. Thus, in Eq. (218) Hillert combines the unknown terms and computes

$$RT \ln(x_\gamma/x_\alpha) = \Delta F_j^{\gamma \to \alpha} + (A - G)$$

Since Hillert's interest is restricted to dilute solution, he evaluates the left-hand side as a function of temperature [and finds that it has a complex behavior (107)], but he does not obtain separate assessments of $\Delta F_j^{\gamma \to \alpha}$, A, and G.

Rudman (115) has recently developed a complete computer-based approach for integrating thermodynamic measurements and phase diagram data. Rudman's description is based on the subregular model as detailed by Eqs. (210)–(218) where the temperature-dependent components of the excess free energy are retained. In addition, the lattice stability terms are included. Thus each phase of i and j is characterized by a heat of fusion and a melting temperature. Rudman has applied his model to analysis of phase diagram and thermodynamic measurements as well as the synthesis of equilibria data from thermodynamic properties and vice versa. A significant number of systems have been analyzed with satisfactory results. The major difference between the present work and Rudman's approach (apart from the use of regular vs. subregular models) is that Rudman avoids defining the lattice stability of unstable forms. In order to avoid defining the lattice stability of the β form of a metal j (which is stable as the α form up to its melting point) in computing β/L equilibrium in the i–j system, Rudman employs Eq. (217) and solves for x_L as if x_β were equal to zero. Thus Eq. (218) is left indeterminate and evaluation of $\Delta F_j^{\beta \to L}$ is circumvented. The only other difference is that Rudman's approach utilizes equilibria and thermodynamic data extensively, while the current approach detailed in Chapters III–IX utilizes empirical models to compute interaction parameters and phase diagrams.

4. Correlations Based on Electron/Atom Ratios or Group-Number Effects

Buckley and Hume-Rothery (118, 119) have considered the relative stability of liquid and solid solution phases to be the combined result of melting point effects (mp effects) and lattice disturbing effects (ΔF_Y). The influence of these factors in controlling the relative stability of liquid and solid solutions [Hume-Rothery (119, pp. 8–13)] has been described as follows:

$$\Delta G_Y = \text{mp effect} + \Delta F_Y \qquad (232)$$

where

$$\Delta G_Y = RT \ln(Y^S/Y^L)$$

and Y^S and Y^L are the atomic fractions of element Y at the solid/(solid + liquid) and (solid + liquid)/liquid boundaries. In the present notation [Eqs. (65)–(81) and Kaufman (33)] one of the equations for equilibrium between a bcc and liquid phase in the i–j system yields

$$RT \ln(x_\beta/x_L) = \Delta F_j^{\beta \to L} + [F_E{}^L + (1 - x)(\partial F_E{}^L/\partial x)]_{x_L}$$
$$- [F_E{}^\beta + (1 - x)(\partial F_E{}^\beta/\partial x)]_{x_\beta}$$

Numerical evaluation of ΔG_Y is performed by determining the slope of the liquids curve $Y^L[T]$ and the solidus curve $Y^S[T]$ at infinite dilution (118):

$$\Delta G_Y = \Delta F_j^{\beta \to L} + (\partial F_E{}^L/\partial x)_{x_L=0} - (\partial F_E{}^\beta/\partial x)_{x_\beta=0} \qquad \text{at} \quad T = \overline{T}_i{}^\beta \qquad (233)$$

Thus, Eq. (233) shows that the quantity ΔG_Y consists of two terms which are taken to be characteristic of the solute ($Y = j$). The first term, $\Delta F_j^{\beta \to L}$, is the difference in free energy between the liquid and bcc forms of element Y (or j) at the melting point of element i. The second term is the difference between the derivative of the excess free energy of the liquid and bcc phases at infinite dilution. If the liquid and the bcc forms were represented by regular solutions, then Eq. (233) would reduce to

$$RT \ln(x_\beta/x_L) = \Delta F_j^{\beta \to L} + L - B$$

The first term in Eq. (232) is set equal to $-2.2(\overline{T}_i{}^\beta - \overline{T}_j{}^\beta)$; hence,

$$\text{mp effect} = -2.2(\overline{T}_i{}^\beta - \overline{T}_j{}^\beta) = \Delta F_j^{\beta \to L} \qquad \text{at} \quad T = \overline{T}_i{}^\beta$$

The second term, ΔF_Y, is then considered to be the sum of size factors and electronic factors (118, 119):

$$\Delta F_Y = \text{size factors} + \text{electronic factors} = L - B$$

By comparison, the difference $(L - B)$ is defined in the present description as $L - B = e_1 + e_2$, where e_1 is dependent on the volumes of i and j, and e_2 is dependent upon the group number of i and j as indicated in Chapter IV, Eqs. (142) and (146). Thus, the approach of Buckley and Hume-Rothery is seen to be identical to the present description. The only difference is involved in the evaluation of the mp effect that has been performed in considering the influence of various solutes on the melting point of bcc iron (118). In dealing with solute elements that are not stable in the bcc form (i.e., Tc, Re, and elements in Groups VIIIA, VIIIB, VIIIC, and 1B), the difference between the melting point of the stable form ($\overline{T}_j^{\epsilon}$ or $\overline{T}_j^{\alpha}$) and the bcc form ($\overline{T}_j^{\beta}$) was neglected as being small (118). This is equivalent to ignoring the lattice stability terms. Recently Sinha, Buckley, and Hume-Rothery have reconsidered this problem taking the difference in melting points into account (120) by using the lattice stability values estimated by the author (16). Their new results indicate that in some cases, the lattice disturbing factors (ΔF_Y) can favor the solid phase rather than the liquid phase. In the case of the iron base alloys considered (120), this result may be due to magnetic complications. However, such a conclusion is not so unusual as it may seem at first glance. Figure 37 shows that the "lattice disturbing effects" favor the ϵ phase over the liquid phase in the W–Rh system.

Equation (232) has been applied by Zener (121) and many others to describe the effect of alloying on the fcc/bcc transformation in iron base alloys. This method has also been applied to consideration of the bcc/hcp transformation in titanium. In all of these instances, the perception of the significance of Eq. (233) that is evidenced by Sinha, Buckley, and Hume-Rothery is absent. Thus, in the analysis of the transitions in iron or titanium base alloys, the logarithm of the ratio of solute solubilities in the bcc and fcc phases (bcc and hcp phases for titanium alloys) multiplied by RT is assumed to be a constant. It is then named the "enthalpy required to transfer one gram atom of solute from the bcc to the fcc phase in iron" (hcp to bcc for titanium). The "constant heat of transformation" so obtained is then examined for many solutes having difference electronic group numbers to show a correlation.

Reference to Eq. (83) shows clearly that RT times the ratio of solubilities is a complex function of temperature and composition that can be a constant only by accident! It is related to the difference in free energy of the solute between the crystal forms under consideration. Moreover, it also depends upon the difference between the partial molar free energies of the solute at the compositions corresponding to the phase boundaries for the temperature in question. Thus, assumption of a "constant heat of transformation" is not justified (33, 122). This conclusion is reinforced by the recent work of Hillert and coworkers (107, 116) which was noted above.

In 1934, Hume-Rothery, Mabbott, and Channell-Evans (*123*) put forth two hypotheses which have served as landmarks in the development of the theory of alloy phases. The first of these stated that where the atomic diameters of solvent and solute differ by more than 15%, the size factor is unfavorable and the solid solution is very restricted. When atomic diameters are within this limit, the size factor is favorable and solid solutions may form subject to other restrictions. This hypothesis is reflected in the size effect term e_1 discussed in Chapter IV.

The second hypothesis concerned the solubility limit of characteristic metallic structures. In particular, consideration of the phase field of stability of the α and β structures in Cu–Zn, Cu–Ga, Cu–Al, and Ag–Cd systems showed that if the equilibrium diagrams are drawn in terms of electron concentration, the solid solubility curves are roughly superposed, yielding a composition limit near electron/atom ratio, $e/a = 1.36$. Subsequently, Jones (*124*) suggested an instability in the α phase at concentrations above $e/a = 1.36$. A similar instability was computed above $e/a = 1.50$ for the β phase (*125, 126*). These instabilities were predicted (*124*) on the basis of Fermi surface–Brillouin zone interactions. Although the model of Jones has been shown to be incorrect (*26*), the concept of electron compounds has been retained. Indeed, characteristic values of the electron/atom ratio were defined at $e/a = 1.50$ (bcc), 21/13 (γ-brass), and 1.75 (hcp). Specification of these characteristic e/a ratios required definition of a system of valences for the transition metals. Nonetheless, many examples of Hume-Rothery electron phases have been identified since the hypothesis of characteristic e/a ratios were formulated.

With reference to the criteria for phase equilibria discussed in Chapter III, the concept of electron compounds or fixed e/a values at α/β or β/ϵ boundaries has little meaning. The "competitive basis" for equilibrium requires that the free energy of each of the competing phases be examined so that the most stable structure can be identified. Figures 25, 27, 77, 92, and 96 show that the phase boundary composition is a sensitive function of the interaction between two or more phases. Thus, even if the model of Jones was correct, the " competitive basis " for equilibrium would require that calculations of the stability be performed for the bcc, hcp, and fcc phases as a function of e/a. The resultant curves would then be examined on the basis of the lowest common tangent rule (Fig. 24) to define the appropriate field of stability. Such a calculation has never been performed. In this regard, it is interesting to note that the hcp phase is observed (*127*) for $1.2 \leqslant e/a \leqslant 1.8$. This observation appears to be inconsistent with the notion that the compositional limit is of necessity to be associated with a discontinuity in the energy–concentration curve [cf. Fig. 3 of Heine (*47*)].

Engel (*128, 129, 130*) has proposed an integral electron concentration rule which is suggested (*131*) as an extension of the $(18 - N)$ rule proposed by Hume-Rothery (*125, 126*). Brewer has applied and broadened the Engel correlation in considering a wide variety of metallic systems (*24, 25, 68*). Characteristic electron/atom ratios have been postulated for the bcc (below 1–1.75 electrons/atom), hcp (1.8–2.2 electrons/atom) and fcc (2.25 to very slightly over 3 electrons/atom) structures (*131*). The methods employed by Engel and Brewer to assign valences or numbers of outer electrons to atoms in various environments differ radically from those employed by Hume-Rothery.

Brewer has performed a detailed calculation of the difference in enthalpy between the bcc and hcp forms of the transition metals (*25*). His values have been averaged (within a given group) and plotted in Fig. 137 for comparison with those derived by the author (*16*). Although the variation of $\Delta H^{\beta \to \epsilon}$ with group number is similar for both sets of results, Brewer's values are larger by a factor of 5. In cases where Brewer's values of $\Delta H^{\beta \to \epsilon}$ lie between zero and -1000 cal/g-atom, polymorphism is predicted (*25*) due to the higher entropy postulated for the bcc phase. Large differences in enthalpy (5000–15,000 cal/g-atom) are computed for the V, Nb, Ta, Cr, Mo, and W group. These values are considered to be appropriate (*132*) on the grounds that they "have only one structure, even though they have been studied over wide ranges of temperature and pressure." In contrast to Brewer's values [which suggest that the author's results (*16*) are *too small* in magnitude], the analysis of Sinha, Buckley, and Hume-Rothery (*120*) applied to the effects of solutes on bcc $\rightleftharpoons$ fcc equilibria in iron alloys suggest that the author's results (*16*) are *too large*.

The recent preparation of fcc forms of these metals by Chopra and co-workers (see Table X) offers an excellent opportunity to check the differences indicated in Fig. 137. In particular, values of $\Delta H^{\beta \to \epsilon}_{Mo} = 2000$ cal/g-atom [Kaufman (*16*)] and 15,000 cal/g-atom [Brewer (*25*)] should be distinguishable experimentally. It should be noted that the preparation of metastable films of Nb, Ta, Mo, and W in the fcc structure by Chopra and co-workers suggests that Brewer's values are too high. Table X indicates that the fcc structure would not form at high pressure because its volume is greater than that of the bcc form. It is difficult to determine the reasons for the large differences between the values suggested by Brewer and those presented by the author for $\Delta H^{\beta \to \epsilon}$ in the case of Nb, Ta, Mo, and W. The origin of these differences is due to assignment of specific electron configurations to the bcc, hcp, and fcc forms in Brewer's calculation. However, it is not clear whether or not relaxation of these restrictions would alter the results.

Recently, Hume-Rothery has presented several reviews of the Engel–Brewer theory (*133–135*). In his most recent review Hume-Rothery points out (*135*, pp. 250, 251, and 263) that many of the predictions cannot be

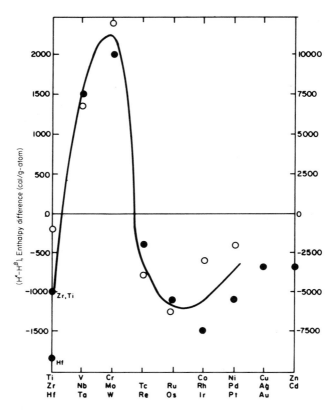

FIG. 137. Calculated enthalpy differences between the hcp (ϵ) and bcc (β) forms of the transition metals after Kaufman (16) (solid circles and left scale) and Brewer (25) (open circles and right scale).

verified by the interested reader and must be "taken on trust." This unfortunate situation makes application of the theory by others difficult, if not impossible. Moreover, Hume-Rothery (135, p. 249) suggests that Brewer's calculation of the energies of sublimation of all the transition metals in both bcc and chp structures permits calculation of the terminal-solution limits in alloys of two metals of different crystal structures based on regular-solution theory. It is also suggested that detailed calculations of the attractive-energy term arising from the mixture of transition metals from the left side of the Periodic Table with those of the right side have been made. The result is proportional to the number of d electrons made available for bonding by transfer from the internally paired condition of the metal to the right because of transfer to the vacant orbitals to the left. An energy repulsive term is also added for phases with atoms of different sizes. Unfortunately, the author has never seen the details of such calculations. In fact, the terminal solubility calculation would be quite interesting in view of the above mentioned large values of $\Delta H^{\beta \to \epsilon}$. Moreover, the calculation of attractive and repulsive energy terms could be directly compared with the e_0 and e_1 values described in Chapter IV.

As indicated above, the Engel correlation is based on a set of postulates (131, 136) that define characteristic values of the e/a ratio for each of the principal phases. These limits appear to be the result of empirical observations by Engel [cf. Fig. 1 of Engel (130)]. Detailed procedures for calculating and comparing free energy–composition curves analogous to those shown in Figs. 24, 25, 27, 77, 92, 96, and 100 have not been presented.

Brewer has outlined the method for defining the free energy–composition curves of the principal solution phases (24). He indicates that the internal pressure term e_p is present in the liquid phase and that substantial size differences between i and j will result in higher (more positive) activity coefficients in the solid. The latter statement is in keeping with the present definition of e_1. Unfortunately, Brewer has not detailed an explicit method for computing the free energy–composition curves for L, β, ϵ, and α structures as shown in Figs. 24, 25, 27, 77, 92, 96, and 100. Development of such a procedure could offer an alternative to the current regular-solution formulation and evaluation of the e_0, e_2, e_3, and e_3 terms. In the absence of a detailed description, the Engel–Brewer correlation appears as a method for describing the composition ranges of stability for various structures in terms of characteristic e/a ratios.

The recent series of exchanges between Hume-Rothery, Brewer, and Engel have provided some insight into major points of disagreement (68, 131, 133–136). However, several comments that have been made are relevant to the present description of phase equilibria. Hume-Rothery (134) suggested that evaluation of the Engel–Brewer correlation be restricted to atmospheric pressure. This eliminated hcp iron (consistent with Engel–Brewer) and fcc Cs, hcp Ba, and other cases (which are inconsistent with the Engel–Brewer correlation) from consideration. In the opinion of the present authors, such a restriction is artificial. The mere fact that most of the information available to us has been gathered at 1 atm should not eliminate relevant data obtained at high pressure. On the other hand, Engel (130) has arbitrarily dismissed the observation of stable bcc form of Be, which is in disagreement with his predictions. Brewer (68) has suggested that many of the sources of disagreement between the Engel–Brewer correlation and observations stem from small differences in energy or weak bonding (i.e., Na, Li, Be, Ca, Sr, etc.). Currently available information on phase stability of metals (20, 21) indicates that small energy differences are the rule. Even if Brewer's calculations of the enthalpy difference between the bcc and hcp forms of the transition metals (25) are accepted as being correct, we would conclude that in half of these cases (14 of 27), energy differences between ±2000 cal/g-atom are characteristic. This is the result of a sampling of "strongly bound" metals. Under these conditions, it appears that specification of the most stable structure under ambient conditions is insufficient. Indeed, the relative stability of competing structures must be described quantitatively!

CHAPTER XI

Regular-Solution Model of Ternary Phase Equilibria

The metals of the second and third transition row form an interesting series from both the practical and theoretical points of view. The high melting points, low vapor pressures, and outstanding oxidation resistance exhibited by metals, alloys, and compounds in this group have led to intensive research designed to describe their properties in recent years.

Since limited information is available concerning phase equilibria in ternary alloy systems composed of the above-mentioned elements, Chapters XI, XII, and XIII describe equilibrium in three-component systems. Accordingly, the ternary version of the regular-solution model has been applied explicitly in terms of the pairwise interaction parameters determined numerically. Computer programs have been developed for solving the equations for equilibrium between the solution phases (L, β, ϵ, and α), miscibility gap formation in the solution phases, and as interactions between the solution phases and the λ and ξ phases. These programs can be employed to compute all of the 286 ternary phase diagrams between the above-mentioned elements. In addition, they can be used to compute tie line compositions in composite structures, activity gradients in single-phase alloys arising from temperature gradients, vapor pressures and vaporization characteristics for component elements in ternary alloys, and to estimate heats of formation of ternary compound phases. Such computations would be based on the regular-solution model and the previously developed methods for estimating the numerical values of lattice stability and interaction parameters. Appendix 2 contains programs TERNRY and TERCP for calculation of phase equilibria between regular solutions in ternary systems and between ternary compounds and ternary solution phases.

The integral free energy of a single-phase alloy solution having the Φ

217

structure and n atomic species may be described by Eqs. (234) and (235) as

$$F^{\Phi}[x_1, x_2 \cdots x_n, T] = \sum_{i=1}^{n} x_i F_i^{\Phi}[T] + RT \sum_{i=1}^{n} x_i \ln x_i + \sum_{i=1}^{n} \sum_{j=i}^{n} x_i x_j E_{ij}^{\Phi} \quad (234)$$

$$\sum_{i=1}^{n} x_i = 1 \tag{235}$$

where x_i is the atomic fraction of element i, T the absolute temperature, $F_i^{\Phi}[T]$ the free energy of the Φ form of pure element i, and E_{ij}^{Φ} the Φ structure interaction parameter between elements i and j (137, 138). This is the extension of the regular-solution model which has been applied to binary systems in Section 1 of Chapter III, inasmuch as the E_{ij}'s are considered to be constants and higher-order interactions are ignored. If I, J, and K denote the three elements comprising a ternary solution at temperature T, then

$$y = \text{atomic fraction of element K}$$
$$x = \text{atomic fraction of element J}$$
$$z = \text{atomic fraction of element I} = (1 - x - y)$$

and Eqs. (234) and (235) yield

$$F^{\Phi}[x, y, T] = zF_i^{\Phi} + xF_j^{\Phi} + yF_k^{\Phi} + RT\{z \ln z + x \ln x + y \ln y\}$$
$$+ xzE_{ij}^{\Phi} + yzE_{ik}^{\Phi} + xyE_{jk}^{\Phi} \tag{236}$$

The partial free energies of I, J, and K in the Φ phase are

$$\overline{F}_i^{\Phi} = F^{\Phi} - x\frac{\partial F^{\Phi}}{\partial x} - y\frac{\partial F^{\Phi}}{\partial y}$$
$$= F_i^{\Phi} + RT \ln z + x^2 E_{ij}^{\Phi} + y^2 E_{ik}^{\Phi} + xy \, \Delta E^{\Phi} \tag{237}$$

$$\overline{F}_j^{\Phi} = F^{\Phi} + (1-x)\frac{\partial F^{\Phi}}{\partial x} - y\frac{\partial F^{\Phi}}{\partial y}$$
$$= F_j^{\Phi} + RT \ln x + (1-x)^2 E_{ij}^{\Phi} + y^2 E_{ik}^{\Phi} - y(1-x) \, \Delta E^{\Phi} \tag{238}$$

$$\overline{F}_k^{\Phi} = F^{\Phi} - x\frac{\partial F^{\Phi}}{\partial x} + (1-y)\frac{\partial F^{\Phi}}{\partial y}$$
$$= F_k^{\Phi} + RT \ln y + x^2 E_{ij}^{\Phi} + (1-y)^2 E_{ik}^{\Phi} - x(1-y)\Delta E^{\Phi} \tag{239}$$

where

$$\Delta E^{\Phi} = E_{ij}^{\Phi} + E_{ik}^{\Phi} - E_{jk}^{\Phi} \tag{240}$$

Equations (237)–(240) reduce to Eqs. (77)–(79) for a binary system.

1. Interactions between Solution Phases

The equilibria between two phases $\Phi 1$ and $\Phi 2$ is described graphically by the "common tangent plane" construction of Fig. 138, which shows the locus of points (x_1, y_1) and (x_2, y_2) corresponding to possible tangent points of a plane tangent to two free energy surfaces $F^1[x, y, T]$ and $F^2[x, y, T]$.

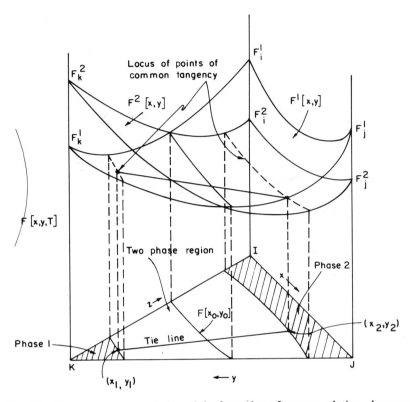

FIG. 138. Free energy representation of the formation of ternary-solution phase equilibria. The phase diagram projected below results from the interaction of free energy surfaces as shown.

Mathematically, this corresponds to the condition of minimum free energy for the system or, equivalently, the equilibration of the partial free energies of I, J, and K across the two-phase field, namely,*

$$G_i = \bar{F}_i{}^2(x_2, y_2) - \bar{F}_i{}^1(x_1, y_1) = 0 \qquad (241)$$

$$G_j = \bar{F}_j{}^2(x_2, y_2) - \bar{F}_j{}^1(x_1, y_1) = 0 \qquad (242)$$

$$G_k = \bar{F}_k{}^2(x_2, y_2) - \bar{F}_k{}^1(x_1, y_1) = 0 \qquad (243)$$

* Note that $F_i^2 \cdots$ and E_{ij}^2 refer to the partial free energy and interaction parameters of phase 2 and *not* to these quantities squared.

Substituting Eqs. (237), (238), and (239) into Eqs. (241), (242), and (243) yields three equations relating x_1, y_1, x_2, y_2, and T, namely,

$$G_i = \Delta F_i^{1 \to 2} + RT \ln(z_2/z_1) + (E_{ij}^2 x_2^2 - E_{ij}^1 x_1^2) + (E_{ik}^2 y_2^2 - E_{ik}^1 y_1^2)$$
$$+ (\Delta E^2 x_2 y_2 - \Delta E^1 x_1 y_1) = 0 \qquad (244)$$

$$G_j = \Delta F_j^{1 \to 2} + RT \ln(x_2/x_1) + [E_{ij}^2(1 - x_2)^2 - E_{ij}^1(1 - x_1)^2]$$
$$+ (E_{ik}^2 y_2^2 - E_{ik}^1 y_1^2) - [\Delta E^2 y_2(1 - x_2) - \Delta E^1 y_1(1 - x_1)] = 0 \qquad (245)$$

$$G_k = \Delta F_k^{1 \to 2} + RT \ln(y_2/y_1) + (E_{ij}^2 x_2^2 - E_{ij}^1 x_1^2) + [E_{ik}^2(1 - y_2)^2$$
$$- E_{ik}^1(1 - y_1)^2] - [\Delta E^2 x_2(1 - y_2) - \Delta E^1 x_1(1 - y_1)] = 0 \qquad (246)$$

where

$$\Delta F_i^{1 \to 2} = F_i^2 - F_i^1 = \Delta H_i^{1 \to 2} - T \Delta S_i^{1 \to 2} \qquad (247)$$

$$\Delta F_j^{1 \to 2} = F_j^2 - F_j^1 = \Delta H_j^{1 \to 2} - T \Delta S_j^{1 \to 2} \qquad (248)$$

$$\Delta F_k^{1 \to 2} = F_k^2 - F_k^1 = \Delta H_k^{1 \to 2} - T \Delta S_k^{1 \to 2} \qquad (249)$$

Values of $\Delta F_i^{1 \to 2}$ and E_{ij}^1 have been derived for cases where i (and/or j) refers to Zr, Nb, Mo, Ru, Rh, Pd, Hf, Ta, W, Re, Os, Ir, and Pt, and phase 1 (and/or 2) refers to the liquid (L), bcc (β), hcp (ϵ), or fcc (α) form. For a given T, any one of the variables x_1, y_1, x_2, or y_2 may be chosen as an independent parameter. If x_1, for example, is selected as the independent parameter, then we seek solutions of the form $y_1[x_1, T]$, $x_2[x_1, T]$, and $y_2[x_1, T]$ from the three simultaneous equations (244), (245), and (246). These solutions may be recast as

$$f_1[x_1, y_1, T] = 0$$
$$f_2[x_2, y_2, T] = 0 \qquad (250)$$

which now represent the curves of common tangency illustrated in Fig. 138. These curves define the limits of the single-phase fields 1 and 2 in thermodynamic equilibrium. Figure 139 illustrates the way in which an isothermal section of a ternary is built up from the three binaries I–J ($y = 0$), J–K ($z = 0$), and K–I ($x = 0$). The binaries provide only the starting points at an edge of the ternary section, such as x_L^0 and x_β^0 or z_L^0 and z_β^0 in Fig. 139. The equilibrium equations (244), (245), and (246) are required to calculate the paths of Eq. (250) between these end points. The straight line connecting corresponding points (x_1, y_1) and (x_2, y_2) is the two dimensional analog of the horizontal tie line in binary diagrams. For every pair of phase limit curves there is an infinite set of tie lines.

Before investigating the numerical solutions of Eqs. (244), (245), and (246) we examine the curve $f_0[x_0, y_0] = 0$ formed by the intersection of the free energy surfaces at some temperature T. These curves are the analogs of the $x_0[T_0]$ curves used extensively in the study of binary equilibria to select the stable phases from among a set of possible phases. As in the binary case, the curve $f_0[x_0, y_0, T] = 0$ must fall within the two-phase field separating the phase pair in question. As shown in Fig. 140, the f_0 curve is determined by constructing the energy difference*:

$$\Delta F^{1 \to 2}[x_0, y_0, T] = F^2[x_0, y_0, T] - F^1[x_0, y_0, T] = 0 \qquad (251)$$

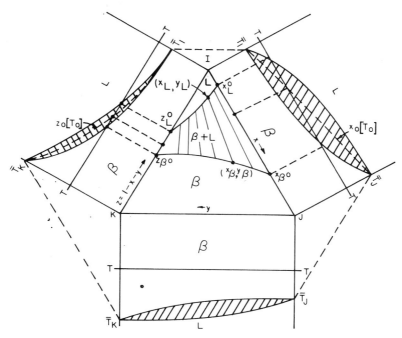

FIG. 139. Illustration of the construction of an isothermal ternary section and the associated binary phase diagrams for the melting region of a single phase β.

Inserting Eq. (236) into Eq. (251) and solving for y_0 as a function of x_0, we find

$$y_0 = A[x_0] \{1 \pm (1 + B[x_0])^{1/2}\} \qquad (252)$$

$$A[x_0] = \frac{D_3 + D_4 x_0}{2D_6} \qquad (253)$$

$$B[x_0] = \frac{4(D_1 + D_2 x_0 - D_5 x_0^2)D_6}{(D_3 + D_4 x_0)^2} \qquad (254)$$

where

$$D_1 = \Delta H_i^{2 \to 1} - T \Delta S_i^{2 \to 1} \qquad (255)$$

* Note that $F^2[x_0, y_0, T]$ is the free energy of phase 2 and not the free energy squared.

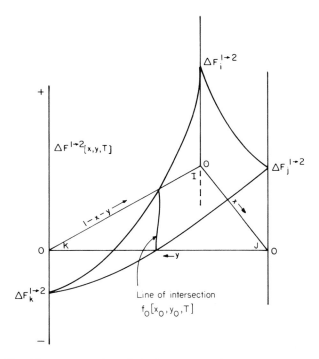

FIG. 140. Graphic representation of the free energy difference function and the intersection curve.

$$D_2 = (\Delta H_j^{2 \to 1} - \Delta H_i^{2 \to 1}) - T(\Delta S_j^{2 \to 1} - \Delta S_i^{2 \to 1}) + (E_{ij}^1 - E_{ij}^2) \tag{256}$$

$$D_3 = (\Delta H_k^{2 \to 1} - \Delta H_i^{2 \to 1}) - T(\Delta S_k^{2 \to 1} - \Delta S_i^{2 \to 1}) + (E_{ik}^1 - E_{ik}^2) \tag{257}$$

$$D_4 = (E_{jk}^1 + E_{ij}^2 + E_{ik}^2) - (E_{jk}^2 + E_{ij}^1 + E_{ik}^1) \tag{258}$$

$$D_5 = E_{ij}^1 - E_{ij}^2 \tag{259}$$

$$D_6 = E_{ik}^1 - E_{ik}^2 \tag{260}$$

The sign of the square root in Eq. (252) is chosen so as to maintain

$$0 \leqslant x_0 \leqslant 1, \qquad 0 \leqslant y_0 \leqslant 1, \qquad 0 \leqslant x_0 + y_0 \leqslant 1 \tag{261}$$

When $E_{ik}^1 = E_{ik}^2$, $(D_6 = 0)$, then

$$y_0 = \frac{D_5 x_0^2 - D_2 x_0 - D_1}{D_3 + D_4 x_0} \tag{262}$$

Also, if $D_3 + D_4 x_0 = 0$, then

$$y_0 = \left(\frac{D_1 + D_2 x_0 - D_5 x_0{}^2}{D_6}\right)^{1/2} \qquad (263)$$

As in the binary case, the calculation of the $y_0[x_0]$ curves is simply that of solving a quadratic equation, because the transcendental logarithm terms arising from configurational entropy cancel out of the energy difference expression.

The equilibria equations (244), (245), and (246) are solved numerically by the Newton–Raphson iteration technique (139) by writing the recursion relations as

$$x_2^{(k+1)} = x_2^{(k)} + \Delta x_2^{(k)}, \qquad y_1^{(k+1)} = y_1^{(k)} + \Delta y_1^{(k)}, \qquad y_2^{(k+1)} = y_2^{(k)} + \Delta y_2^{(k)} \quad (264)$$

Here k represents the kth iteration after initial guesses of $x_2^{(0)}$, $y_1^{(0)}$, and $y_2^{(0)}$. The correctors Δx_2, Δy_1, Δy_2 to the previous iterates x_2, y_1, y_2 in determinant form are

$$J\,\Delta x_2 = \begin{vmatrix} -G_i & \dfrac{\partial G_i}{\partial y_1} & \dfrac{\partial G_i}{\partial y_2} \\[2mm] -G_j & \dfrac{\partial G_j}{\partial y_1} & \dfrac{\partial G_j}{\partial y_2} \\[2mm] -G_k & \dfrac{\partial G_k}{\partial y_1} & \dfrac{\partial G_k}{\partial y_2} \end{vmatrix} \qquad (265)$$

$$J\,\Delta y_1 = \begin{vmatrix} \dfrac{\partial G_i}{\partial x_2} & -G_i & \dfrac{\partial G_i}{\partial y_2} \\[2mm] \dfrac{\partial G_j}{\partial x_2} & -G_j & \dfrac{\partial G_j}{\partial y_2} \\[2mm] \dfrac{\partial G_k}{\partial x_2} & -G_k & \dfrac{\partial G_k}{\partial y_2} \end{vmatrix} \qquad (266)$$

$$J\,\Delta y_2 = \begin{vmatrix} \dfrac{\partial G_i}{\partial x_2} & \dfrac{\partial G_i}{\partial y_1} & -G_i \\[2mm] \dfrac{\partial G_j}{\partial x_2} & \dfrac{\partial G_j}{\partial y_1} & -G_j \\[2mm] \dfrac{\partial G_k}{\partial x_2} & \dfrac{\partial G_k}{\partial y_1} & -G_k \end{vmatrix} \qquad (267)$$

where J, the Jacobian determinant, is

$$
J = \begin{vmatrix}
\dfrac{\partial G_i}{\partial x_2} & \dfrac{\partial G_i}{\partial y_1} & \dfrac{\partial G_i}{\partial y_2} \\[2mm]
\dfrac{\partial G_j}{\partial x_2} & \dfrac{\partial G_j}{\partial y_1} & \dfrac{\partial G_j}{\partial y_2} \\[2mm]
\dfrac{\partial G_k}{\partial x_2} & \dfrac{\partial G_k}{\partial y_1} & \dfrac{\partial G_k}{\partial y_2}
\end{vmatrix}
\tag{268}
$$

and the derivatives are explicitly

$$
\frac{\partial G_i}{\partial x_2} = -\frac{RT}{z_2} + \Delta E^2 y_2 + 2E_{ij}^2 x_2
\tag{269}
$$

$$
\frac{\partial G_j}{\partial x_2} = \frac{RT}{x_2} - \Delta E^2 y_2 - 2E_{ij}^2 (1 - x_2)
\tag{270}
$$

$$
\frac{\partial G_k}{\partial x_2} = -\Delta E^2 (1 - y_2) + 2E_{ij}^2 x_2
\tag{271}
$$

$$
\frac{\partial G_i}{\partial y_1} = \frac{RT}{z_1} - \Delta E^1 x_1 - 2E_{ik}^1 y_1
\tag{272}
$$

$$
\frac{\partial G_j}{\partial y_1} = \Delta E^1 (1 - x_1) - 2E_{ik}^1 y_1
\tag{273}
$$

$$
\frac{\partial G_k}{\partial y_1} = -\frac{RT}{y_1} - \Delta E^1 x_1 + 2E_{ik}^1 (1 - y_1)
\tag{274}
$$

$$
\frac{\partial G_i}{\partial y_2} = -\frac{RT}{z_2} + \Delta E^2 x_2 + 2E_{ik}^2 y_2
\tag{275}
$$

$$
\frac{\partial G_j}{\partial y_2} = -\Delta E^2 (1 - x_2) + 2E_{ik}^2 y_2
\tag{276}
$$

$$
\frac{\partial G_k}{\partial y_2} = \frac{RT}{y_2} + \Delta E^2 x_2 - 2E_{ik}^2 (1 - y_2)
\tag{277}
$$

The iteration process described by Eq. (264) converges when the relative error between successive iterates falls below some tolerable limit, i.e.,

$$
|\Delta x_2^{(k)}| \leqslant \epsilon |x_2^{(k)}|, \qquad |\Delta y_1^{(k)}| \leqslant \epsilon |y_1^{(k)}|, \qquad |\Delta y_2^{(k)}| \leqslant \epsilon |y_2^{(k)}|
\tag{278}
$$

It is emphasized that the choice of x_1 as the independent variable is arbitrary and that any of three other possible configurations are equally valid. The computer program developed to perform ternary calculations does, in fact, make provision for choosing any one of the configurations, depending upon the particular case being solved. The only change in the Newton–Raphson equations, Eqs. (264)–(276), is a matter of subscripts or labels. In essence, one fixes a temperature T and chooses a starting value x_1 for the independent variable, then chooses starting values of $x_2{}^0$, $y_1{}^0$, and $y_2{}^0$ on which to begin the iteration. The iteration is continued until it converges on a solution x_2, y_1, y_2. Then x_1 is changed by some small Δx_1 and the procedure

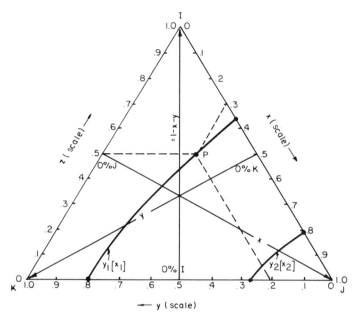

FIG. 141. Conventions for graphing isothermal ternary sections.

repeated using the previous solutions found for x_1 to start the new solutions at $x_1 + \Delta x_1$. In this way a pair of isothermal phase-limit curves $y_2[x_2]$ and $y_1[x_1]$ is generated for each phase pair considered.

Figure 141 illustrates a case in point. For the curve $y_1[x_1]$ it is seen that x_1 varies between 0.2 and 0.35. Any value of x_1 outside this range leads, mathematically, to negative values of x_2, y_1, and y_2; i.e., points outside the triangle. To calculate phase-limit curves on an isothermal section, then, it is important to know the range over which the independent variable exists for real solutions. In Fig. 141, one could start either along the I–J edge $x_1 = 0.35$, $x_2 = 0.8$, $y_1 = y_2 = 0$ or alternatively along the J–K edge $x_1 = 0.2$, $x_2 = 0.73$, $y_1 = 0.8$, $y_2 = 0.27$. Starting along the I–K edge is of little value because neither curve intersects this edge. The edge points can be found from a solution of the binaries I–J or J–K using the program TRSE described in Appendix 1. The program TRSE is of great aid in the ternary problem, because, in addition to providing numerical starting points, it automatically sorts out the stable from the unstable phase equilibrium pairs and their temperature–composition ranges of stability. This avoids the problem of mathematically possible but physically unacceptable solutions to the equilibria equations that arise when the interaction parameters are positive.

2. FORMATION OF MISCIBILITY GAPS

A special case of Eq. (236) arises when one or more of the interaction parameters E_{ij}^{Φ}, E_{ik}^{Φ}, or E_{jk}^{Φ} is positive and large compared with $2RT$. Positive interaction parameters represent repulsive interactions among atoms in solution, and we may therefore expect to find, under certain conditions, the formation of a ternary miscibility gap. In particular when a binary miscibility gap is found on two edges of a ternary, say I–J and I–K with both E_{ij} and $E_{ik} > 2RT$, then the gap will extend across the ternary triangle between the I–J and I–K edges. We label the composition limits of the gap as curves 1 and 2, which represent now the locus of points (x_1, y_1) and (x_2, y_2) formed by the two contacts of a common tangent plane on a single free energy surface. Figure 142 illustrates the graphical construction of a typical miscibility gap when E_{ij} and E_{ik} are both positive. The problem of finding the isothermal phase-limit curves $f_1(x_1, y_1)$ and $f_2(x_2, y_2)$ is identical to that of the previous two-solution phase interaction, except that now a single phase, rather than two different phases, is involved.

The mathematical criterion for forming a miscibility gap is, as before, the equilibration of the partial molar free energies across the gap, i.e.,

$$\bar{F}_{\nu}^{\Phi}[x_1, y_1, T] = \bar{F}_{\nu}^{\Phi}[x_2, y_2, T], \qquad \nu = i, j, k \qquad (279)$$

Inserting Eqs. (237), (238), and (239) into (279) generates the three equations

$$G_i = RT \ln(z_2/z_1) + E_{ij}(x_2{}^2 - x_1{}^2) + E_{ik}(y_2{}^2 - y_1{}^2) + \Delta E(x_2 y_2 - x_1 y_1) = 0 \quad (280)$$

$$G_j = RT \ln(x_2/x_1) + E_{ij}[(1 - x_2)^2 - (1 - x_1)^2] + E_{ik}(y_2{}^2 - y_1{}^2) \\ - \Delta E[y_2(1 - x_2) - y_1(1 - x_1)] = 0 \quad (281)$$

$$G_k = RT \ln(y_2/y_1) + E_{ij}(x_2{}^2 - x_1{}^2) + E_{ik}[(1 - y_2)^2 - (1 - y_1)^2] \\ - \Delta E[x_2(1 - y_2) - x_1(1 - y_1)] = 0 \quad (282)$$

These equations are identical in form to Eqs. (244), (245), and (246) except that $\Delta F_i^{1 \rightarrow 2} = 0$, $E_{ij}^1 = E_{ij}^2 = E_{ij}$, etc. Selecting x_1 as the independent parameter in Eqs. (280), (281), and (282), the numerical solution of these equations proceeds by the identical iteration process as was used in the two-solution phase interaction, namely Eqs. (264)–(268). The superscripts 1 and 2 in the partial derivative expressions in Eqs. (269)–(277) are dropped for the present case.

Numerical solutions to Eqs. (280), (281), and (282) are started on the edge

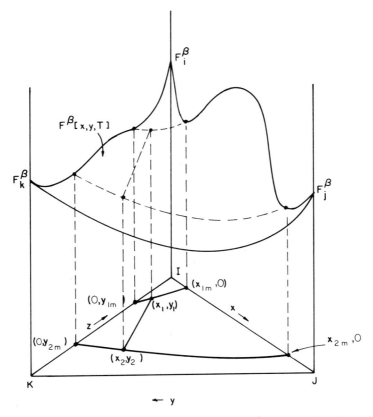

FIG. 142. Illustration of the formation of a ternary miscibility gap. The phase diagram projected below stems from the free energy surface shown.

of the ternary. For example on the I–J edge ($y = 0$), the solutions x_{1m} and x_{2m} follow immediately from the Newton–Raphson equations:

$$H[x_{1m}] = RT \ln\left(\frac{x_{1m}}{1 - x_{1m}}\right) + E_{ij}(1 - 2x_{1m}) = 0 \qquad (283)$$

$$\frac{\partial H}{\partial x_{1m}} = \frac{RT}{x_{1m}(1 - x_{1m})} - 2E_{ij} \qquad (284)$$

$$x_{1m}^{(k+1)} = x_{1m}^{(k)} - \left(\frac{H}{\partial H/\partial x_{1m}}\right)_{x_{1m}^{(k)}}, \qquad x_{1m}^{(0)} = \delta \qquad (285)$$

$$x_{2m} = 1 - x_{1m} \qquad (286)$$

Similarly, on the I–K edge $(x = 0)$, the solutions y_{1m} and y_{2m} are derived from

$$H[y_{1m}] = RT \ln\left(\frac{y_{1m}}{1 - y_{1m}}\right) + E_{ik}(1 - 2y_{1m}) = 0 \qquad (287)$$

$$\frac{\partial H}{\partial y_{1m}} = \frac{RT}{y_{1m}(1 - y_{1m})} - 2E_{ik} \qquad (288)$$

$$y_{1m}^{(k+1)} = y_{1m}^{(k)} - \left(\frac{H}{\partial H/\partial y_{1m}}\right)_{y_{1m}^{(k)}}, \qquad y_{1m}^{(k)} = \delta \qquad (289)$$

$$y_{2m} = 1 - y_{1m} \qquad (290)$$

It then follows that the limits on the variables are $0 \leqslant x_2 \leqslant x_{2m}$, $0 \leqslant y_1 \leqslant y_{1m}$, $0 \leqslant y_2 \leqslant y_{2m}$. Starting the general solution on the I–K edge and advancing x_1 from 0 to Δx_1, the initial estimates to the solution are $x_1 = x_2 = \Delta x_1$, $y_1 = y_{1m} - \delta$, $y_2 = y_{2m} - \delta$. In these equations δ is any reasonable small number such as 0.0001 which ensures that the iteration will converge and that logarithmic terms in the G and H functions above will not diverge. Binary miscibility gap curves are symmetrical about 0.5, i.e., $x_{2m} = 1 - x_{1m}$ as in Eq. (286); however, no such symmetry can be demonstrated in the general ternary case. When miscibility gaps are absent in the component binary system at the temperature under consideration, the free energy of mixing of the ternary solution phase given by Eq. (234) must be examined along the lines indicated by Meijering (137) in order to determine if isolated gaps exist. Such gaps do not touch the binary edges. Program MIGAP contained in Appendix 2 incorporates the development of Meijering (137) to compute ternary miscibility gaps for all values of E_{ij}, E_{ik}, and E_{kj} (positive and negative) in Eq. (234) since isolated ternary gaps can exist in special cases even when all the interaction parameters are negative.

CHAPTER XII

Description of Ternary Compound Phases

Intrusion of compound phases in binary phase diagrams was considered in Chapters VI and IX by idealizing the behavior of compounds and treating them as line compounds. On this basis, hexagonal Laves phases (λ) and fcc $AuCu_3$ phases (ξ) were considered. Formulation of the free energy of the λ phases was based on the ϵ phase for the pure metals while the α (fcc) form of the pure metals served as the basis for the description of the ξ phase. In the ternary case, if the I–J binary contains the compound $I_{1-x_*}J_{x_*}$ (compound A) and the J–K binary system contains $K_{1-x_*}J_{x_*}$ (compound B), then $F^\Psi[y, T]$ is defined as follows [see Eq. (91)]:

$$F^\Psi[y, T] = (1 - x_* - y)F_i^\theta + x_*F_j^\theta + yF_k^\theta + \Delta F^\Psi[y, T] \tag{291}$$

$$\Delta F^\Psi[y, T] = \left(1 - \frac{y}{1 - x_*}\right)\Delta F_A^\Psi + \left(\frac{y}{1 - x_*}\right)\Delta F_B^\Psi$$
$$+ RT\{y \ln y + (1 - x_* - y)\ln(1 - x_* - y) - (1 - x_*)\ln(1 - x_*)\} \tag{292}$$

where ΔF_A^Ψ and ΔF_B^Ψ are the free energies of formation of compounds A and B based on Eq. (91), $\theta = \epsilon$ (hcp) for Laves phases and $\theta = \alpha$ (fcc) for $AuCu_3$-type phases (i.e., for $\Psi = \xi$). In this case, the composition variable y is constrained to be within the range $0 \leqslant y \leqslant 1 - x_*$. The partial molar free energies are found to be

$$\bar{F}_i^\Psi = \bar{F}_j^\Psi = (1 - x_*)F_i^\theta + x_*F_j^\theta + \Delta F_A^\Psi + (1 - x_*)RT \ln\left(1 - \frac{y}{1 - x_*}\right) \tag{293}$$

$$\bar{F}_k^\Psi = F_k^\theta + x_*(F_j^\theta - F_i^\theta) + \left(\frac{1}{1 - x_*}\right)(\Delta F_B^\Psi - x_* \Delta F_A^\Psi)$$
$$+ RT\left[\ln\left(\frac{y}{1 - x_*}\right) - x_* \ln\left(1 - \frac{y}{1 - x_*}\right)\right] \tag{294}$$

229

The equivalence of $\bar{F}_i{}^{\Psi}$ and $\bar{F}_j{}^{\Psi}$ follows from the fact that F^{Ψ} is independent of the composition x.

To determine the phase-limit equilibrium between the solution phase Φ and the compound phase Ψ, the usual equilibration at constant T of the partial molar free energies across the two-phase field is now invoked, i.e.,

$$\bar{F}_\nu{}^{\Phi}[x_\Phi, y_\Phi, T] = \bar{F}_\nu{}^{\Psi}[x_*, y_\Psi, T], \qquad \nu = i; j, k \tag{295}$$

where x_Φ and y_Φ denote a point on the phase-limit curve of the Φ phase in equilibrium with Ψ. Figure 143 illustrates the tangent plane construction corresponding to the three relations of Eq. (295). Two cases are considered, one in which the compound phase is stable across the entire ternary field, as in Fig. 143, and one in which the compound phase is stable only over a portion of the ternary field, as shown in Fig. 144. Figure 145 illustrates other possible cases of a partially stable compound interacting with a solution phase. These cases are distinguished by possible intersections of the free energy surfaces F^{Ψ} and F^{Φ}.

Equation (295) for $\nu = i$ and $\nu = j$ requires that $\bar{F}_i{}^{\Phi}[x_\Phi, y_\Phi] = \bar{F}_j{}^{\Phi}[x_\Phi, y_\Phi]$, since $\bar{F}_i{}^{\Psi} = \bar{F}_j{}^{\Psi}$. Mathematically this is equivalent to the condition

$$\left|\frac{\partial F^{\Phi}}{\partial x}\right|_{x_\Phi, y_\Phi} = 0 \tag{296}$$

$$F_j{}^{\Phi} - F_i{}^{\Phi} + RT \ln(x_\Phi/z_\Phi) + E_{ij}(1 - 2x_\Phi) - \Delta E y_\Phi = 0 \tag{297}$$

where

$$z_\Phi = 1 - x_\Phi - y_\Phi \tag{298}$$

Equation (295) applied to $\nu = k$ using Eqs. (239) and (294) generates the following relation:

$$(F_k{}^{\Phi} - F_k{}^{\theta}) + x_*(F_i{}^{\theta} - F_j{}^{\theta}) + \frac{1}{1 - x_*}(x_* \Delta F_A{}^{\Psi} - \Delta F_B{}^{\Psi})$$

$$+ RT\left\{\ln\left[(1 - x_*)\frac{y_\Phi}{y_\Psi}\right] + x_* \ln\left(1 - \frac{y_\Psi}{1 - x_*}\right)\right\}$$

$$+ x_\Phi{}^2 E_{ij} + (1 - y_\Phi)^2 E_{ik} - x_\Phi(1 - y_\Phi) \Delta E = 0 \tag{299}$$

Defining $\Delta F_\nu{}^{\theta \to \Phi} = F_\nu{}^{\Phi} - F_\nu{}^{\theta}$ $(\nu = i, j, k)$ as the free energy difference of I, J, or K between states Φ and θ, then Eq. (299) may be manipulated to read

$$F_i{}^{\theta} - F_j{}^{\theta} = \Delta F_j{}^{\theta \to \Phi} - \Delta F_i{}^{\theta \to \Phi} + RT \ln(x_\Phi/z_\Phi) + E_{ij}(1 - 2x_\Phi) - \Delta E y_\Phi \tag{300}$$

Upon inserting Eq. (300) into the second term in Eq. (299) we arrive at

$$G_1[x_\Phi, y_\Phi, y_\Psi] = \Delta F_k{}^{\theta \to \Phi} + x(\Delta F_j{}^{\theta \to \Phi} - \Delta F_i{}^{\theta \to \Phi}) + \frac{1}{1 - x_*}(x_* \Delta F_A{}^{\Psi} - \Delta F_B{}^{\Psi})$$

$$+ RT\left[(1 - x_*)\ln(1 - x_*) + \ln\left(\frac{y_\Phi}{y_\Psi}\right) + x_* \ln\left(x_\Phi \frac{z_\Psi}{z_\Phi}\right)\right]$$

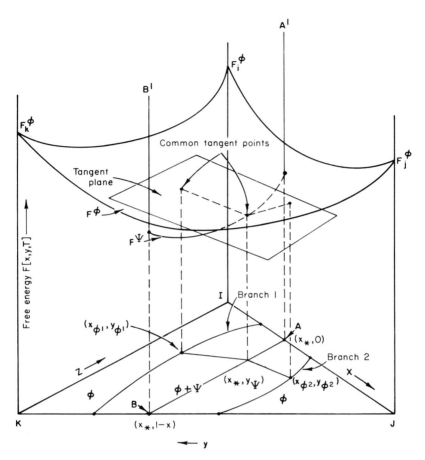

FIG. 143. Illustration of the tangent plane construction for the equilibrium between a line compound Ψ and a solution phase Φ. Case I: Ψ phase stable for all compositions.

$$+ E_{ij}[x_\Phi^2 + x_*(1 - 2x_\Phi)] + E_{ik}(1 - y_\Phi)^2$$
$$+ \Delta E \, [y_\Phi(x_\Phi - x_*) - x_\Phi] = 0 \tag{301}$$

where

$$z_\Psi = 1 - x_* - y_\Psi \tag{302}$$

Equation (301) provides one relation between x_Φ, y_Φ, and y_Ψ. To provide another we make use of the fact that

$$\frac{\partial F^\Phi}{\partial y}\bigg|_{x_\Phi, \, y_\Phi} = \frac{\partial F^\Psi}{\partial y}\bigg|_{x_\Psi, \, y_\Psi} \tag{303}$$

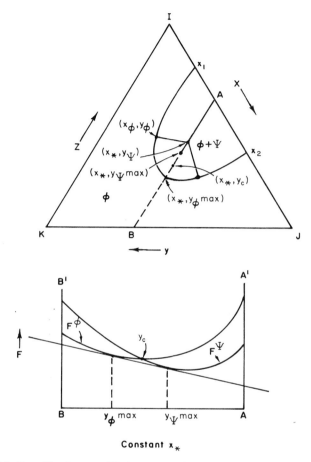

FIG. 144. Case II: phase equilibria construction when Ψ phase is partially stable.

Operating on Eqs. (236) for F^{Φ} and (291) and (292) for F^{Ψ}, condition (303) leads to

$$G_2[x_{\Phi}, y_{\Phi}, y_{\Psi}] = \Delta F_k^{\theta \to \Phi} - \Delta F_i^{\theta \to \Phi} + \frac{1}{1-x_*}(\Delta F_A^{\Psi} - \Delta F_B^{\Psi}) - \Delta E\, x_{\Phi}$$

$$+ RT \ln\left(\frac{y_{\Phi}\, z_{\Psi}}{z_{\Phi}\, y_{\Psi}}\right) + E_{ik}(1 - 2y_{\Phi}) = 0 \qquad (304)$$

An alternative relation follows from the condition

$$z_{\Psi}\, \bar{F}_i{}^{\Phi}[x_{\Phi}, y_{\Phi}] + x_*\, \bar{F}_j{}^{\Phi}[x_{\Phi}, y_{\Phi}] + y_{\Psi}\, \bar{F}_k{}^{\Phi}[x_{\Phi}, y_{\Phi}] = F^{\Psi}[x_*, y_{\Psi}] \qquad (305)$$

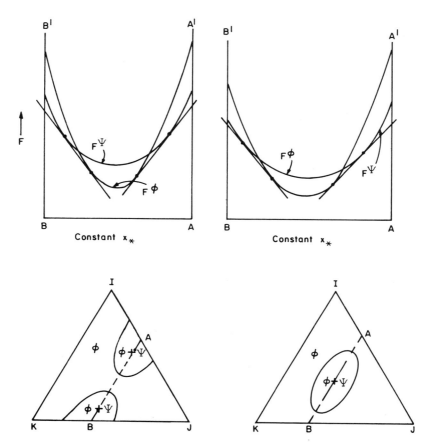

FIG. 145. Other possible phase equilibria constructions when Ψ phase is partially stable.

which upon expanding becomes

$$z_{\Psi'} \Delta F_i^{\theta \to \Phi} + x_* \Delta F_j^{\theta \to \Phi} + y_{\Psi'} F_k^{\theta \to \Phi} - \Delta F^{\Psi}[x_*, y_{\Psi'}]$$
$$+ RT\{z_{\Psi'} \ln z_{\Phi} + x_* \ln x_{\Phi} + y_{\Psi'} \ln y_{\Phi}\}$$
$$+ E_{ij}\{x_{\Phi}^2 + x_*(1 - 2x_{\Phi})\} + E_{ik}\{y_{\Phi}^2 + y_{\Psi'}(1 - 2y_{\Phi})\}$$
$$+ \Delta E\{(x_{\Phi} - x_*)y_{\Phi} - x_{\Phi}y_{\Psi'}\} = 0 \qquad (306)$$

and $\Delta F^{\Psi}[x_*, y_{\Psi'}]$ is defined by Eq. (292) with $y = y_{\Psi'}$.

If we consider $y_{\Psi'}$ as an independent parameter, then Eq. (301) and (304) provide two equations in the two unknowns x_{Φ} and y_{Φ}. For each value $y_{\Psi'}$ there will in general be two tie lines from $(x_*, y_{\Psi'})$ to points on the Φ-phase limit curve on either side of the Ψ phase (see Fig. 143). We designate these two branches as 1 and 2 with points $(x_{\Phi 1}, y_{\Phi 1})$ and $(x_{\Phi 2}, y_{\Phi 2})$ corresponding to independent solutions of the simultaneous equations (301) and (304) given y_Ψ. The Newton–Raphson equations for this numerical problem appear below:

$$G_1[x_\Phi, y_\Phi] = 0 \tag{307}$$

$$G_2[x_\Phi, y_\Phi] = 0 \tag{308}$$

$$x_\Phi^{(n+1)} = x_\Phi^{(n)} + \Delta x_\Phi^{(n)} \tag{309}$$

$$y_\Phi^{(n+1)} = y_\Phi^{(n)} + \Delta y_\Phi^{(n)} \tag{310}$$

$$\Delta x_\Phi = \frac{G_2(\partial G_1/\partial y_\Phi) - G_1(\partial G_2/\partial y_\Phi)}{(\partial G_1/\partial x_\Phi)(\partial G_2/\partial y_\Phi) - (\partial G_1/\partial y_\Phi)(\partial G_2/\partial x_\Phi)} \tag{311}$$

$$\Delta y_\Phi = \frac{G_1(\partial G_2/\partial x_\Phi) - G_2(\partial G_1/\partial x_\Phi)}{(\partial G_1/\partial x_\Phi)(\partial G_2/\partial y_\Phi) - (\partial G_1/\partial y_\Phi)(\partial G_2/\partial x_\Phi)} \tag{312}$$

$$x_\Phi = x_\Phi^{(m)} \quad \text{when} \quad |\Delta x_\Phi^m| \leqslant \epsilon|x_\Phi^{(m)}| \tag{313}$$

$$y_\Phi = y_\Phi^{(m)} \quad \text{when} \quad |\Delta y_\Phi^m| \leqslant \epsilon|y_\Phi^{(m)}| \tag{314}$$

$$\frac{\partial G_1}{\partial x_\Phi} = \frac{RTx_*(1 - y_\Phi)}{x_\Phi z_\Phi} + 2E_{ij}(x_\Phi - x_*) - \Delta E\,(1 - y_\Phi) \tag{315}$$

$$\frac{\partial G_1}{\partial y_\Phi} = \frac{RT(z_\Phi + x_* y_\Phi)}{y_\Phi z_\Phi} - 2E_{ik}(1 - y_\Phi) + \Delta E\,(x_\Phi - x_*) \tag{316}$$

$$\frac{\partial G_2}{\partial x_\Phi} = \frac{RT}{z_\Phi} - \Delta E \tag{317}$$

$$\frac{\partial G_2}{\partial y_\Phi} = \frac{RT(1 - x_\Phi)}{y_\Phi z_\Phi} - 2E_{ik} \tag{318}$$

To start the numerical solution of Eqs. (301) and (304) we require that initial estimates for x_Φ and y_Φ be provided at any specified value of y_Ψ. Once a solution is found on either branch of the phase-limit curve, the solution on that branch at the next step in y_Ψ is easily generated by using as initial estimates the solution at the previous y_Ψ. An initial solution can be found on the I–J ternary edge when $y_\Psi = y_\Phi = 0$. Equation (306) in this limit reduces to

$$H[x_\Phi] = RT \ln\{(1 - x_\Phi)^{1-x} * x_\Phi^x *\} + E_{ij}\{x_\Phi^2 + x_*(1 - 2x_\Phi)\}$$
$$+ (1 - x_*)\,\Delta F_i^{\theta \to \Phi} + x_*\,\Delta F_j^{\theta \to \Phi} - \Delta F_A^\Psi = 0 \tag{319}$$

The two solutions of Eq. (319) are the terminal points x_1 and x_2 calculated by iterating on

$$x_1^{(n+1)} = x_1^{(n)} - \left(\frac{H}{dH/dx}\right)_{x_1^{(n)}}, \qquad x_1^{(0)} = x_* - \delta \tag{320}$$

$$x_2^{(n+1)} = x_2^{(n)} - \left(\frac{H}{dH/dx}\right)_{x_2^{(n)}}, \qquad x_2^{(0)} = x_* + \delta \tag{321}$$

$$\frac{\partial H}{\partial x} = (x - x_*)\left(2E_{ij} - \frac{RT}{x(1-x)}\right) \tag{322}$$

A similar degenerate solution occurs at the J–K edge when $y_{\Psi} = 1 - x_*$, $x_\Phi = 1 - y_\Phi$. Regardless of which edge is used, the solution can be propagated along each branch by stepping y_{Ψ} between the limits $0 \leqslant y_{\Psi} \leqslant 1 - x_*$.

The previous discussion assumes that the Ψ phase is stable across the entire ternary field, i.e., F^{Ψ} is always more negative than F^{Φ} over the allowed range of y. If, however, F^{Ψ} is less negative than F^{Φ} anywhere in that range, then the partial stability of the Ψ phase arises as illustrated in Figs 144 and 145. To determine when this occurs, we examine solutions y_c to the equation determined by the intersection of the free energy surfaces, namely,

$$F^{\Phi}[x_*, y_c] = F^{\Psi}[x_*, y_c] \tag{323}$$

When Eqs. (236) and (291) are employed, we arrive at the quadratic form

$$ay_c^2 + by_c - c = 0 \tag{324}$$

$$a = E_{ik} \tag{325}$$

$$\begin{aligned} b = x_* \, \Delta E - E_{ik} + \Delta F_i^{\theta \to \Phi} - \Delta F_k^{\theta \to \Phi} \\ - [1/(1 - x_*)](\Delta F_A^{\Psi} - \Delta F_B^{\Psi}) \end{aligned} \tag{326}$$

$$\begin{aligned} c = (1 - x_*) \, \Delta F_i^{\theta \to \Phi} + x_* \, \Delta F_j^{\theta \to \Phi} - \Delta F_A^{\Psi} + E_{ij}^{\Phi} x_*(1 - x_*) \\ + RT \ln x_*^{x_*}(1 - x_*)^{1-x_*} \end{aligned} \tag{327}$$

When one root y_c of Eq. (324) lies within the range $0 < y_c < 1 - x_*$, then the terminal composition $y^{\Psi}(\max)$ of the Ψ phase must occur at $0 < y_{\Psi}(\max) < y_c$ while the maximum composition of the two-phase field $y_\Phi(\max)$ must occur at $y_c < y_\Phi(\max) < 1 - x_*$. This situation is depicted in Fig. (144), while other situations that could arise when both roots of Eq. (324) fall within range are illustrated in Fig. (145).

Both $y_{\Psi}(\max)$ and $y_\Phi(\max)$ can be calculated from Eqs. (301) and (304) by setting $x_\Phi = x_*$ and solving the equations simultaneously. The relationship between $y_{\Psi}(\max)$ and $y_\Phi(\max)$ is given by

$$y_{\Psi\max} = (1 - x_*) - z_{\Phi\max} u_{\Phi\max} \tag{328}$$

where

$$u_{\Phi\max} = \exp\left(\frac{c + E_{ik}^{\Phi} y_{\Phi\max}^2}{(1 - x_*)RT}\right) \tag{329}$$

$$z_{\Phi\max} = 1 - x_* - y_{\Phi\max} \tag{330}$$

and $y_{\Phi max}$ itself is extracted from the equations

$$Q[y_{\Phi max}] = RT \ln\left(\frac{y_{\Phi max}}{y_{\Psi max}}\right) + \frac{E_{ik}^{\Phi} y_{\Phi max}}{1 - x_*}[y_{\Phi max} - 2(1 - x_*)] - b \qquad (331)$$

$$\frac{dQ}{dy_{\Phi max}} = -\frac{2E_{ik}^{\Phi} z_{\Phi max}}{1 - x_*}$$

$$+ \frac{RT}{y_{\Phi max} y_{\Psi max}}\left[y_{\Psi max} + u_{\Phi max}\left(\frac{2z_{\Phi max} E_{ik}^{\Phi}}{(1 - x_*)RT} - 1\right)\right] \qquad (332)$$

$$y_{\Phi max}^{(n+1)} = y_{\Phi max}^{(n)} - \left(\frac{Q}{dQ/dy_{\Phi max}}\right)_{y_{\Phi max}^{(n)}}, \qquad y_{\Phi max}^{(0)} = y_c + \delta \qquad (333)$$

The procedure, then, for any case of a ternary compound-solution phase interaction problem is to first investigate the roots y_c of Eq. (324). If neither root falls within the limits $0 \leqslant y_c \leqslant 1 - x_*$, the compound phase is fully stable across the ternary and $y_\Psi(\text{max}) = 1 - x_*$. If, however, either root does fall within these limits, then the compound phase is only partially stable and $y_\Psi(\text{max})$ is calculated as in Eqs. (328)–(333). The equation used to estimate $\Delta F_A{}^\Psi$ and $\Delta F_B{}^\Psi$ for a binary compound A or B is Eq. (91):

$$\Delta F_A{}^\Psi = x_*(1 - x_*)(L_{ij} - C) \qquad (334)$$

$$\Delta F_B{}^\Psi = y_*(1 - y_*)(L_{jk} - C) \qquad (335)$$

where C is a constant whose value is 18,000 cal/g-atom for stable Laves phases and fcc $AuCu_3$ phases, and $C = 0$ for unstable phases.

It is reemphasized that the roles of elements I, J, and K are interchangeable by rotation of the ternary diagram about its center, or equivalently the interchange of composition variables x, y, and z in the preceding analyses. Thus, superpositions of several calculated equilibria curves on a common ternary diagram is easily facilitated by renaming the corners of the triangular grids when necessary to preserve the correct sense of I, J, and K as dictated by the equations.

The numerical ternary equilibria problems previously discussed have been programmed for an IBM 1130 computer system, with automatic plotting output of the two regular-solution calculation. FORTRAN listings and descriptions of the programs appear in Appendix 2.

CHAPTER XIII

Calculation of Ternary Phase Diagrams

Several ternary refractory metal systems appearing in the literature have been selected to provide a comparison between the analytical methods described in Chapters XII and XIII and experimental results. In particular Mo–W–Os has been investigated by Taylor and Doyle (140), Zr–Ta–W by Pease and Brophy (141), Re–W–Ta by Wulff and Brophy (142), and Re–Hf–Mo by Taylor and Doyle (143). Calculations of the isothermal ternary sections of these systems at several temperatures has proceeded along identical lines to the construction of the binary phase diagrams. First, any solution phase interactions appearing on the calculated constituent binary diagrams are recalculated for the ternary with tie lines across the two-phase fields. In the cases considered, the only solid solution phases appearing are β (bcc), ϵ (hcp), and L (liquid). Second, any miscibility gaps that might form are calculated and superimposed on the solution phase diagram; this occurs in the Zr–Ta–W system. Finally, two possible compound phases λ [Laves phases having a hexagonal structure with $x_* = \frac{2}{3}$ and ξ (fcc $AuCu_3$) phases with $x_* = \frac{3}{4}$] are investigated for stability. When they exist in the systems considered, their interactions with the stable solution phases are calculated and superimposed on the solution phase diagram to yield the final ternary diagram, which may then be compared with the experimental diagrams. Compound phases other than λ and ξ have been observed experimentally, but are not taken into account in the present calculations.

1. THE MOLYBDENUM–TUNGSTEN–OSMIUM SYSTEM

The calculated Mo–W–Os ternary is a simple system consisting of two solution phases ϵ (Os) and β (Mo and W) with a single interaction ϵ/β. The calculated isothermal sections at 1873°K and 2648°K appear in Figs. 146 and 148 and are to be compared with the observed (*140*) diagrams of Figs. 147 and 149.

Explicit application of Eqs. (251) and (252) in order to generate the $y_0[x_0]$ and the $[x_1, y_1]$, $[x_2, y_2]$ pairs (which define the tie lines across two-phase fields at any temperature) were generated for L/β, L/ϵ, L/α, and ϵ/β equilibrium. The appropriate numerical values for $\Delta F_i^{1 \to 2} \cdots$ and $E_{ij}^1 \cdots$ where I = Mo, J = W, and K = Os and phases 1 and 2 refer to L (liquid), β (bcc), ϵ (hcp), and α (fcc) are given in Tables VIII and XIII of Chapters III and IV. This procedure was employed to choose the most stable equilibrium and perform the required calculations. For example, in the case of β/ϵ equilibrium at 1873°K, $\Delta F_i^{1 \to 2} = \Delta F_{Mo}^{\beta \to \epsilon} = 2000$ cal/mole, $\Delta F_j^{1 \to 2} = \Delta F_W^{\beta \to \epsilon} = 2000$ cal/mole, and $\Delta F_k^{1 \to 2} = \Delta F_{Os}^{\beta \to \epsilon} = -2618$ cal/mole. In addition, for the $\beta = 1$ phase, $E_{ij}^1 = B(\text{Mo–W}) = 1296$ cal/g-atom, $E_{ik}^1 = B(\text{Mo–Os}) = -5904$ cal/g-atom, and $E_{jk}^1 = B(\text{W–Os}) = -5860$ cal/g-atom. Similarly, for the $\epsilon = 2$ phase $E_{ij}^2 = E(\text{Mo–W}) = 1296$ cal/g-atom, $E_{ik}^2 = E(\text{Mo–Os}) = -7584$ cal/g-atom, and $E_{jk}^2 = E(\text{W–Os}) = -7540$ cal/g-atom. The values of $E_{ij}^1 = B(\text{Mo–W})$ and $E_{ij}^2 = E(\text{Mo–W})$ are not contained in Table XIII of Chapter IV. Numerical values are obtained following the procedure shown in Tables XI and XII, where e_0 is zero for Mo–W and $e_p = L = 1280$ cal/g-atom. Since $e_1 = 16$ cal/g-atom and $e_2 = e_3 = 0$, $B = E = 1296$ cal/g-atom for Mo–W.

The calculated and observed Mo–Os and W–Os binaries are compared in Fig. 72, Chapter V. (Mo–W is a trivial system consisting of a single β field that melts above about 3000°K.) The observed (*140*) diagrams contain compound phases θ at 1873°K and σ at 1873 and 2648°K, which are not considered in the calculation.* The θ phase is a narrow, partially stable phase observed as Mo_3Os in the Mo–Os system below 2480°K and has a cubic β-W (A-15) structure. A broad σ phase having a complex tetragonal structure similar to the σ phase in Fe–Cr alloys extends across the ternary triangle between Mo–Os and W–Os. The absence of the phase in the calculated diagrams permits substantially greater extension of the ϵ and β solution fields into the ternary than has been observed.

2. The Zirconium–Tantalum–Tungsten System

At 1873°K, the only solution appearing in the calculated Zr–Ta–W ternary is a bcc solution phase. The calculated interaction parameters B(Zr–Ta), B(Zr–W), and B(Ta–W) are sufficiently large and positive, however, that a ternary miscibility gap is expected to form at that temperature. Figure 150 is the calculated β miscibility gap ($\beta + \beta'$) extending from the Zr–W edge

---

* All observed compound phases are indicated as shaded areas.

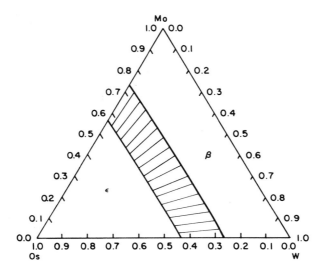

FIG. 146. Calculated ternary diagram of Mo–W–Os at 1873 °K.

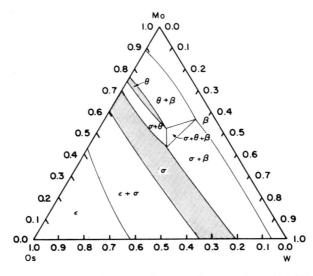

FIG. 147. Observed ternary diagram of Mo–W–Os at 1873 °K.

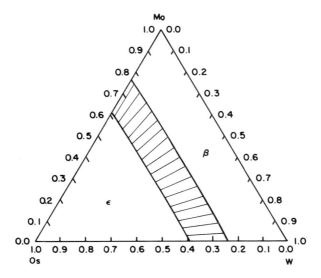

FIG. 148. Calculated ternary diagram of Mo–W–Os at 2648 °K.

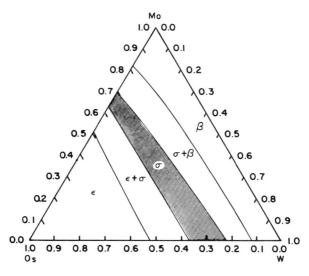

FIG. 149. Observed ternary diagram of Mo–W–Os at 2648 °K.

across the ternary to the Zr–Ta edge. Comparison should be made with the calculated binaries of Zr–Ta, Zr–W, and Ta–W appearing in Figs. 52, 54, and 63, respectively, of Chapter V. Appendix 2 illustrates the calculations of the gap compositions. A Laves phase ZrW_2 is found to be stable in the Zr–W binary (Fig. 91 of Chapter VI), but the corresponding $ZrTa_2$ counterphase system is not stable. Consequently, the computed ternary λ phase is a partially stable compound phase in equilibrium with β. Appendix 2 shows the details for the computation of λ (Laves phase) vs. β (bcc) equilibrium shown in Fig. 151. As indicated earlier, following Eq. (335), a value of $C = 18,000$ was employed in estimating the free energy of the ZrW_2 phase, while a value of $C = 0$ was employed for the unstable $ZrTa_2$ counterphase. Thus, the free energy of the ternary $Zr(W, Ta)_2$ phase was estimated by Eq. (336), which follows directly from Eqs. (291), (292), and (335) and Table XIII. For $Zr(W, Ta)_2$

$$F^\lambda[y, T] = \tfrac{1}{3}F^\epsilon_{Zr} + (\tfrac{2}{3} - y)F^\epsilon_W + yF^\epsilon_{Ta} - (1 - \tfrac{3}{2}y)(1976)$$
$$+ \tfrac{3}{2}y(1398) + RT[y \ln y + (\tfrac{2}{3} - y)\ln(\tfrac{2}{3} - y) - \tfrac{2}{3} \ln \tfrac{2}{3}] \qquad (336)$$

where y is the atomic fraction of tantalum.

The computed phase limits are shown in Fig. 151, where it is seen that the $\lambda + \beta$ field extends part way into the ternary. When the miscibility gap equilibrium of Fig. 150 and the β/λ equilibrium of Fig. 151 are combined, the result is Fig. 152, which compares the calculated phase diagram (thick lines) with the observed (141) phase diagram (thin lines). The positions of the three-phase field ($\beta + \beta' + \lambda$) in the computed and observed diagrams are in relatively good accord. The miscibility gap of the observed Zr–Ta system is not symmetric about the composition $x = 0.50$ (Fig. 52 of Chapter V compares the computed and observed binary phase diagram). The present calculations are based on symmetrical heats of mixing. Consequently, the calculated intersections of the gap phase limit curves and the λ/β phase limit curves occurs at slightly larger tungsten compositions than observed.

3. THE RHENIUM–TUNGSTEN–TANTALUM SYSTEM

The W–Re, W–Ta, and Ta–Re binary phase diagrams are computed and compared with observed binary diagrams in Figs. 81, 64, 80, and 99 of Chapters V and VI. These calculations showed the stability of the β, ϵ, and L phases in various temperature and compositional ranges. Subsequently, a ξ phase was inserted in the compositional range near $TaRe_3$. Although the free energy of this phase was computed on the basis of an $AuCu_3$-type (ξ) phase, it actually exhibits an α-Mn structure (see Chapter VI, discussion of Ta–Re). Reference to the above-mentioned W–Re binary (Fig. 81 in Chapter V) shows the occurrence of a compound phase near WRe_3. The latter also

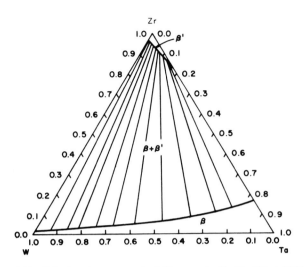

Fig. 150. Calculated ternary miscibility gap in Zr–Ta–W at 1873 °K.

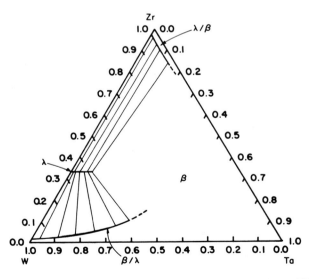

Fig. 151. Calculated compound phase equilibria in Zr–Ta–W at 1873 °K.

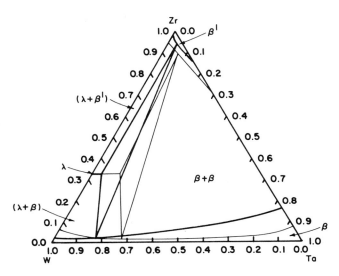

FIG. 152. Comparison of observed and calculated ternary diagrams of Zr–Ta–W at 1873 °K. Thick lines are calculated and thin lines are observed.

exhibits the α-Mn structure. Finally, the W–Re binary system contains a broad σ phase field, which has not been considered. Wulff and Brophy (142) have presented ternary sections for Re–W–Ta at 2293 and 2953°K. These sections have been computed considering β, ε, α, L, and ξ phases. In line with the procedure described earlier the free energy of the ξ phase $F^{\xi}[y, T]$ is given by Eq. (337) for $(Ta, W)Re_3$:

$$F^{\xi}[y, T] = \tfrac{3}{4}F^{\alpha}_{Re} + (\tfrac{1}{4} - y)F^{\alpha}_{Ta} + yF^{\alpha}_{W} - (1 - 4y)(8032) - 4y(5625)$$
$$+ RT\{y \ln y + (\tfrac{1}{4} - y)\ln(\tfrac{1}{4} - y) - \tfrac{1}{4}\ln\tfrac{1}{4}\} \tag{337}$$

This free energy is used to compute the stability of the ξ phase which is to be associated with the α-Mn(Ta, W)Re$_3$ phase. Equation (337) has been applied for the Ta–Re binary (i.e., $y = 0$) to compute the required interactions between the TaRe$_3$ phase and the β and ε phases. The results, which are shown in Fig. 99 of Chapter VI, indicate that observed TaRe$_3$ phase is more stable than the computed TaRe$_3$ phase. Application of Eq. (337) to compute the stability of the WRe$_3$ phase (i.e., $y = \tfrac{1}{4}$) was not performed earlier (Fig. 81 of Chapter V shows the computed β/ε interaction). If Eq. (337) is applied to TaRe$_3$, the temperature limit of stability of the WRe$_3$ with respect to the ε phase is computed to be 2260°K. This value is somewhat lower than observed. Finally, since formulation of the free energy of the σ phase has not been considered in the binary diagrams, it is not included in the computation of the ternary cases.

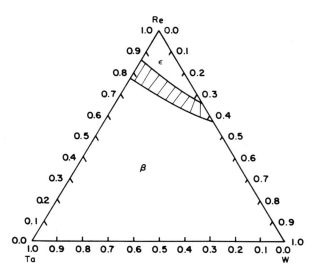

Fig. 153. Calculated solution phase equilibria in Re–W–Ta at 2293 °K.

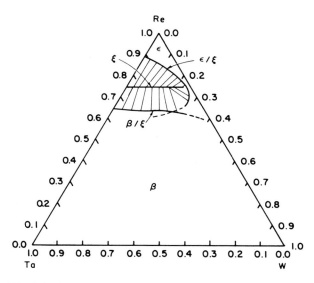

Fig. 154. Calculated compound phase equilibria in Re–W–Ta at 2293 °K.

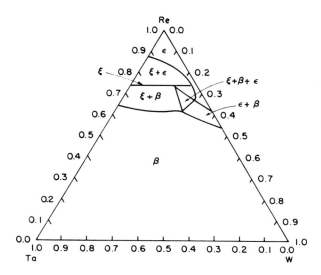

FIG. 155. Calculated ternary diagram of Re–W–Ta at 2293 °K.

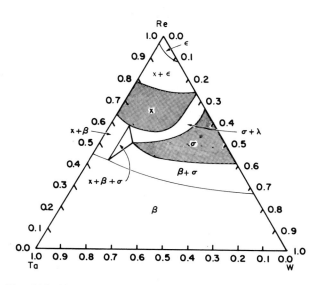

FIG. 156. Observed ternary diagram of Re–W–Ta at 2293 °K.

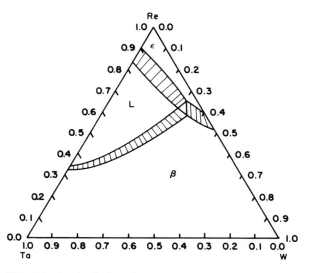

FIG. 157. Calculated solution phase equilibria in Re–W–Ta at 2953 °K.

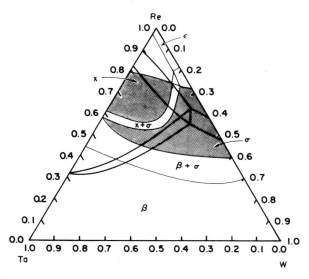

FIG. 158. Comparison of observed and calculated ternary diagrams of Re–W–Ta at 2953 °K. Thick lines are calculated and thin lines are observed.

Computation of the ternary sections is performed along the lines described earlier. This includes initial identification of the stable solution phases (β, ϵ, α, L), followed by superposition of the compound intrusion equilibria. The results are presented and compared with the observed diagrams in Figs. 153–158. Thus, Fig. 153 presents the computed β/ϵ two-phase field and tie lines at 2293°K; Fig. 154 shows the computed interaction of the (Ta, W)Re$_3$ phase with the β and ϵ phases; and Fig. 155 contains the superposition of the β, ϵ, and (Ta, W)Re$_3$ phase–solution phase interactions. This is to be compared with the observed phase diagram shown in Fig. 156. The latter exhibits a broad σ phase (shaded) and a (Ta, W)Re$_3$ χ phase (α-Mn structure). This phase is stable across the entire ternary section, while the computed (Ta, W)Re$_3$ (ξ) phase terminates just short of the W–Re edge. As indicated earlier, this phase is computed to become unstable at 2260°K and would appear on computed isothermal sections below this temperature. As noted earlier, the present description of the ξ phase indicates a level of stability that is slightly less than that of the χ phase, which exhibits the α-Mn structure for the (Ta, W)Re$_3$ compositions. At 2953°K, the computed phase equilibria (Fig. 157) shows β, L, and ϵ interactions, while the observed phase diagrams show that the χ and σ phases actually dominate the equilibria and eliminate the liquid field. Alteration of the value of C in Eqs. (334) and (335), which are currently employed to describe the ξ phase from $C = 18,000$ to $C = 25,000$, would yield better agreement between observed and computed equilibria. Such a change would convert the fourth and fifth terms of Eq. (337) to $-(1-4y)(9345)$ and $-4y(6938)$, respectively. These changes would stabilize the computed (Ta, W)Re$_3$ phase to higher temperatures, and the computed diagram of Fig. 158 would resemble that of Fig. 155 more closely. The recomputed phase diagrams are shown in Figs. 159 and 160.

4. THE RHENIUM–HAFNIUM–MOLYBDENUM SYSTEM

Computed isothermal sections through the Re–Hf–Mo ternary system at 1873, 2273, and 2673°K are displayed and compared with observed sections in Figs. 161–172. The calculated phase diagrams in Figs. 91, 99, and 81 of Chapters V and VI for Hf–Mo, Hf–Re, and Mo–Re indicate the existence of β (Mo) and ϵ (Hf, Re) solid-solution phases at the lower temperature. At the higher temperature, portions of the Hf–Re and Hf–Mo binaries are liquid. With the addition of Laves phases HfRe$_2$ and HfMo$_2$, the final computed binary diagrams take the forms indicated in Figs. 91 and 99 of Chapter VI.

In line with the present development [i.e., Eqs. (291), (292), (335), and Table XIII], the free energy of the Hf(Re, Mo)$_2$ Laves phase is given by Eq. (338) as

$$F^\lambda[y, T] = (\tfrac{2}{3} - y)F^\epsilon_{Re} + \tfrac{1}{3}F^\epsilon_{Hf} + yF^\epsilon_{Mo} - (1 - \tfrac{3}{2}y)(8273) - \tfrac{3}{2}y(3929)$$
$$+ RT\{y \ln y + (\tfrac{2}{3} - y) \ln(\tfrac{2}{3} - y) - \tfrac{2}{3} \ln \tfrac{2}{3}\} \tag{338}$$

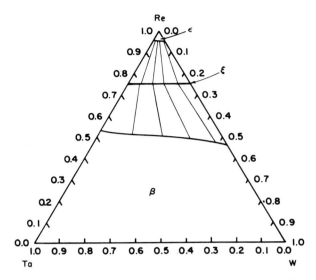

FIG. 159. Recalculation of ternary sections Re–W–Ta system at 2293 °K with $C = 25,000$ for (Ta, W)Re$_3$ Phase. [Compare with Fig. 154 computed with $C = 18,000$ for (Ta, W)Re$_3$ and observed diagram in Fig. 156.]

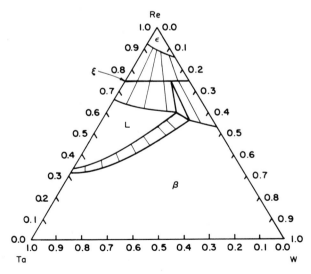

FIG. 160. Recalculation of ternary sections Re–W–Ta system at 2953 °K with $C = 25,000$ for (Ta, W)Re$_3$ phase. [Compare with Fig. 157 computed with $C = 18,000$ for (Ta, W)Re$_3$ and observed diagram in Fig. 158.]

where y is the atomic fraction of Mo. At 1873°K, the β/ϵ equilibrium is calculated in Fig. 161, showing the tie lines across the $\beta + \epsilon$ field. Figure 162 displays the β/λ and ϵ/λ interactions of a fully stable λ phase Hf(Mo, Re)$_2$. Superposition of the last two figures yields that of Fig. 163, the final calculated isothermal, which should now be compared with the observed (*142*) diagram of Fig. 164. Comparison with the observed diagram is complicated by the existence of three observed compounds χ, σ, and Φ, which are not taken into account in the computations. The χ and σ phases are the same phases observed in the Re–W–Ta system, but the Φ phase, essentially HfRe, has a pseudotetragonal structure. However, the observed λ phase, which dominates the system, is stable across the ternary triangle in agreement with the present calculation. The Mo–Hf edges of Figs. 163 and 164 are in general agreement, in spite of the intrusion of the Φ compound phase, but the upper portion of the observed diagram differs from that computed due to a relatively broad χ phase.

Figures 165, 166, and 167 represent the development of the 2273°K isothermal section. Figures 167 and 163 differ due to the liquid phase which enters at the higher temperature and replaces the right hand branch of the λ/β equilibria with a λ/L interaction. The same situation occurs in the observed diagram of Fig. 166, the λ phase remaining fully stable and the ϵ phase of Hf transforming into β on heating from 1873 to 2273°K in agreement with the calculation.

At 2673°K, the computed liquid phase occupies a larger portion of the ternary as Hf melts and the ϵ phase of Re becomes narrower in Fig. 169. In Figs. 170 and 171 the λ phase is no longer fully stable, since HfMo$_2$ melts and the λ/L interaction dominates the lower half of the ternary. The observed and calculated diagrams of Figs. 171 and 172 show the same tendency. However, the observed λ phase extends further into the triangle to a Mo concentration of about 50 at.%, while the calculated λ phase terminates at about 15 at.% Mo. Thus, the observed Mo β phase remains relatively narrow. The major discrepancy between Figs. 171 and 172 is the extent of the λ field. This discrepancy is due to the fact that the value of $C = 18,000$ employed for both the HfMo$_2$ and HfRe$_2$ phases yields an accurate calculation of the melting point of HfMo$_2$ (Fig. 91 of Chapter VI), but yields a computed melting point of HfRe$_2$ that is some 500°K lower than observed (Fig. 99 of Chapter VI). If the value of C were increased from 18,000 to 21,000 for HfRe$_2$, the computed melting point of the binary compound would agree with the observed value and the computed stability of the ternary compound Hf(Mo, Re)$_2$ in the 2673°K ternary section would be in close agreement with the observed result. This change would convert the fourth term on the right of Eq. 338 to $-(1 - \frac{3}{2}y)$ (8938).

(text continues on page 256)

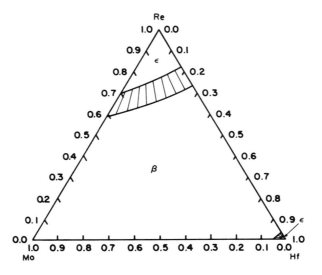

FIG. 161. Calculated solution phase equilibria in Re–Hf–Mo at 1873 °K.

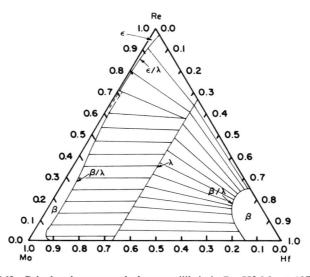

FIG. 162. Calculated compound phase equilibria in Re–Hf–Mo at 1873 °K.

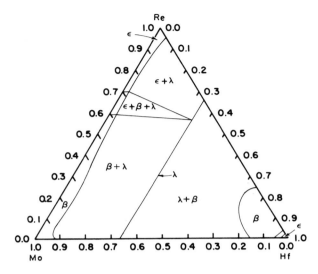

FIG. 163. Calculated ternary diagram of Re–Hf–Mo at 1873 °K.

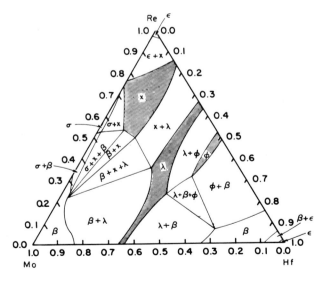

FIG. 164. Observed ternary diagram of Re–Hf–Mo at 1873 °K.

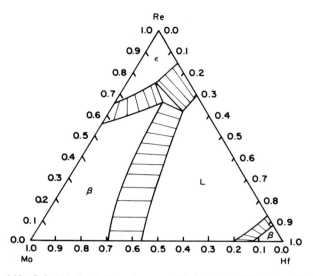

FIG. 165. Calculated solution phase equilibria in Re–Hf–Mo at 2273 °K.

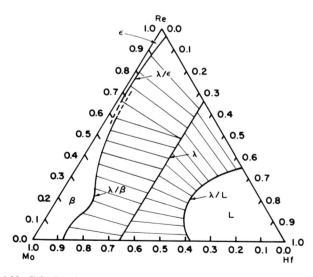

FIG. 166. Calculated compound phase equilibria in Re–Hf–Mo at 2273 °K.

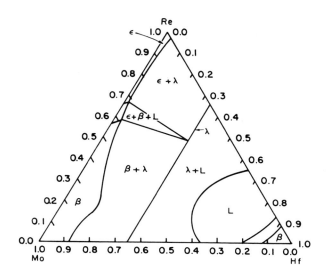

FIG. 167. Calculated ternary diagram of Re–Hf–Mo at 2273 °K.

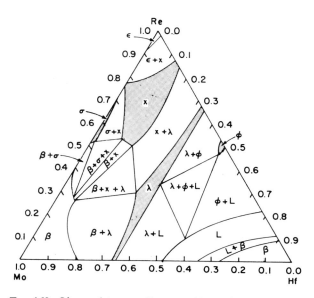

FIG. 168. Observed ternary diagram of Re–Hf–Mo at 2273 °K.

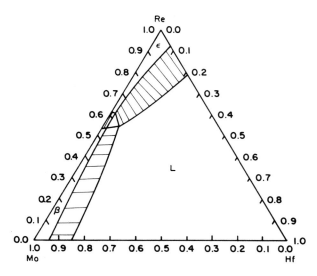

FIG. 169. Calculated solution phase equilibria in Re–Hf–Mo at 2673 °K.

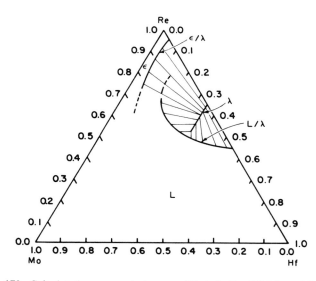

FIG. 170. Calculated compound phase equilibria in Re–Hf–Mo at 2673 °K.

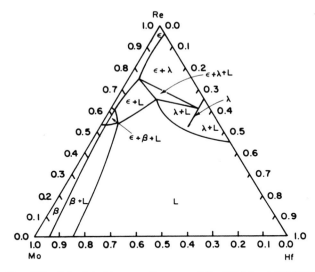

FIG. 171. Calculated ternary diagram of Re–Hf–Mo at 2673 °K.

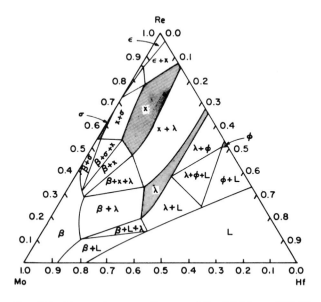

FIG. 172. Observed ternary diagram of Re–Hf–Mo at 2673 °K.

5. DISCUSSION OF RESULTS

The framework presented in Chapters XI–XIII covers development of computational techniques for dealing with ternary interactions between solution phases (L, β, ϵ, and α) and the formation of miscibility gaps in these phases, as well as interactions between these solution phases and the Laves (λ) and AuCu$_3$ (ξ) phases treated earlier. Generalized and detailed descriptions of these methods are contained in the above-mentioned chapters. Application of these methods to prediction of ternary sections in the Mo–W–Os, Zr–Ta–W, Re–W–Ta, and Re–Hf–Mo systems at selected temperatures was limited by the availability of experimental data for comparison. The present calculations could be applied directly to 286 ternary systems containing Zr, Nb, Mo, Ru, Rh, Pd, Hf, Ta, W, Re, Os, Ir, and Pt at all temperatures. Numerical input parameters required for such computations are contained in Tables VIII and XIII. In addition, computations of component vapor pressures and vaporization rates for all of the components can be obtained along the lines indicated earlier by application of Eqs. (237)–(239). Similarly, the heats of formation of ternary compounds can be estimated as before by application of Eqs. (291), (292), (334), and (335). The results obtained by direct computation of ternary sections for four systems of refractory transition metals constituted a direct evaluation of the simple description of the liquid, bcc, hcp, fcc, Laves, and AuCu$_3$-type phases developed earlier.

Evaluation of the present calculations is best accomplished by considering the computed and observed phase equilibria for the above-mentioned systems. Figures 148 and 149 compared the Mo–W–Os sections at 2468°K. Here the chief difference is the presence of a σ phase that intrudes on the β/ϵ equilibria. The earlier study did not present a formulation of the free energy of the σ phase. If this were done, the effect of σ phase intrusion could be computed. Nevertheless, the computed section (Fig. 148), is consistent with the observed section (Fig. 149) in that the computed β/ϵ equilibria is completely within the observed $\beta/\sigma/\epsilon$ fields. Comparison of the computed and observed Zr–Ta–W section at 1873°K (Fig. 152) is quite satisfactory. The principal source of discrepancy here is due to the assumed symmetrical form of the free energy of mixing, and the crude estimate of the free energy of formation of ZrTa$_2$ (λ), which is not stable [Eq. (336)]. In addition, realistic treatment of the λ phase is a refractory compound of variable composition rather than a " line compound " would improve agreement.

Figures 156–160 compare the observed and computed equilibria in the Re–W–Ta system. Here, discrepancies arise because the (Ta, W)Re$_3$ phase has the α-Mn structure (χ) rather than the ξ form. However, this can be eliminated by setting $C = 25,000$ (Figs. 159 and 160). As in the case of the

Mo–W–Os system, intrusion of the σ phase is important, though it is not included in the calculations. Finally, consideration of the compounds as variable-composition phases rather than line compositions would improve the overall agreement. With these allowances, it can be seen that the computed equilibria (Figs. 159 and 160) are consistent with the observed sections (Figs. 156 and 158).

Finally, inclusion of the χ and σ phases in the Re–Hf–Mo system along with a modified constant for $HfRe_2$ ($C = 21,000$) and consideration of variable compound compositions would improve the agreement between computed and observed sections in the Re–Hf–Mo ternary system (Figs. 158–171).

In conclusion, the present comparisons show no obvious inconsistencies between computed and observed equilibria that cannot be corrected by (a) inclusion of σ and χ phase intrusion, (b) treatment of the compounds as phases of variable composition, and (c) adjustment of the characteristic compound constant C from the fixed value of $C = 18,000$ to a specific value that optimizes agreement in the binary. Each of these procedures could be applied to predict ternary phase diagrams when data on the component binary diagrams were limited or completely absent.

CHAPTER XIV

Summary

As indicated at the outset of this monograph, quantitative descriptions of phase stability require specification of free energy–composition curves for competing phases over wide ranges of temperature and pressure. Consequently, simultaneous consideration of lattice stability, internal pressure, size, electronic, vibrational, magnetic, and volumetric contributions in addition to those depending on group number, valence, or electron/atom ratios is necessary. The literature abounds with theoretical descriptions of phase stability in which emphasis is placed upon electronic factors, Fermi surface effects, crystal chemical factors, size and valence effects, etc. In most instances, these descriptions treat the stability (or solubility limit) of a given phase as if it were a property of the phase itself, rather than the result of competition with other phases that can form in a given system. Such approaches are not based on the concept of phase competition, which offers the broadest framework for explaining and predicting the stability of metallic phases in quantitative terms.

By contrast, the present work attempts to describe all of the principal phases quantitatively. A variety of factors (valence, size effects, lattice stability) are included in the evaluation of properties, and phase competition is chosen as the basis for equilibrium. Simplified models for phase equilibria have been presented which include the "competitive aspect" and utilize computer techniques to examine simultaneous interactions among many phases. These methods have been applied to approximately 100 binary transition metal systems. The format for extending this approach to ternary systems has been developed and applied to several systems. This description includes explicit representation of the activity of component metals and the heats of formation of compound and solution phases. At present, accuracy

258

is limited by the representation of the excess free energy of solution phases (the regular-solution model is employed) and the procedures for calculating numerical values for lattice stability and interaction parameters. Thus, the present calculations must be considered as approximate, limited by the simplifications inherent in the regular-solution model and the line compound phase restriction. At the same time, it should be noted that both these restrictions could be lifted without encountering difficulties in computer programing.

In spite of these drawbacks, the present framework provides the most realistic picture of phase competition available. Since the phase competition concept is acknowledged to lie at the base of phase equilibria, the general implications of the present work must be judged accordingly. The interaction parameters discussed in Chapter IV are merely *estimated descriptions* of an *approximate* model (regular solutions) of metallic phases. Nevertheless, when coupled with the lattice stability parameters presented in Chapter III, they provide a *complete and explicit* description of the solution phases in the systems involved. Although this description may be inaccurate to a certain degree because of inadequacies in the model or the estimation procedures employed, it represents the first attempt to specify the stability of these solution phases in quantitative terms. Refinements of the current solution model along the lines indicated in Chapter X or modification of the lattice stability values could provide an improved but more complex description of phase stability.

Although fundamental calculations of the relative energies of metallic structures are presently incapable of providing more accurate estimates of these quantities at present, measurements of high-pressure equilibria and of the properties of metastable phases can provide relevant data. Future studies along these lines offer a logical means for improving the current description of metallic phases and relaxation of the artificial constraints imposed by the regular-solution model that is employed at present. Notwithstanding the simplicity of the present models, the agreement between computed and observed phase equilibria and thermodynamic properties is surprisingly good. As a consequence, the scope of the present treatment qualifies it as a valuable tool for guiding research on thermodynamic properties and phase equilibria in complex systems. In addition, it should provide substantial assistance in the development of new alloy and composite systems, since it provides a rapid and relatively inexpensive method for considering the effects of simultaneously varying several alloying ingredients.

References

1. J. J. Van Laar, *Z. Physik. Chem.* **63**, 216 (1908); **64**, 257 (1908).
2. M. Hansen and K. Anderko, " Constitution of Binary Alloys. " McGraw-Hill, New York, 1958.
3. R. P. Elliot, " Constitution of Binary Alloys, " First Supplement. McGraw-Hill, New York, 1965.
4. C. H. M. Jenkins, *J. Inst. Metals* **36**, 63 (1926).
5. D. Stockdale, *J. Inst. Metals* **44**, 75 (1930).
6. J. L. Meijering, *Philips Res. Rept.* **3**, 281 (1948).
7. R. Hultgren, R. L. Orr, P. D. Anderson, and K. K. Kelley, " Selected Values of Thermodynamic Properties of Metals and Alloys. " Wiley, New York, 1963.
8. O. Kubaschewski, " Phase Stability in Metals and Alloys " (P. S. Rudman, J. Stringer, and R. I. Jaffe , eds.), p. 63. McGraw-Hill, New York, 1967. See also O. Kubaschewski and W. Slough, *Progr. Mater. Sci.* **14**, No. 1, 1 (1969).
9. W. Hume-Rothery, " Phase Stability in Metals and Alloys " (P. S. Rudman, J. Stringer, and R. I. Jaffee, eds.), p. 3. McGraw-Hill, New York, 1967.
10. R. Oriani, " The Physical Chemistry of Metallic Solutions and Intermetallic Compounds, " Vol. I, Chapter 2A (*Natl. Phys. Lab. Symp.* No. 9). HM Stationery Office, London, 1958.
11. J. Friedel, *Advan. Phys.*, **3**, 446 (1954).
12. O. J. Kleppa, " Metallic Solid Solutions " (J. Friedel and A. Guinier, eds.), Chapter XXXIII. Benjamin, New York, 1963.
13. R. J. Weiss, " Solid State Physics for Metallurgists, " p. 300. Pergamon Press, London, and Addison-Wesley, Reading, Mass., 1963.
14. J. H. Hildebrand and R. L. Scott, " Solubility of Nonelectrolytes," 3rd ed. Van Nostrand (Rheinhold), Princeton, New Jersey, 1950.
15. L. R. Bidwell and R. Speiser, *Acta Met.*, **13**, 61 (1965).
16. L. Kaufman, " Phase Stability in Metals and Alloys " (P. S. Rudman, J. Stringer, and R. I. Jaffee, eds.), p. 125. McGraw-Hill, New York, 1967.
17. W. Lomer, " Phase Stability in Metals and Alloys " (P. S. Rudman, J. Stringer, and R. I. Jaffee, eds.), p. 569. McGraw-Hill, New York, 1967.
18. J. Lumsden, " Thermodynamics of Alloys " (*Inst. Metals Monogr. Rept. Ser.* No. 11), p. 354. Inst. Metals, London, 1952.

19. L. Kaufman, *Acta Met.* **9**, 861 (1961).
20. L. Kaufman, " Solids Under Pressure " (W. Paul and D. Warschauer, eds.), p. 303. McGraw-Hill, New York, 1963.
21. L. Kaufman, " Energetics in Metallurgical Phenomena III " (W. Mueller, ed.), p. 53. Gordon and Breach, New York, 1967.
22. O. Kubaschewski and T. G. Chart, *J. Inst. Metals* **93**, 329 (1964).
23. J. L. Meijering, " The Physical Chemistry of Metallic Solutions and Intermetallic Compounds, " Vol. II, p. 5A.7 (*Natl. Phys. Lab. Symp.* No. 9). H.M. Stationery Office, London, 1958.
24. L. Brewer, " High Strength Materials " (V. F. Zackay, ed.), p. 12. Wiley, New York, 1965.
25. L. Brewer, " Phase Stability in Metals and Alloys " (P. S. Rudman, J. Stringer, and R. I. Jaffee, eds.), p. 39. McGraw-Hill, New York, 1967.
26. J. Stringer, " Phase Stability in Metals and Alloys " (P. S. Rudman, J. Stringer, and R. I. Jaffee, eds.), p. 165. McGraw-Hill, New York, 1967.
27. W. Klement and A. Jayaraman, *Progr. Solid State Chem.* **3**, 289 (1966).
28. W. Klement, A. Jayaraman, and G. C. Kennedy, *Phys. Rev.* **131**, 632 (1963).
29. L. Kaufman and E. V. Clougherty, " Metallurgy at High Pressures and High Temperatures " (K. A. Gschneider, M. T. Hepworth, and N. A. D. Parlee, eds.), p. 322. Gordon and Breach, New York, 1964.
30. C. H. Bates, F. Dachille, and R. Roy, *Science* **147**, 860 (1965).
31. C. H. Bates, doctoral dissertation, Dept. of Geochemistry, Pennsylvania State University, March, 1966; Tech. Rept. 22 Contract NONR 656(00), Office Naval Res. (1966).
32. L. Kaufman, *Progr. Mater. Sci.* **14**, No. 2, 55 (1969).
33. L. Kaufman, *Acta Met.* **7**, 575 (1959).
34. L. Kaufman, E. V. Clougherty, and R. J. Weiss, *Acta Met.* **11**, 323 (1963).
35. L. D. Blackburn, L. Kaufman, and M. Cohen, *Acta Met.* **13**, 533 (1965).
36. G. S. Stepakoff and L. Kaufman, *Acta Met.* **16**, 13 (1968).
37. K. J. Tauer and R. J. Weiss, *J. Phys. Chem. Solids* **4**, 135 (1958).
38. M. M. Rao, R. J. Russell, and P. G. Winchell, *Trans. Met. Soc. AIME* **239**, 634 (1967).
39. F. Heiniger, E. Bucher, and J. Muller, *J. Condensed Matter* **5**, 243 (1966).
40. H. Claus, *J. Phys. Chem. Solids* **28**, 2449 (1967).
41. K. Andres, E. Bucher, J. P. Maita, and R. C. Sherwood, *Phys. Rev.* **178**, 702 (1969).
42. H. Claus, *J. Phys. Chem. Solids* **30**, 782 (1969).
43. L. Kaufman, Physical properties of martensite and bainite, Special Rept. No. 93, p. 48, Iron and Steel Inst., London (1965).
44. A. N. Holden, " Physical Metallurgy of Uranium," pp. 36–49. Addison-Wesley, Reading, Mass., 1958.
45. W. B. Pearson, " Lattice Spacing of Metals and Alloys." Pergamon Press, London, 1958.
46. W. A. Harrison, " Pseudo Potentials in the Theory of Metals " (D. Pines, ed.), pp. 193–195. Benjamin, New York, 1966.
47. V. Heine, " Phase Stability in Metals and Alloys " (P. S. Rudman, J. Stringer, and R. I. Jaffee, eds.), p. 103. McGraw-Hill,, New York, 1967.
48. R. M. Pick, " Phase Stability in Metals and Alloys " (P. S. Rudman, J. Stringer, and R. I. Jaffee, eds.), p. 219. McGraw-Hill, New York, 1967.
49. A. Seegar, " Phase Stability in Metals and Alloys " (P. S. Rudman, J. Stringer, and R. I. Jaffee, eds.), p. 249. McGraw-Hill, New York, 1967.
50. W. A. Harrison, *Phys. Rev.* **129**, 2512 (1963).

51. W. A. Harrison, *Phys. Rev.* **136**, A1107 (1964).
52. A. Taylor and N. J. Doyle, *J. Less Common Metals* **7**, 38 (1964).
53. S. J. Michalik and J. H. Brophy, *Trans. AIME* **227**, 1047 (1963).
54. E. J. Rapperport and M. F. Smith, Refractory metal constitution diagrams, WADD-TR-132, Part II (March, 1966).
55. L. Kaufman and H. Bernstein, " Anisotropy in Single Crystal Refractory Compounds " (F. W. Vahldiek and S. Mersol, eds.), Vol. I, p. 269. Plenum Press, New York, 1968.
56. L. Kaufman, *Bull. Am. Phys. Soc.* **4**, 181 (1959); Office Tech. Services Rept. PB144220 (1959).
57. G. F. Hurley and J. H. Brophy, *J. Less Common Metals* **7**, 267 (1964).
58. J. M. Dickenson and L. S. Richardson, *Trans. Am. Soc. Metals* **51**, 1055 (1959).
59. A. Taylor, N. J. Doyle, and B. Kagle, *J. Less Common Metals* **4**, 436 (1962).
60. E. J. Rapperport and M. F. Smith, *Trans. AIME* **230**, 6 (1964).
61. E. A. Anderson, *J. Less Common Metals* **6**, 81 (1964).
62. W. K. Goetz and J. H. Brophy, *J. Less Common Metals* **6**, 345 (1964).
63. M. A. Tylkina, I. A. Tsyganova, and E. M. Savitskii, *Russian J. Inorg. Chem. (English Transl.)* **7**, No. 8, 990 (1962).
64. M. A. Tylkina, V. P. Polyakova, and O. Kh. Khamidov, *Russian J. Inorg. Chem. (English Transl.)* **8**, No. 3, 395 (1963).
65. E. Raub, *Z. Metallk.* **55**, 316 (1964).
66. R. D. Reiswig and J. M. Dickenson, *Trans. AIME* **230**, 469 (1964).
67. K. L. Chopra, M. R. Randlett, and R. J. Duff, *Phil. Mag.* **16**, 261 (1967).
68. L. Brewer, *Acta Met.* **15**, 553 (1967).
69. R. A. Buckley and W. Hume-Rothery, *J. Iron Steel Inst. (London)* **201**, 228 (1963).
70. O. J. Kleppa, Thermodynamics and properties of liquid solutions, "Liquid Metals and Solidifications." Am. Soc. Metals, Cleveland, Ohio, 1957.
71. B. W. Mott, *Phil. Mag.* **2**, 259 (1957).
72. K. Furakawa, *J. Japan. Inst. Metals* **23**, A322 (1959).
73. H. Wada, *Trans. Japan. Natl. Res. Inst. Metals* **6**, No. 3, 96 (1964).
74. D. R. Stull and G. C. Sinke, " Thermodynamic Properties of the Elements." Am. Chem. Soc., Washington, D.C., 1956.
75. P. T. B. Shaffer, " High Temperature Materials." Plenum Press, New York, 1964.
76. J. M. Criscione, R. A. Mercuri, E. P. Schram, A. W. Smith, and H. F. Valk, High temperature protective coatings for graphite, TDR-64-173, Part II, Air Force Mater. Lab. (October, 1964).
77. J. F. Elliot and M. Gleiser, " Thermochemistry for Steelmaking," Vol. I. Addison-Wesley, Reading, Mass., 1960.
78. L. Kaufman and A. Sarney, Compounds of Interest in Nuclear Reactor Technology, *in* " Nuclear Metallurgy " (J. T. Waber, P. Chiotti, and W. Miner, eds.), Vol. X, p. 267. AIME, New York, 1964.
79. B. C. Giessen, R. H. Kane, and N. J. Grant, *Trans. AIME* **233**, 855 (1965).
80. P. S. Rudman, *Trans. Met. Soc. AIME* **239**, 64 (1967).
81. Cottrell, A. H., " Theoretical Structural Metallurgy." St. Martin's Press, New York, 1960.
82. A. G. Knapton, *J. Less Common Metals* **2**, 113 (1960).
83. P. S. Rudman, *Trans. AIME* **233**, 864 (1965).
84. E. Raub and E. Roschel, *Z. Metallk.* **54**, 455 (1963).
85. H. Kato and M. Copeland, *U.S. Bur. Mines Rept.* BMU 1031, BMU 1059, and BMU 1082 (1963).

86. D. L. Ritter, B. C. Giessen, and N. J. Grant, *Trans. AIME* **230**, 1250 (1964).
87. B. C. Giessen, R. Koch, and N. J. Grant, *Trans. AIME* **230**, 1268 (1964).
88. W. H. Ferguson, Jr., B. C. Giessen, and N. J. Grant, *Trans. AIME* **227**, 1401 (1963).
89. B. C. Giessen, H. Ibach, and N. J. Grant, *Trans. AIME* **230**, 113 (1964).
90. E. M. Savitskii, V. P. Polyakova, M. A. Tylkina, and G. S. Burkanov, *Russian J. Inorg. Chem. (English Transl.)* **9**, No. 7, 890 (1964).
91. M. A. Tylkina, I. A. Tsyganova, and E. M. Savitskii, *Russian J. Inorg. Chem. (English Transl.)* **7**, No. 8, 994 (1962).
92. L. Kaufman and E. V. Clougherty, Thermodynamic factors controlling the stability of solid phases at high temperatures and pressures, *in* " Metallurgy at High Pressures and High Temperatures " (K. A. Geschneider, M. T. Hepworth, and N. A. D. Parlee, eds., Met. Soc. Conf. AIME), Vol. 22, p. 322. Gordon and Breach, New York, 1964.
93. L. Kaufman, Thermodynamic properties of transition metal diborides, "Compounds of Interest in Nuclear Reactor Technology " (J. T. Waber, P. Chiotti, and W. Miner, eds., Nuclear Metallurgy Ser.), Vol. X, p. 193. AIME, New York, 1964.
94. L. Kaufman and E. V. Clougherty, Investigations of boride compounds for high temperature applications, " Materials for the Space Age " (F. Benesovsky, ed.), p. 722. Metallwek Plansee, Reutte, Austria, 1964.
95. P. S. Rudman, *Trans. AIME* **233**, 872 (1965).
96. L. Kaufman and H. Bernstein, " The Science, Technology and Application of Titanium " (R. I. Jaffee and N. Promisel, eds.), p. 349. Pergamon Press, New York, 1969.
97. E. Rudy and St. Windisch, *Trans. Met. Soc. AIME* **242**, 953 (1968).
98. M. Hoch, J. V. Hackworth, and R. J. Usell, " The Science, Technology and Application of Titanium " (R. I. Jaffee and N. Promisel, eds.), p. 347. Pergamon Press, New York, 1969.
99. M. J. Pool, R. Speiser, and G. R. St. Pierre, *Trans. Met. Soc. AIME* **239**, 1180 (1967).
100. W. Johnson, K. Komarek, and E. Miller, Thermodynamic properties of the Cr–Al and V–Al Systems, Contract NONR 285(64), Office Naval Res., Washington, D.C., Department of Metallurgy, New York University, New York (August, 1967).
101. C. Zener, *Trans. Met. Soc. AIME* **203**, 619 (1955).
102. R. J. Weiss and K. J. Tauer, *Phys. Rev.* **102**, 1495 (1956).
103. J. A. Hoffman, A. Paskin, K. J. Tauer, and R. J. Weiss, *J. Phys. Chem. Solids* **1**, 45 (1956).
104. K. J. Tauer and R. J. Weiss, *Phys. Rev.* **100**, 1223 (1955).
105. R. J. Weiss and K. J. Tauer, *Proc. Seminar Theory Alloy Phases*, p. 290. Am. Soc. Metals, Cleveland, Ohio (1956).
106. L. Kaufman and M. Cohen, *Progr. Metal Phys.* **7**, 165 (1958).
107. M. Hillert, T. Wada, and H. Wada, *J. Iron Steel Inst. (London)* **205**, 539 (1967).
108. T. Wada, *Trans. Iron Steel Inst. Japan* **8**, 1 (1968).
109. R. J. Weiss and K. J. Tauer, *J. Phys. Chem. Solids* **7**, 249 (1959).
110. J. L. Meijering, *J. Phys. Chem. Solids* **18**, 267 (1961).
111. H. K. Hardy, *Acta Met.* **1**, 202 (1953).
112. H. A. Wreidt, *Trans. Met. Soc. AIME* **377**, 221 (1961).
113. M. Hillert, " Metallic Solid Solutions " (J. Friedel and A. Guinier, eds.), p. XLVII-8. Benjamin, New York, 1963.
114. B. E. Sundquist, *Trans. Met. Soc. AIME*, **236**, 1111 (1966).
115. P. S. Rudman, *J. Mater. Sci. Eng.* to be published.

116. M. Hillert, " Phase Transformations " (M. Cohen, ed.). Am. Soc. Metals, Cleveland, Ohio, 1969.
117. R. W. Krenzer and M. J. Pool, *Trans. Met. Soc. AIME* **245**, 91 (1969), Appendix by L. Kaufman.
118. R. A. Buckley and W. Hume-Rothery, *J. Iron Steel Inst. (London)* **201**, 227 (1963).
119. W. Hume-Rothery, " Phase Stability in Metals and Alloys " (P. S. Rudman, J. Stringer, and R. I. Jáffe, eds.), p. 1. McGraw-Hill, New York, 1967.
120. A. K. Sinha, R. A. Buckley, and W. Hume-Rothery, *J. Iron Steel Inst. (London)* **207**, 36 (1969).
121. C. Zener, *Trans. AIME* **167**, 513 (1946).
122. L. Kaufman, *Trans. AIME* **215**, 218 (1959).
123. W. Hume-Rothery, G. W. Mabbott, and K. M. Channel-Evans, *Phil. Trans. Roy. Soc. (London)* **A233**, 1 (1934).
124. H. Jones, *Proc. Phys. Soc. (London)* **49**, 243, 250 (1937).
125. W. Hume-Rothery, " The Structure of Metals and Alloys " (Monogr. Rept. Ser. No. 1). Inst. Metals, London, 1936.
126. W. Hume-Rothery, " Atomic Theory for Students of Metallurgy (Monogr. Rept. Ser. No. 3). Inst. Metals, London, 1946.
127. T. B. Massalski, "Phase Stability in Metals and Alloys " (P. S. Rudman, J. Stringer, and R. I. Jaffe, eds.), p. 243. McGraw-Hill, New York, 1967.
128. N. Engel, *Kem. Maanedsbl.* **30**, 53 (1949).
129. Engel, N., *Kem. Maanedsbl.* **30**, 97, 105, 113 (1949).
130. Engel, N., *Trans. Am. Soc. Metals* **57**, 611 (1964).
131. N. Engel, *Acta Met.* **15**, 565 (1967).
132. L. Brewer, " Phase Stability in Metals and Alloys " (P. S. Rudman, J. Stringer, and R. I. Jaffe, eds.), p. 246. McGraw-Hill, New York, 1967.
133. W. Hume-Rothery, *Acta Met.* **15**, 1039 (1965).
134. W. Hume-Rothery, *Acta Met.* **15**, 567 (1967).
135. W. Hume-Rothery, *Progr. Mater. Sci.* **13**, No. 5, p. 229 (1967).
136. N. Engel, *Acta Met.* **15**, 557 (1967).
137. J. L. Meijering, Segregation in regular ternary solutions, *Philips Res. Rept.* **5**, 333 (1950); **6**, 183 (1951).
138. J. L. Meijering, The physical chemistry of metallic solutions and intermetallic compounds .*Natl. Phys. Lab. Symp.* Chapter 5A. HM Stationery Office, London (1958); *Acta Met.* **5**, 257 (1957).
139. J. B. Scarborough, " Numerical Mathematical Analysis," p. 213. John Hopkins Press, Baltimore, 1962.
140. A. Taylor and N. J. Doyle, *J. Less Common Metals* **9**, 190 (1965).
141. L. F. Pease and J. H. Brophy, *J. Less Common Metals* **6**, 118 (1964).
142. J. Wulff and J. H. Brophy, Refractory metal constitution diagrams, WADD TR-60-132, Massachusetts Institute of Technology, pp. 84–85. Wright-Patterson Air Force Base, Ohio (June, 1960).
143. A. Taylor and N. J. Doyle, Refractory metal constitution diagrams, ASD TDR-62-132, Westinghouse Research Laboratories, pp. 107–109. Wright-Patterson Air Force Base, Ohio (March, 1962).

Computer Programs TRSE and LCRSE

A computer program for calculating the phase-boundary curves as a function of temperature for regular-solution phases is detailed below. The main program is called TRSE (Two Regular-Solution Equilibria) and contains the subroutines ZEROS, AMAX1, AMIN1, XAXB, and FABMI. Although this program is written specifically for the IBM 1130 computer, it is readily adaptable to any machine using FORTRAN IV language. The input data consist of the free energy difference parameters $\Delta H_i^{2 \to 1}$ (DHA), $\Delta S_i^{2 \to 1}$ (DSA), $\Delta H_j^{2 \to 1}$ (DHB), and $\Delta S_j^{2 \to 1}$ (DSB); the interaction parameters Z_1 and Z_2 (i.e., L, B, E, or A); the highest temperature T_i (TI), the lowest temperature T_f (TF), and the temperature increments ΔT (DT) specifying the temperature range over which the curves $x_1[T]$ and $x_2[T]$ will be computed; and finally a comment called CASE, which identifies the system and equilibria pair being computed, e.g., W–IR (B–L) which signifies the β:L equilibrium in the tungsten–iridium system. The printed output consists of a listing of the input data and the results x_1, x_2, and x_0 vs. T in the single-solution case or x_1, x_2, $\bar{x}_2$, $\bar{x}_1$, x_0, and $\bar{x}_0$ vs. T in the double-solution case. The latter occurs when the equilibria curves exhibit maxima or minima as in the β/L for Zr–Mo (Fig. 56). The message "NONCONVERGENT FOR ABOVE INPUT" is displayed when the iteration procedure cannot converge for any particular temperature.

A program for computing the $x_0[T]$ or $T_0[x]$ curves follows the TRSE program and is called XOTO. This information is part of the TRSE output. There are two options specified by INDEX within this program, depending upon whether a solid/liquid equilibrium or a solid/solid equilibrium is being calculated. The solid/liquid curves run between $x = 0$ and $x = 1$ and require the first option (INDEX = 1) in the form of $T_0[x]$, while the solid/solid curves vary with temperature and require the second option (INDEX = 2) in the form of $x_0[T]$. Input data, as for TRSE, consist of DHI, DSI, DHJ, DSJ, E1, E2, CASE, and the option INDEX. Output consists of T_0 vs. x from $x = 0$ to 1.0 in steps of 0.005 for option 1 or x_0 vs. T from $T = 100°$ to $4000°$K in steps of $100°$.

Finally a program called LCRSE (Line Compound Regular-Solution Equilibria) is detailed which calculates the boundary of the two-phase field between a regular-solution phase Φ (L, β, ϵ, or α) and a general line compound (Ψ) having a stoichiometric composition x_*. The curve $T[x_{\Phi\Psi}]$ for the phase boundary $x_{\Phi\Psi}$ has the form

$$T[x_{\Phi\Psi}] = \frac{(C_\Psi - L)x_*(1-x_*) + \Delta H_i^{\theta \to \Phi}(1-x_*) + \Delta H_j^{\theta \to \Phi}x_* + \Phi[x_*(1-2x_{\Phi\Psi}) + x^2_{\Phi\Psi}]}{\Delta S_i^{\theta \to \Phi}(1-x_*) + \Delta S_j^{\theta \to \Phi}x_* - R[(1-x_*)\ln(1-x_{\Phi\Psi}) + x_*\ln x_{\Phi\Psi}]}$$

which is a generalization Eq. (94) of Chapter III and of Eqs. (153)–(164) of Chapter VI. Here C_Ψ is a constant that is equal to 18,000 for $\Psi = \lambda$ ($x_* = 0.667$) compounds such as HfRe$_2$ as well as for $\Psi = \xi$ ($x_* = 0.750$) compounds such as NbRh$_3$; L is the usual interaction parameter for the liquid phase; Φ is the interaction parameter (L, B, E, or A) for the particular solution phase examined; and the symbol θ in the superscripts of ΔH_i, ΔS_i, ΔH_j and ΔS_j is the base phase and refers to the hexagonal form ϵ of the pure elements in the case when Ψ is the Laves phase ($x_* = \frac{2}{3}$). When Ψ is the ξ phase ($x_* = \frac{3}{4}$), then θ in the superscript refers to the fcc form α of the elements. Input required is C, L, PHI, DHI, DSI, DHJ, DSJ, XO, and CASE [e.g., ZR–W (ZRW$_2$–L) meaning the phase boundary curve of the liquid phase in equilibrium with the compound ZrW$_2$ within the system Zr–W]. The output again lists the input data and yields T vs. x in steps of $dx = 0.05$.

(Computer software for all programs is available from ManLabs.)

TRSE

```
// JOB                 TWO REGULAR SOLUTION INTERACTION PROBLEM
// FOR
*IOCS(CARD,TYPEWRITER,KEYBOARD,DISK)
*ONE WORD INTEGERS
*EXTENDED PRECISION
*LIST SOURCE PROGRAM
*NAME   TRSE
      DIMENSION CASE(3),X(6)
   10 FORMAT(3A4,2(F7.0,F6.2),2F8.0,2F6.0)
   20 FORMAT(//,3A4,2(F7.0,F6.2),2F8.0,2F6.0,//)
   30 FORMAT(F5.0,'  NON CONVERGENT')
   40 FORMAT(F5.0,3F9.4)
   50 FORMAT(F5.0,6F9.4)
    1 READ(2,10) CASE,DHA,DSA,DHB,DSB,E1,E2,TI,TF
      WRITE(1,20) CASE,DHA,DSA,DHB,DSB,E1,E2,TI,TF
      R=1.987
      DT=(TF-TI)/20.
      T=TF+DT
      DO 9 I=1,21
      T=T-DT
      RT=R*T
      A=(DHA-T*DSA)/RT
      B=(DHB-T*DSB)/RT
      Z1=E1/RT
      Z2=E2/RT
      IF(A) 200,100,100
  100 A=-A
      B=-B
      ZZ=Z1
      Z1=Z2
      Z2=ZZ
  200 CALL XAXB(A,B,Z1,Z2,N,X,IND)
      IF(IND) 3,4,4
    3 WRITE(1,30) T
      GO TO 8
    4 IF(N) 8,8,5
    5 IF(B) 7,6,6
    6 WRITE(1,40) T,X(1),X(5),X(2)
      GO TO 8
    7 WRITE(1,50) T,X(1),X(5),X(2),X(3),X(6),X(4)
    8 IF(DT) 9,1,9
    9 CONTINUE
      GO TO 1
      END
```

```
// FOR
*EXTENDED PRECISION
*LIST SOURCE PROGRAM
      SUBROUTINE ZEROS(A,B,SA,SB,X,Y,XLO,XHI,YLO,YHI,IND)
      ITERS = 100
      XMID = XHI+XLO
      XDIF = XHI-XLO
      YMID = YHI+YLO
      YDIF = YHI-YLO
      IND = 7
      DO 3 I=1,ITERS
      OLDX = X
      OLDY = Y
      T1=1.-X
      T2=1.-Y
      T3 = X-Y
      IF(T1) 3, 10, 10
   10 IF (T2)3,11,11
   11 IF(X) 3,12,12
   12 IF (Y) 3,13,13
   13 IF (T3)14,3,14
   14 GA=A+ALOG( T1 )+SA*X**2-ALOG( T2 )-SB*Y**2
      GB=B+ALOG(X)+SA*( T1 )**2-ALOG(Y)-SB*( T2 )**2
      DX=X*T1*(Y*GB+T2*GA)/(1.-2.*SA*X*T1)/T3
      DY=Y*T2*(X*GB+T1*GA)/(1.-2.*SB*Y*T2)/T3
      X=X+DX
      Y=Y+DY
      IF (ABS(DX)-.0001)1,1,3
    1 IF (ABS(DY)-.0001)2,2,33
   33 IF (X-XLO) 35,35,34
   34 IF (X-XHI) 36,35,35
   35 DX = DX/2.
      IF ((X-DX)-X) 355,365,355
  355 X = X-DX
      GO TO 33
  365 X = (2.*OLDX+XMID+SIGN(XDIF,DX))/4.
   36 IF (Y-YLO) 38,38,37
   37 IF (Y-YHI) 3,38,38
   38 DY = DY/2.
      IF ((Y-DY)-Y) 39,300,39
   39 Y = Y-DY
      GO TO 36
  300 Y = (2.*OLDY+YMID+SIGN(YDIF,DY))/4.
    3 CONTINUE
      IND = -7
    2 CONTINUE
      RETURN
      END
```

```
// FOR
*EXTENDED PRECISION
*LIST SOURCE PROGRAM
      FUNCTION AMAX1(A,B)
      IF(A-B)1,1,2
   1  AMAX1=B
      RETURN
   2  AMAX1=A
      RETURN
      END
// FOR
*EXTENDED PRECISION
*LIST SOURCE PROGRAM
      FUNCTION AMIN1(A,B)
      IF(A-B)1,1,2
   1  AMIN1=A
      RETURN
   2  AMIN1=B
      RETURN
      END



// FOR
*EXTENDED PRECISION
*LIST SOURCE PROGRAM
      SUBROUTINE XAXB(A,B,SA,SB,N,X,IND)
      DIMENSION X(6)
      XA=0.
      XB=0.
      XAB=0.
      XBB=0.
      AMB=A-B
      ATB=A*B
      SAMSB=SA-SB
      IF(SAMSB)1000,400,1000
 1000 ATBD=AMB*SAMSB
      ABQM=AMB/SAMSB
      ABQP=(A+B)/SAMSB
      ABQMS=ABQM**2
      ROOT = SQRT(ABS(1.+2.*ABQP+ABQMS))
      X0 = 1. -ABQM
      IF (ATB) 101,6,6
```

```
C      A AND B OPPOSITE SIGNS
C      CASE I.
  101  IF(ROOT-.0005*ABS(X0)) 13,1,1
   13  XA = X0/2.
       XB = XA
       N= 1
       GO TO 300
    1  IF (ATBD) 2,12,12
C      DIFFERENT
    2  X0=(X0-ROOT)/2.
       GO TO 11
C      SAME
   12  X0=(X0+ROOT)/2.
   11  CONTINUE
       XI=.5
       TA = SA-2.
       IF (TA) 1130,1130,1105
 1105  XI = (1.-SQRT(TA/SA))/2.
 1130  XAM = XI/10.
       CALL FABMI (A,B,SA,XAM,XI,IND)
       IF (IND) 1110,1110,1120
 1110  XAM = 0.
       XA = AMIN1(1.E-8,XI/1000.)
 1120  IF (XI-.5) 11210,11200,11210
11200  XI = X0
11210  XBM = .5
       XIB = .5
       TB = SB-2.
       IF (TB) 1160,1160,1140
 1140  XIB = (1.-SQRT(TB/SB))/2.
       XBM = XIB/10.
       CALL F BMI (0.,0.,SB,XBM,XIB,IND)
       XIB = 1.-XIB
       IF (IND) 1150,1150,1160
 1150  WRITE (1,1151)
 1151  FORMAT (1H0,30HERROR - CANT FIND BETA MINIMUM)
       GO TO 200
 1160  TA = 1.-XAM
       TB = 1.-XBM
       DIFF = A*TA+ALOG(TA/TB)-(SB*TB+ALOG(XBM/TB))*XBM
       IF (XAM) 1180,1180,1170
 1170  DIFF = DIFF+(B+SA*TA+ALOG(XAM/TA))*XAM
       GO TO 1185
 1180  XAM = XA
 1185  XA = .99999999*AMIN1(XAM,X0/2.)
       XB = .99999999*AMIN1(XBM,(XBM+X0)/2.)
       IF (DIFF) 1190,1195,1200
 1195  DIFF = -0.
 1190  XB = AMAX1(1.-XB,AMAX1((X0+1.)/2.,(XIB+1.)/2.))
       XA = AMAX1(1.00000001*XAM,(XAM+X0)/2.)
```

```
1200 CONTINUE
     IF (A)5,5,4
   4 XX=XA
     XA=XB
     XB=XX
   5 CALL ZEROS(A,B,SA,SB,XA,XB,0.,AMIN1(X0,XI),AMAX1(X0,(.5-SIGN(.5,DI
    1FF))*XIB),1.-(.5+SIGN(.5,DIFF))*XIB,IND)
     N=2
     GO TO 300
C                    CASE II
   6 C=ABS(A)+ABS(B)+SQRT(4.*.7B)
     XC=(ABS(A)+SQRT(ATB))/C
     IF(A) 7,7,8
   7 IF (ABS(ABS(SAMSB)-C)-.0005) 100,7000,7000
7000 IF(ABS(SAMSB)-C)200,100,8
   8 X0 = (X0-ROOT)/2.
     X0B = X0+ROOT
     IF (SB-2.) 2000,2000,200
2000 XI = X0
     XIR=X0B
     TA = SA-2.
     IF (TA) 2020,2020,2010
2010 XI = (1.-SQRT(TA/SA))/2.
     XIR=1.-XI
2020 XA = AMIN1(X0/2.,XI/2.)
     XB=(X0+XC)/2.
     IF (A) 10,9,9
   9 XX=XB
     XB=XA
     XA=XX
  10 CALL ZEROS(A,B,SA,SB,XA,XB,0.,AMIN1(X0,XI),X0,X0B,IND)
     XAB = AMAX1(1.-XA,AMAX1((X0B+1.)/2.,(1.+XIR)/2.))
     XBB = (XB+X0B)/2.
     CALL ZEROS(A,B,SA,SB,XAB,XBB,AMAX1(X0B,XIR)   ,1.,X0,X0B,IND)
     N=4
     IF (XB-XBB) 300,300,3000
3000 XAM = XI/10.
     CALL FABMI (0.,0.,SA,XAM,XI,IND)
     XA = XAM
     XAB = 1.-XA
     XB = 1.E35
     XBB = XB
     N = 4
     GO TO 300
 100 TA = SA-2.
     IF (TA) 21000,21000,21100
21100 XI = (1.-SQRT(TA/SA))/2.
```

```
           GO TO 3000
  21000 XA=XC
           XB = XC
           XAB = XC
           XBB = XC
           N=1
           GO TO 300
   200   N=0
    300  X(1)=XA
           X(2)=XB
           X(3)=XBB
           X(4)=XAB
           X(5)=X0
           X(6)=X0B
           RETURN
    400   X0=ABS(A)/(ABS(A)+ABS(B))
           GO TO 11
           END



// FOR
*EXTENDED PRECISION
*LIST SOURCE PROGRAM
      SUBROUTINE FABMI (A,B,SAB,X,XI,IND)
      IND = 7
      DO 50 ITER=1,100
      OLDX = X
      T = 1.-X
      DX = -(ALOG(X/T)+SAB*(T-X)-A+B)*X*T/(1.-2.*SAB*X*T)
      IF (ABS(DX)-5.E-4*ABS(X)) 100,100,10
   10 X = X+DX
   20 IF (X) 40,40,30
   30 IF (X-XI) 50,40,40
   40 DX = DX/2.
      IF ((X-DX)-X) 45,90,45
   45 X = X-DX
      GO TO 20
   50 CONTINUE
      IND = -7
   90 X = OLDX
  100 RETURN
      END
```

XOTO


```
// FOR
*IOCS(CARD,TYPEWRITER,KEYBOARD,DISK)
*LIST SOURCE PROGRAM
*NAME  XOTO
       DIMENSION CASE(3)
    1 FORMAT(3A4,I2,2(F7.0,F6.2),4F8.0)
    2 FORMAT(/,'SOLID-LIQUID INTERACTION',/)
    3 FORMAT(/,'SOLID-SOLID INTERACTION',/)
    4 FORMAT(F8.4,5X,F5.0)
    5 FORMAT('NEGATIVE ARGUMENT')
    6 FORMAT(F8.4,5X,F8.4,5X,F5.0)
    7 FORMAT(//,3A4,I2,2(F7.0,F6.2),2F8.0)
   99 READ(2,1) CASE,KODE,DHA,DSA,DHB,DSB,E1,E2,DE1,DE2
      E1=E1+DE1
      E2=E2+DE2
      WRITE(1,7) CASE,KODE,DHA,DSA,DHB,DSB,E1,E2
      H=DHB-DHA
      S=DSB-DSA
      E=E1-E2
      IF(KODE) 10,10,20
   10 WRITE(1,2)
      X=-0.05
      DO 100 I=1,21
      X=X+.05+1.0E-06
      T=(DHA+X*(H+E)-X*X*E)/(DSA+X*S)
  100 WRITE(1,4) X,T
      GO TO 99
   20 WRITE(1,3)
      T=500.
      DO 200 I=1,15
      T=T+200.
      Q=E+H-T*S
      IF(E) 40,30,40
   30 X=(T*DSA-DHA)/(H-T*S)
      WRITE(1,4) X,T
      GO TO 200
   40 ARG=Q*Q+4.*E*(DHA-T*DSA)
      IF(ARG) 50,60,60
   50 WRITE(1,5)
      GO TO 200
   60 ROOT=SQRT(ARG)
      X1=(Q-ROOT)/(2.*E)
      X2=(Q+ROOT)/(2.*E)
      WRITE(1,6) X1,X2,T
  200 CONTINUE
      GO TO 99
      END
```

```
// JOB                   LINE COMPOUND REGULAR SOLUTION EQUILIBRIA
// FOR
*ONE WORD INTEGERS
*IOCS(CARD,TYPEWRITER,KEYBOARD,DISK)
*EXTENDED PRECISION
*LIST SOURCE PROGRAM
*NAME   LCRSE
        DIMENSION CASE(5)
   10 FORMAT(5A4,2(F7.0,F6.2),F6.4,2F8.0,F7.0,F5.0)
   20 FORMAT(//,'CASE   ',5A4,5X,'DHA=',F6.0,2X,'DSA=',F5.2,2X,'DHB=',F6.
      10,2X,'DSB=',F5.2,2X,'XO=',F6.4,/,'PHI=',F7.0,2X,'L=',F7.0,2X,'C=',
      2F6.0,2X,'DF=',F7.0,2X,'TM=',F5.0,//)
   30 FORMAT('X=',F7.4,3X,'T=',F5.0)
    1 READ(2,10) CASE,DHA,DSA,DHB,DSB,XO,PHI,EL,C,TMIN
      XMO=1.-XO
      S=XMO*DSA+XO*DSB
      DF=XO*XMO*(EL-C)
      Q=XMO*DHA+XO*DHB+XO*XMO*PHI-DF
      R=1.987
      TERM=R*(XMO*EALOG(XMO)+XO*EALOG(XO))
      TM=Q/(S-TERM)
      WRITE(1,20) CASE,DHA,DSA,DHB,DSB,XO,PHI,EL,C,DF,TM
      ARG=Q/PHI
      IF(ARG) 2,3,3
    2 ARG=-ARG
    3 ROOT=ESQRT(ARG)
      XS=XO-ROOT
      XF=XO+ROOT
      IF(XS) 4,4,5
    4 XS=0.
    5 IF(1.-XF) 6,6,7
    6 XF=1.00
    7 DX=(XF-XS)/20.
      X=XS
      DO 100 I=1,20
      X=X+DX+1.0E-06
      XM=1.-X
      TERMX=R*(XMO*EALOG(XM)+XO*EALOG(X))
      T=(Q+PHI*(X-XO)**2)/(S-TERMX)
      WRITE(1,30) X,T
  100 CONTINUE
      GO TO 1
      END
```

Computer Programs TERNRY, MIGAP, and TERCP

1. TERNARY EQUILIBRIA

A computer mainline program TERNRY has been developed to solve the ternary equilibria Eqs. (244)–(246), for refractory metal solutions of Zr, Hf, Nb, Ta, Mo, W, Re, Ru, Os, Rh, Ir, Pd, and Pt. The entire contents of Table VIII (free energy parameters) and Table XIII (regular-solution parameters), Chapters III and IV, for these elements have been stored for convenience. One chooses a ternary system consisting of elements I, J, and K and a temperature T. From visual observation of the binary phase diagrams, Figs. 52–105 in Chapters V and VI, the number and kind of stable phase pairs that are to be calculated are noted. Input to the program consists of the following:

Card 1. Table of elements and phases:

ZRHFNBTAMO WRERUOSRHIRPDPT LBEA

This is a fixed card accompanying every calculation. The last four symbols denote possible phases L (liquid), B (bcc), E (hcp), and A (fcc).

Card 2. Chemical symbols for elements I, J, and K; N the number of curve pairs to be plotted; T the temperature of the isothermal section.

Format: $\langle (3(A2, 1X), I2, 1X, F5.0) \rangle$

Cards 3 through $2 + N$. Chemical symbols for elements I–J, J–K, K–I of the binary mixture at the starting edge; names of the phases being studied; option for selecting which side of a minimum or maximum to operate upon.

Format: $\langle (A2, 1X, A2, 1X, 2A1, 1X, I2) \rangle$

There may be as many sets of Cards 3 through $2 + N$ as desired. Each set will generate one graph and a corresponding table of data. The phase limit curves are started on one edge selected by the operator after inspecting the binary phase diagrams. The order of the phases must be maintained in the proper sense so that the signs of ΔF_i, ΔF_j, and ΔF_k are properly assigned. The edge solutions are generated by the subroutine XAXB (a part of TRSE), and the ternary solution is generated from these starting values until another edge is reached. The temperature or phase pair can then be changed and plots made either on the same graph or another graph. A table of the points (x_1, y_1, z_1) and (x_2, y_2, z_2) is printed as well as the points (x_0, y_0) to provide additional accuracy to the graphical output.

Subroutines called by TERNRY3 are:

INIT	This subroutine reads in two tables—table of elements and table of phases L, B, E, and A.
TRIGRD	This subroutine draws and labels the triangular grid.
GTINX	This subroutine does a table lookup to get the index of each of the elements of the ternary.
PHNMP	This subroutine does a table lookup to get the index of the two-phase system being examined.
PROP, PROP1, PROP2	These subroutines calculate the free energy parameters and interaction parameters divided by RT needed to evaluate the functions G_i.
XAXB	This subroutine solves the binary-solution problem and is used to compute starting edge values of the ternary problem.
TRIRO	This subroutine rotates the triangle so that y_1 or z_1 can be independent variables.
SOL3	This routine steps the independent variable (x_1) and calculates guesses to the other three variables (x_2, y_1, y_2). It then calls subroutine THREE which calculates the solution. It also calls the routines that plot the solution.
UNRO	This routine undoes the work of TRIRO, returning the triangle to its original state so that the values can be written and plotted with the proper labels.
SCLTRG	This subroutine scales the values of x, y, and z so they may be plotted correctly on the triangular grid.
THREE	This subroutine (see Fig. 174) uses Newton's method to find the roots of the three equations

$$G_i[x_1, x_2, y_1, y_2, T] = 0, \qquad G_j[x_1, x_2, y_1, y_2, T] = 0,$$
$$G_k[x_1, x_2, y_1, y_2, T] = 0$$

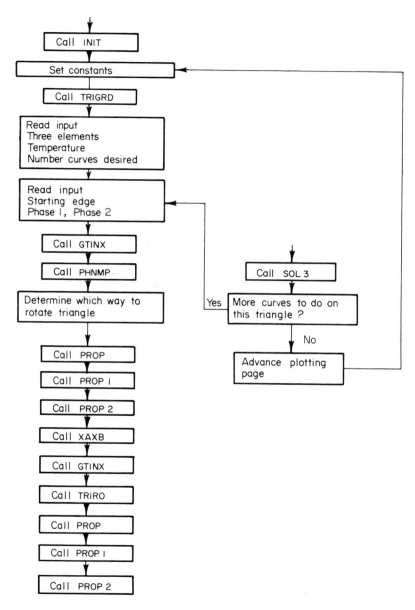

FIG. 173. Overall flow chart (TERNRY).

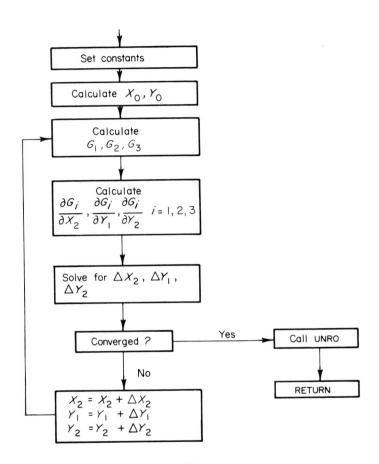

FIG. 174. Overall flow chart (THREE).

```
       PROGRAM TERNRY
       DIMENSION IZZ7(2)
       DIMENSION X(10),XX(4),XXX(10)
       DIMENSION BIN(2),KAK(2),IBN(2),IAN(3)
       DIMENSION AN(3),ANM(13),PH(4)
       COMMON I,J,K,IP1,IP2,X,XX,ICX,I1,I2,S,T,ISW
       COMMON        DF(13,4,4),EL(13,13,4),RT
       COMMON ISOL,FACTOR
       COMMON /DPSI/ITERS
       INTEGER XLR
       EQUIVALENCE(BIN,IBN),(IAN,AN)
       ISW=1
       ITERS=100
       S=-1.
       CALL DPSIV(8H MANLABS)
       CALL INIT(ANM,PH)
  111  CONTINUE
       DO 20 II=1,4
       DO 20 KI=1,4
       DO 20 KK=1,13
   20  DF(KK,KI,II)=0.
       DO 21 II=1,4
       DO 21 KI=1,13
       DO 21 KK=1,13
   21  EL(KK,KI,II)=0.
       READ (5,901) (AN(K),K=1,3),NSOL,T
  901  FORMAT(A2,1XA2,1XA2,1X,I2,1X,F5.0)
       IF(EOF,5)1234,1235
 1234  CALL PLOTND(0)
       CALL EXIT
 1235  CONTINUE
       CALL TRIGRD(AN)
       RT=1.987*T
       DO 100 ISOL=1,NSOL
       CALL GTINX(I,J,K,AN,ANM)
       READ(5,900) BIN,P1,P2,XLR
  900  FORMAT(A2,1XA2,1X2A1,1XI2)
       CALL PHNMP(P1,P2,IP1,IP2,PH)
       DO 71 LI=1,2
       IB=IBN(LI)
       DO 70 LJ=1,3
       IF(IB .NE. IAN(LJ)) GO TO 70
       KAK(LI)=LJ
       GO TO 71
   70  CONTINUE
   71  CONTINUE
       KCH=KAK(1)+1
       IF(KCH.EQ.KAK(2).OR.KCH.EQ.KAK(2)+3) GO TO 72
       ITEMP=KAK(1)
       KAK(1)=KAK(2)
       KAK(2)=ITEMP
       TEMP=BIN(1)
       BIN(1)=BIN(2)
       BIN(2)=TEMP
```

```
   72 KSUM=KAK(1)+KAK(2)
      GO TO (73,73,53,54,55),KSUM
   53 I1=1
      I2=3
      GO TO 73
   54 I1=2
      I2=3
      GO TO 73
   55 I1=1
      I2=2
   73 CONTINUE
      WRITE(6,950) AN(2),AN(3),AN(1),P1,P2,T
  950 FORMAT(10XA2,6XA2,6XA2,9X6HPHASE A1,1H,,A1,5X6HTEMP
      DO 400 INDX=1,2                                        =F7.0,//)
      JMP=KAK(INDX)
      GO TO (401,402,403),JMP
  401 IZZZ(INDX)=I
      GO TO 400
  402 IZZZ(INDX)=J
      GO TO 400
  403 IZZZ(INDX)=K
  400 CONTINUE
      I=IZZZ(1)
      J=IZZZ(2)
      J1=I
      J2=J
      CALL PROP
      CALL PROP1
      CALL PROP2
      A=DF(J1,IP2,IP1)/RT
      B=DF(J2,IP2,IP1)/RT
      E2=EL(J1,J2,IP2)/RT
      E1=EL(J1,J2,IP1)/RT
      WRITE(6,1000) A,B,F1,E2,I,J,K,IP1,IP2
 1000 FORMAT(4E16.8/5I10)
      IF(A)40,40,30
   30 A=-A
      B=-B
      EE=E1
      E1=E2
      E2=EE
      IPP=IP1
      IP1=IP2
      IP2=IPP
   40 CALL XAXB(A,B,F1,E2,N,X,IND)
      WRITE (6,1001) X
 1001 FORMAT(5E16.8)
      IF(IND.LT.0) GO TO 100
      XX(2)=0.
      XX(4)=0.
      IF(N .LT. 4) GO TO 33
      IF(XLR .EQ. 0 ) GO TO 33
      XX(1)=X(3)
      XX(3)=X(4)
```

```
      GO TO 31
  33 XX(1)=X(1)
     XX(3)=X(2)
  31 CONTINUE
     X(1)=XX(1)+S*.0001
     X(2)=.0001
     X(3)=XX(3)+S*.0001
     X(4)=.0001
     FACTOR=.1
     CALL GTINX(I,J,K,AN,ANM)
     CALL TRIRO(I1,I2,I,J,K,XXX,ICX)
     CALL PROP
     CALL PROP1
     CALL PROP2
     CALL SOL3
 100 CONTINUE
     CALL FRAMEV(2)
     GO TO 111
     END
     SUBROUTINE FARMIN(A,B,SAB,X,XI,IND)
     IND = 7
     DO 50 ITER=1,100
     OLDX = X
     T = 1.-X
     DX = -(ALOG(X/T)+SAB*(T-X)-A+B)*X*T/(1.-2.*SAB*X*T)
     IF (ABS(DX)-5.E-4*ABS(X)) 100,100,10
  10 X = X+DX
  20 IF (X) 40,40,30
  30 IF (X-XI) 50,40,40
  40 DX = DX/2.
     IF ((X-DX)-X) 45,90,45
  45 X = X-DX
     GO TO 20
  50 CONTINUE
     IND = -7
  90 X = OLDX
 100 RETURN
     END
     SUBROUTINE GTINX(I,J,K,IX,NM)
     DIMENSION NM(13),IX(3),II(3)
     DO 1 L=1,3
     II(L)=0
     DO 1 LL=1,13
     IF(IX(L)- NM(LL))1,2,1
   2 II(L)=LL
   1 CONTINUE
     I=II(1)
     J=II(2)
     K=II(3)
     RETURN
     END
     SUBROUTINE INIT (ANM,PH)
     DIMENSION ANM(13),PH(4)
     READ (5,900) (ANM(I),I=1,13),(PH(I),I=1,4)
```

```
  900    FORMAT(13A2,4A1)
         RETURN
         END
         SUBROUTINE PHNMP(IA,IB,IP1,IP2,IPH)
         DIMENSION II(2),IPH(4) ,IX(2)
         IX(1)=IA
         IX(2)=IB
         DO 1 L=1,2
         II(L)=0
         DO 1 LL=1,4
         IF(IX(L)-IPH(LL))1,2,1
    2    II(L)=LL
    1    CONTINUE
         IP2=II(1)
         IP1=II(2)
         RETURN
         END
         SUBROUTINE PROP
CPROPITIES OF ELEMENTS AND COMPOUNDS
         DIMENSION X(10),XX(4)
         COMMON I,J,K,IP1,IP2,X,XX,ICX,I1,I2,S,T,ISW
         COMMON          DF(13,4,4),EL(13,13,4),RT
         COMMON ISOL
         T24=T*2.4
         T28=T*2.8
         T305=T*3.05
         T8=T*.8
         T9=T*.9
         T2=T+T
         T29= 2.9*T
         T215=2.15*T
         T23=2.3*T
         II=MINO(IP1+IP2,8)
         GO TO (10,10,3,4,55,8,7,10),II
   10    RETURN
    3 DF( 1,2,1)=4250.-T2
      DF( 2,2,1)=4990.-T2
      DF( 3,2,1)=5480.-T2
      DF( 4,2,1)=6540.-T2
      DF( 5,2,1)=5800.-T2
      DF( 6,2,1)=7300.-T2
      DF( 7,2,1)=6500.-T24
      DF( 8,2,1)=3980.-T28
      DF( 9,2,1)=5480.-T28
      DF(10,2,1)=2830.-T305
      DF(11,2,1)=3850.-T305
      DF(12,2,1)=2290.-T28
      DF(13,2,1)=2730.-T28
      GO TO 100
    4 DF( 1,3,1)=5280.-T29
      DF( 2,3,1)=6820.-T29
      DF( 3,3,1)=3980.-T28
      DF( 4,3,1)=5040.-T28
      DF( 5,3,1)=3800.-T2
```

```
   DF ( 6,3,1)=5300.-T2
   DF ( 7,3,1)=6900.-T2
   DF ( 8,3,1)=5100.-T2
   DF ( 9,3,1)=6600.-T2
   DF (10,3,1)=4300.-T215
   DF (11,3,1)=5350.-T215
   DF (12,3,1)=3390.-T23
   DF (13,3,1)=3830.-T23
   GO TO 100
55 IF(IP1/4+IP2/4)10,6,5
 5 T285=2.85*T
   DF ( 1,4,1)=4480.-T29
   DF ( 2,4,1)=6820.-T29
   DF ( 3,4,1)=3330.-T285
   DF ( 4,4,1)=4390.-T285
   DF ( 5,4,1)=3300.-T215
   DF ( 6,4,1)=4800.-T215
   DF ( 7,4,1)=6650.-T23
   DF ( 8,4,1)=4980.-T28
   DF ( 9,4,1)=6480.-T28
   DF (10,4,1)=4480.-T2
   DF (11,4,1)=5500.-T2
   DF (12,4,1)=3640.-T2
   DF (13,4,1)=4080.-T2
   GO TO 100
 6 DF ( 1,2,3)= -1030.+T9
   DF ( 2,2,3)= -1830.+T9
   DF ( 3,2,3)=  1500.+T8
   DF ( 4,2,3)= DF(3,2,3)
   DF ( 5,2,3)=  2000.
   DF ( 6,2,3)=  2000.
   DF ( 7,2,3)=  -400.-.4*T
   DF ( 8,2,3)= -1120.-T8
   DF ( 9,2,3)= DF(8,2,3)
   DF (10,2,3)= -1500.-T9
   DF (11,2,3)= DF(10,2,3)
   DF (12,2,3)= -1100.-.5*T
   DF (13,2,3)= DF(12,2,3)
   GO TO 100
 7 DF ( 1,4,3)= -800.
   DF (2,4,3)= -800.
   DF ( 3,4,3)= -650.-T*.05
   DF ( 4,4,3)= DF(3,4,3)
   DF ( 5,4,3)= -500.-.15*T
   DF ( 6,4,3)= DF(5,4,3)
   DF ( 7,4,3)= -250.-.3*T
   DF ( 8,4,3)= -120.-T8
   DF ( 9,4,3)= DF(8,4,3)
   DF (10,4,3)=  150.+.15*T
   DF (11,4,3)= DF(10,4,3)
   DF (12,4,3)= -DF(7,4,3)
   DF (13,4,3)= DF(12,4,3)
   GO TO 100
 8 DF ( 1,4,2)= 230.-T9
```

```
      DF(  2,4,2)=  1030.-T9
      DF(  3,4,2)= -2150.-T*.85
      DF(  4,4,2)=  DF(3,4,2)
      DF(  5,4,2)= -2500.-.15*T
      DF(  6,4,2)=  DF(5,4,2)
      DF(  7,4,2)=  150.+.1*T
      DF(  8,4,2)=  1000.
      DF(  9,4,2)=  1000.
      DF(10,4,2)=  1650.+1.05*T
      DF(11,4,2)=  DF(10,4,2)
      DF(12,4,2)=  1350.+T8
      DF(13,4,2)=  DF(12,4,2)
100   IF(ISW) 101,110,101
101   ISW=0
110   RETURN
      END
      SUBROUTINE PROP1
      DIMENSION X(10),XX(4)
      COMMON I,J,K,IP1,IP2,X,XX,ICX,I1,I2,S,T,ISW
      COMMON        DF(13,4,4),EL(13,13,4),RT
      COMMON ISOL
      EL(  1,  3,1)=  4297.
      EL(  1,  6,1)=  9109.
      EL(  3,  6,1)=  2213.
      EL(  6,  9,1)= -5806.
      EL(  6,  9,2)= -5860.
      EL(  6,  9,3)= -7540.
      EL(  6,  9,4)= -7660.
      EL(  9,11,1)=    46.
      EL(  9,11,2)=    55.
      EL(  9,11,3)=    55.
      EL(  9,11,4)=    55.
      EL(  6,11,1)= -5560.
      EL(  6,11,2)= -5894.
      EL(  6,11,3)= -7914.
      EL(  6,11,4)= -7894.
      EL(  1,  3,2)=  6934.
      EL(  1,  3,3)=  6934.
      EL(  1,  3,4)=  6934.
      EL(  1,  4,1)=  6299.
      EL(  1,  4,2)=  8920.
      EL(  1,  4,3)=  8920.
      EL(  1,  4,4)=  8920.
      EL(  1,  5,1)=  1512.
      EL(  1,  5,2)=  6551.
      EL(  1,  5,3)=  8981.
      EL(  1,  5,4)=  8981.
      EL(  1,  6,2)= 14425.
      EL(  1,  6,3)= 16855.
      EL(  1,  6,4)= 16855.
      EL(  1,  7,1)= 17298.
      EL(  1,  7,2)=-10732.
      EL(  1,  7,3)= -5402.
      EL(  1,  7,4)= -5302.
```

```
EL( 1, 8,1)=-53356.        EL( 2,10,3)=-69201.
EL( 1, 8,2)=-46642.        EL( 2,10,4)=-69101.
EL( 1, 8,3)=-40092.        EL( 2,11,1)=-73638.
EL( 1, 8,4)=-40012.        EL( 2,11,2)=-70462.
EL( 1, 9,1)=-52773.        EL( 2,11,3)=-65932.
EL( 1, 9,2)=-46887.        EL( 2,11,4)=-65832.
EL( 1, 9,3)=-40337.        EL( 2,12,1)=-73893.
EL( 1, 9,4)=-40257.        EL( 2,12,2)=-73605.
EL( 1,10,1)=-76135.        EL( 2,12,3)=-71875.
EL( 1,10,2)=-71560.        EL( 2,12,4)=-71925.
EL( 1,10,3)=-67030.        EL( 2,13,1)=-72041.
EL( 1,10,4)=-66930.        EL( 2,13,2)=-71551.
EL( 1,11,1)=-72162.        EL( 2,13,3)=-69821.
EL( 1,11,2)=-67776.        EL( 2,13,4)=-69871.
EL( 1,11,3)=-63246.        EL( 3, 5,1)=    97.
EL( 1,11,4)=-63146.        EL( 3, 5,2)=   931.
EL( 1,12,1)=-73993.        EL( 3, 5,3)=   931.
EL( 1,12,2)=-72840.        EL( 3, 5,4)=   931.
EL( 1,12,3)=-71110.        EL( 3, 6,2)=  2920.
EL( 1,12,4)=-71160.        EL( 3, 6,3)=  2920.
EL( 1,13,1)=-71224.        EL( 3, 6,4)=  2920.
EL( 1,13,2)=-69719.        EL( 3, 7,1)=-23867.
EL( 1,13,3)=-67989.        EL( 3, 7,2)=-23054.
EL( 1,13,4)=-68039.        EL( 3, 7,3)=-20154.
EL( 2, 3,1)=  3219.        EL( 3, 7,4)=-20054.
EL( 2, 3,2)=  5137.        EL( 3, 8,1)=-43178.
EL( 2, 3,3)=  5137.        EL( 3, 8,2)=-41788.
EL( 2, 3,4)=  5137.        EL( 3, 8,3)=-40568.
EL( 2, 4,1)=  4948.        EL( 3, 8,4)=-40588.
EL( 2, 4,2)=  6839.        EL( 3, 9,1)=-43023.
EL( 2, 4,3)=  6839.        EL( 3, 9,2)=-42205.
EL( 2, 4,4)=  6839.        EL( 3, 9,3)=-40985.
EL( 2, 5,1)=   300.        EL( 3, 9,4)=-41005.
EL( 2, 5,2)=  4319.        EL( 3,10,1)=-38000.
EL( 2, 5,3)=  6749.        EL( 3,10,2)=-38002.
EL( 2, 5,4)=  6749.        EL( 3,10,3)=-38802.
EL( 2, 6,1)=  7072.        EL( 3,10,4)=-38802.
EL( 2, 6,2)= 11258.        EL( 3,11,1)=-37418.
EL( 2, 6,3)= 13688.        EL( 3,11,2)=-37820.
EL( 2, 6,4)= 13688.        EL( 3,11,3)=-38620.
EL( 2, 7,1)=-19264.        EL( 3,11,4)=-38620.
EL( 2, 7,2)=-13931.        EL( 3,12,1)=-34306.
EL( 2, 7,3)= -8601.        EL( 3,12,2)=-35172.
EL( 2, 7,4)= -8501.        EL( 3,12,3)=-37092.
EL( 2, 8,1)=-54911.        EL( 3,12,4)=-37122.
EL( 2, 8,2)=-49457.        EL( 3,13,1)=-37904.
EL( 2, 8,3)=-42907.        EL( 3,13,2)=-38926.
EL( 2, 8,4)=-42827.        EL( 3,13,3)=-40846.
EL( 2, 9,1)=-54394.        EL( 3,13,4)=-40876.
EL( 2, 9,2)=-49771.        EL( 4, 5,1)=     6.
EL( 2, 9,3)=-43221.        EL( 4, 5,2)=   957.
EL( 2, 9,4)=-43141.        EL( 4, 5,3)=   957.
EL( 2,10,1)=-77111.        EL( 4, 5,4)=   957.
EL( 2,10,2)=-73731.        EL( 4, 6,1)=  1205.
```

```
EL(  4,  6,2)=   2026.        EL( 5,12,4)= -4832.
EL(  4,  6,3)=   2026.        EL( 5,13,1)= -5645.
EL(  4,  6,4)=   2026.        EL( 5,13,2)= -6008.
EL(  4,  7,1)=-24838.        EL( 5,13,3)= -9148.
EL(  4,  7,2)=-23843.        EL( 5,13,4)= -9158.
EL(  4,  7,3)=-20943.        EL( 6,  7,1)=-12000.
EL(  4,  7,4)=-20843.        EL( 6,  7,2)=-11704.
EL(  4,  8,1)=-43720.        EL( 6,  7,3)=-11704.
EL(  4,  8,2)=-42074.        EL( 6,  7,4)=-11704.
EL(  4,  8,3)=-40854.        EL( 6,  8,1)= -5739.
EL(  4,  8,4)=-40874.        EL( 6,  8,2)= -5454.
EL(  4,  9,1)=-43629.        EL( 6,  8,3)= -7134.
EL(  4,  9,2)=-42571.        EL( 6,  8,4)= -7254.
EL(  4,  9,3)=-40351.        EL( 6,10,1)= -4061.
EL(  4,  9,4)=-41371.        EL( 6,10,2)= -4115.
EL(  4,10,1)=-37856.        EL( 6,10,3)= -6135.
EL(  4,10,2)=-37640.        EL( 6,10,4)= -6115.
EL(  4,10,3)=-38440.        EL( 6,12,1)=  4706.
EL(  4,10,4)=-38440.        EL( 6,12,3)=  4530.
EL(  4,11,1)=-37954.        EL( 6,12,3)=  1390.
EL(  4,11,2)=-38034.        EL( 6,12,4)=  1380.
EL(  4,11,3)=-38834.        EL( 6,13,1)= -3038.
EL(  4,11,4)=-38834.        EL( 6,13,2)= -3328.
EL(  4,12,1)=-37466.        EL( 6,13,3)= -6468.
EL(  4,12,2)=-38178.        EL( 6,13,4)= -6478.
EL(  4,12,3)=-40098.        EL( 7,  8,1)=   250.
EL(  4,12,4)=-40128.        EL( 7,  8,2)=   513.
EL(  4,13,1)=-37514.        EL( 7,  8,3)=   513.
EL(  4,13,2)=-38394.        EL( 7,  8,4)=   513.
EL(  4,13,3)=-40314.        EL( 7,  9,1)=   187.
EL(  4,13,4)=-40344.        EL( 7,  9,2)=   296.
EL(  5,  7,1)=-10767.        EL( 7,  9,3)=   296.
EL(  5,  7,2)=-10617.        EL( 7,  9,4)=   296.
EL(  5,  7,3)=-10617.        EL( 7,10,1)=  1859.
EL(  5,  7,4)=-10617.        EL( 7,10,2)=  1923.
EL(  5,  8,1)= -5662.        EL( 7,10,3)=  1583.
EL(  5,  8,2)= -5721.        EL( 7,10,4)=  1723.
EL(  5,  8,3)= -7401.        EL( 7,11,1)=   424.
EL(  5,  8,4)= -7521.        EL( 7,11,2)=   372.
EL(  5,  9,1)= -5567.        EL( 7,11,3)=    32.
EL(  5,  9,2)= -5904.        EL( 7,11,4)=   172.
EL(  5,  9,3)= -7584.        EL( 7,12,1)= 10300.
EL(  5,  9,4)= -7704.        EL( 7,12,2)= 10800.
EL(  5,10,1)= -5915.        EL( 7,12,3)=  9680.
EL(  5,10,2)= -6267.        EL( 7,12,4)=  9650.
EL(  5,10,3)= -8287.        EL( 7,13,1)=  2851.
EL(  5,10,4)= -8267.        EL( 7,13,2)=  3379.
EL(  5,11,1)= -5806.        EL( 7,13,3)=  2259.
EL(  5,11,2)= -6357.        EL( 7,13,4)=  2229.
EL(  5,11,3)= -8377.        EL( 8,10,1)=   713.
EL(  5,11,4)= -8357.        EL( 8,10,2)=   719.
EL(  5,12,1)= -1391.        EL( 8,10,3)=   719.
EL(  5,12,2)= -1682.        EL( 8,10,4)=   719.
EL(  5,12,3)= -4822.        EL( 8,11,1)=    20.
```

```
EL(  8,11,2)=      95.
EL(  8,11,3)=      95.
EL(  8,11,4)=      95.
EL(  8,12,1)=    7009.
EL(  8,12,2)=    7800.
EL(  8,12,3)=    7020.
EL(  8,12,4)=    6850.
EL(  8,13,1)=    1330.
EL(  8,13,2)=    2327.
EL(  8,13,3)=    1547.
EL(  8,13,4)=    1377.
EL(  9,10,1)=     848.
EL(  9,10,2)=     856.
EL(  9,10,3)=     856.
EL(  9,10,4)=     856.
EL(  9,12,1)=    7494.
EL(  9,12,2)=    8173.
EL(  9,12,3)=    7393.
EL(  9,12,4)=    7223.
EL(  9,13,1)=    1520.
EL(  9,13,2)=    2335.
EL(  9,13,3)=    1555.
EL(  9,13,4)=    1385.
EL(10,12,1)=    3219.
EL(10,12,2)=    3330.
EL(10,12,3)=    3330.
EL(10,12,4)=    3330.
EL(10,31,1)=      84.
EL(10,13,2)=     355.
EL(10,13,3)=     355.
EL(10,13,4)=     355.
EL(11,12,1)=    6406.
EL(11,12,2)=    6440.
EL(11,12,3)=    6440.
EL(11,12,4)=    6440.
EL(11,13,1)=    1039.
EL(11,13,2)=    1179.
EL(11,13,3)=    1179.
EL(11,13,4)=    1179.
      RETURN
      END
      SUBROUTINE PROP2
      DIMENSION X(10),XX(4)
      COMMON I,J,K,IP1,IP2,X,XX,ICX,I1,I2,S,T,ISW
      COMMON         DF(13,4,4),EL(13,13,4),RT
      COMMON ISOL
      IF(I-J) 41,41,40
40    EL(I,J,IP1)=EL(J,I,IP1)
      EL(I,J,IP2)=EL(J,I,IP2)
41    IF(J-K)43,43,42
42    EL(J,K,IP1)=EL(K,J,IP1)
      EL(J,K,IP2)=EL(K,J,IP2)
43    IF(I-K)45,45,44
44    EL(I,K,IP1)=EL(K,I,IP1)
```

```
      EL(I,K,IP2)=EL(K,I,IP2)
 45   CONTINUE
      IF(IP1+IP2-5)47,50,47
 50   IF((IP1-3)*(IP1-1))48,49,48
 47   IF(IP1-IP2)49,49,48
 48   DF(I,IP2,IP1)=-DF(I,IP1,IP2)
      DF(J,IP2,IP1)=-DF(J,IP1,IP2)
      DF(K,IP2,IP1)=-DF(K,IP1,IP2)
 49   CONTINUE
      RETURN
      END
      SUBROUTINE SCLTRG(X,Y,IIX,IIY)
      IIX= 365.*(X-Y)+515.
      IIY= -630.*(X+Y)+830.
       RETURN
      END
      SUBROUTINE SOL3
      DIMENSION X(10),XS(4),XX(4),SL(4)
      DIMENSION IPT(7)
      DIMENSION XXX(4)
      COMMON I,J,K,IP1,IP2,X,XX,ICX,I1,I2,S,T,ISW
      COMMON      DF(13,4,4),EI(13,13,4),RT
      COMMON ISOL,FACTOR
      EQUIVALENCE(ICX,IC)
      FACTOR=.01
      IPT(1)=27
      IPT(2)=16
      IPT(3)=44
      IPT(4)=55
      IPT(5)=38
      IPT(6)=63
      IPT(7)=11
      DX=X(1)-XX(1)
      STDX=DX
      LLO=-1
      IISW=0
      IAX=0
      IRW=0
      NNO=3
      DO 1 LL=1,200
      DO 25 L=1,4
 25   XXX(L)=XX(L)
      DO 2II=1,4
 2    XS(II)=X(II)
      CALL THREE(NG1,NG2,NG3,N)
      NN=NG1+NG2+NG3
      Z1=1.-X(1)-X(2)
      Z2=1.-X(3)-X(4)
      WRITE(6,900) X(7),X(5),X(9),X(8),X(6),X(10),X(1),X(2),Z1,X(3),X(4)
     1   ,Z2,NN
900   FORMAT(6H X0,Y03F8.3/6H X0,Y03F8.3/6H   1   3F8.3/6H   2   3F8.3,
     18XI2,1HK///)
      IPP=IPT(ISOL)
      IF(NN .NE. 0) GO TO 2222
```

```
      CALL SCLTRG(X(1),X(2),IX,IY)
      CALL PLOTV(IX,IY,IPP)
      CALL SCLTRG(X(3),X(4),IX,IY)
      CALL PLOTV(IX,IY,IPP)
      CALL TRIRO(I1,I2,I,J,K,X,IC)
2222 CONTINUE
      IF(NG1+NG2+NG3) 11,4,11
   11 IF(FACTOR-.01) 120,120,40
   40 FACTOR=FACTOR*.5
      DO 41 II=2,4
   41 X(II)=XS(II)-SL(II)*STDX*.5
      X(1)=X(1)-STDX*.5
      GO TO 370
  120 IF(LL-LLO-1) 12,13,12
   12 LLO=LL
   22 IF(IISW)21,20,21
   20 S=-S
      IISW=1
      TSTDX=2.*STDX
      DO 3 II=2,4
    3 X(II)=XS(II)-SL(II)*TSTDX
      X(1)=X(1)-TSTDX
      GO TO 370
   21 CONTINUE
      IF(IAX)230,27,230
  230 IF(LL-3) 130,20,130
   27 IAX=1
      FACTOR=.01
      IBW=0
      CALL UNROX(IC,XXX)
      CALL UNROX(IC,XX)
      CALL UNROX(IC,XS)
      CALL UNRO(IC,X)
      IISW=0
      LLO=-1
      I1=I1+1-(I1/3)*3
      I2=I2+1-(I2/3)*3
      CALL TRIRO(I1,I2,3,3,3,XXX,IC)
      CALL TRIRO(I1,I2,I,J,K,X,IC)
      CALL TRIRO(I1,I2,3,3,3,XX,IC)
      CALL TRIRO(I1,I2,3,3,3,XS,IC)
      DO 26 L=1,4
      X(L)=XX(L)
   26 XX(L)=XXX(L)
      S=SIGN(1.,X(1)-XX(1))
      GO TO 36
    4 CONTINUE
      IF(LL-20) 44,44,63
   63 IF(X(1)-.01)130,130,14
   14 IF(1.-X(1)-X(2)-.01)130,130,131
  131 IF(X(2)-.01)130,130,44
   44 CONTINUE
      IAX=0
      IF(IBW)36,37,36
```

```
   37 FACTOR=1.
      IRW=1
   36 NNO=NN
      IISW=1
      DXX=X(1)-XX(1)
      DO 6II=2,4
      SL(II)=(X(II)-XX(II))/DXX
    6 XX(II)=X(II)
      XX(1)=X(1)
      DX=.1-.0258012*(ATAN(ABS(SL(2))))**3
      FACT=FACTOR
      IF(S)15,15,16
   15 FACT=AMAX1(X(1),.001)
   16 STDX=S*DX*FACT
      X(1)=X(1)+STDX
      DO 7II=2,4
    7 X(II)=X(II)+SL(II)*STDX
  370 CONTINUE
      DO 35 L=2,4
      IF(X(L) .LT.0.) X(I)=.0001
   35 CONTINUE
      IF(X(1)+X(2) .GT. 1.) X(2)=.99 -X(1)
      IF(X(3)+X(4) .GT. 1.) X(4)=.99-X(3)
    5 CONTINUE
      GO TO 10
   13 IF(LL-4)22,22,130
  130 RETURN
   10 CONTINUE
    1 CONTINUE
      RETURN
      END
      SUBROUTINE THREE(NG1,NG2,NG3,NG4)
      DIMENSION X(10),G(3),PG(3,3),DELX(3),A(3,3),XX(4)
      COMMON I,J,K,IP1,IP2,X,XX,ICX,I1,I2,S,T,ISW
      COMMON        DF(13,4,4),EL(13,13,4),RT
      COMMON ISOL
      A1=EL(I,J,IP1)
      A2=EL(I,J,IP2)
      B1=EL(I,K,IP1)
      B2=EL(I,K,IP2)
      C1=EL(J,K,IP1)
      C2=EL(J,K,IP2)
  132 CONV=.001
      NG4=0
      D5=A1-A2
      D6=B1-B2+.001
      D4=-D5-D6+C1-C2
      D1=DF(I,IP2,IP1)
      D2=DF(J,IP2,IP1)-DF(I,IP2,IP1)+D5
      D3=DF(K,IP2,IP1)-DF(I,IP2,IP1)+D6
      GG= (D3+D4*X(1))/(2.*D6)
      H= 4.*(D1+D2*X(1)-D5*X(1)**2)*D6/(D3+D4*X(1))**2
      ROOT=SQRT(1.+H)*GG
      X(5)=GG+ROOT
```

```
          X(6)=GG-ROOT
          DO 104 M=5,6
          IF(X(M))104,102,102
  102     IF(X(1)+X(M)-1.)103,103,104
  103     NG4=NG4+1
  104     CONTINUE
  200     DO 13 IJK=1,20
          X1=X(1)
          X2=X(3)
          Y1=X(2)
          Y2=X(4)
          Z1=1.-X1-Y1
          Z2=1.-X2-Y2
          IF(ABS(Z1)-1.E-04)35,36,36
   35     Z1=1.E-04
   36     IF(ABS(Z2)-1.E-04)37,38,38
   37     Z2=1.E-04
   38     G(1)=-DF(I,IP2,IP1)+RT*ALOG(Z2/Z1)+A2*X2*(X2+Y2)-A1*X1*(X1+Y1)
          G(1)=G(1)+B2*Y2*(X2+Y2)-B1*Y1*(X1+Y1)-C2*X2*Y2+C1*X1*Y1
          G(2)=-DF(J,IP2,IP1)+RT*ALOG(X2/X1)+A2*(1.-X2)*Z2-A1*(1.-X1)*Z1
          G(2)=G(2)-B2*Y2*Z2+B1*Y1*Z1+C2*Y2*(1.-X2)-C1*Y1*(1.-X1)
          G(3)=-DF(K,IP2,IP1)+RT*ALOG(Y2/Y1)-A2*X2*Z2+A1*X1*Z1+B2*(1.-Y2)*Z2
          G(3)=G(3)-B1*(1.-Y1)*Z1+C2*X2*(1.-Y2)-C1*X1*(1.-Y1)
C
C
C         PG(X,1) ARE PARTIALS OF G WRT Y1
C         PG(X,2) ARE PARTIALS OF G WRT X2
C         PG(X,3) ARE PARTIALS OF G WRT Y2
C
          PG(1,1)=RT/Z1            -B1*(X1+2.*Y1)+X1*(C1-A1)
          PG(2,1)=B1*(1.-X1-2.*Y1)-(C1-A1)*(1.-X1)
          PG(3,1)=-RT/Y1+B1*(2.-X1-2.*Y1)+X1*(C1-A1)
          PG(1,2)=-RT/Z2            +A2*(2.*X2+Y2)-(C2-B2)*Y2
          PG(2,2)=RT/X2-A2*(2.-2.*X2-Y2)-(C2-B2)*Y2
          PG(3,2)=-(1.-2.*X2-Y2)*A2+(C2-B2)*(1.-Y2)
          PG(1,3)=-RT/Z2            +B2*(X2+2.*Y2)-(C2-A2)*X2
          PG(2,3)=-(1.-X2-2.*Y2)*B2+(C2-A2)*(1.-X2)
          PG(3,3)=RT/Y2-B2*(2.-X2-2.*Y2)-(C2-A2)*X2
          DO 2 L=1,3
          A(L,1)=PG(L,2)
          A(L,2)=PG(L,1)
    2     A(L,3)=PG(L,3)
          KRET=1
          GO TO 10
    3     DET=D
          DO 4 L=1,3
    4     A(L,2)=-G(L)
          KRET=2
          GO TO 10
    5     DELX(1)=D/DET
          DO 6 L=1,3
          A(L,1)=-G(L)
    6     A(L,2)=PG(L,1)
          KRET=3
          GO TO 10
```

```
      7 DELX(2)=D/DET
        DO 8 L=1,3
        A(L,1)=PG(L,2)
      8 A(L,3)=-G(L)
        KRET=4
        GO TO 10
      9 DELX(3)=D/DET
        IC=0
        DO 11 L=1,3
        LL=L+1
        X(LL)=X(LL)+DELX(L)
     32 IF(ABS(DELX(L))-ABS(X(LL))*CONV)   11,11,20
     20 IC=1
     11 CONTINUE
        IF(X(1)+X(2)-1.)51,51,52
     51 IF(X(3)+X(4)-1.)53,53,52
     52 NG1=1
        NG2=1
        NG3=1
        GO TO 201
     53 CONTINUE
        IF(X(2))52,54,54
     54 IF(X(3))52,55,55
     55 IF(X(4))52,56,56
     56 CONTINUE
        IF(IC)13,12 ,13
     13 CONTINUE
        NG1=1
        NG2=1
        NG3=1
        GO TO 201
     12 NG3=0
        NG1=0
        NG2=0
C
C
C       CHECK THAT POINTS SATISFY THE CONSTRAINTS
C
        DO 14 L=1,2
        IF(X(L))15,14,14
     14 CONTINUE
        DO 16 L=3,4
        IF(X(L))17,16,16
     16 CONTINUE
        IF(X(1)+X(2)-1.)18,18,15
     18 IF(X(3)+X(4)-1.) 201,201,17
     15 NG1=1
        GO TO 201
     17 NG2=1
    201 CONTINUE
     50 CALL UNRO(ICX,X)
        WRITE (6,10001) A1,A2,B1,B2,C1,C2
  10001 FORMAT (21H1A1,A2,B1,B2,C1,C2 = 6(   E15.8,1H,))
        WRITE (6,10002) I,J,K,IP1,IP2
  10002 FORMAT (17H0I,J,K,IP1,IP2 = 5(I15,1H,))
```

```
      WRITE(6,10005) G
10005 FORMAT(3H0G=3E20.8)
      WRITE (6,10003)    X1,X2,Y1,Y2,Z1,Z2
10003 FORMAT (21H0X1,X2,Y1,Y2,Z1,Z2 =        6(E15.8,1H, ))
      WRITE (6,10004) IC,DELX,CONV
10004 FORMAT(23H0IC,DELX,CONV          =        I15,1H,,4(E15.8,1H, ))
      RETURN
C
C     SUBPROG. TO CALCULATE DETERMINANTS
C
   10 D=A(1,1)*(A(2,2)*A(3,3)-A(2,3)*A(3,2))
      D=D-A(1,2)*(A(2,1)*A(3,3)-A(2,3)*A(3,1))
      D=D+A(1,3)*(A(2,1)*A(3,2)-A(2,2)*A(3,1))
      GO TO (3,5,7,9),KRET
      END
      SUBROUTINE TRIGRD(C)
      DIMENSION  CHR(11),C(3)
      DIMENSION IYS(11),IXS(11),IXB(11)
      DATA (CHR(1)=3H0.0),(CHR(2)=3H0.1),(CHR(3)=3H0.2),(CHR(4)=3H0.3),
     1(CHR(5)=3H0.4),(CHR(6)=3H0.5),(CHR(7)=3H0.6),(CHR(8)=3H0.7),
     2(CHR(9)=3H0.8),(CHR(10)=3H0.9),(CHR(11)=3H1.0)
      DATA(IYS=200,263,326,389,452,515,578,641,704,767,830)
      DATA(IXS=150,186,223,259,296,332,369,405,442,478,515)
      DATA(IXB=150,223,296,369,442,515,588,661,734,803,880)
      DO  10 I=1,3
      CALL LINEV(150,200,880,200)
      CALL LINEV(150,200,515,830)
   10 CALL LINEV(515,830,880,200)
      DO  5 I=1,11
      IM12 = 12-I
      IX=IXS(I)
      IY=IYS(I)
      IX3=1030-IX
      IX1=IXB(I)
      IX2=365+IX
      IY2=1030-IY
      CALL LINEV (IX,IY,IX3,IY)
      CALL LINEV(IX1,200,IX2,IY2)
      CALL LINEV (IX,IY,IX1,200)
      CALL RITE2V(IX-79,IY+4,1023,90,2,3,1,CHR(I   ),ERR)
      CALL RITE2V(IX3+25,IY+4,1023,90,2,3,1,CHR(IM12),ERR)
      CALL RITE2V(IX1-27,175,1023,90,2,3,1,CHR(IM12),ERR)
    5 CONTINUE
      CALL RITE2V (497,895,1023,90,2,2,1,C(1),ERR)
      CALL RITE2V (930,150,1023,90,2,2,1,C(2),ERR)
      CALL RITE2V ( 60,150,1023,90,2,2,1,C(3),ERR)
      RETURN
      END
      SUBROUTINE TRIRO(I1,I2,I,J,K,X,IC)
      DIMENSION X(10),XX(4)
      II=I1+I2
      X1=X(1)
      Y1=X(2)
      X2=X(3)
```

```
      Y2=X(4)
      GO TO (50,50,3,1,2),II
   1  IC=1
      GO TO 100
   2  IS=I
      I=K
      K=J
      J=IS
      X(1)=1.-X1-Y1
      X(3)=1.-X2-Y2
      X(2)=X1
      X(4)=X2
      IC=2
      GO TO 100
   3  KS=K
      K=I
      I=J
      J=KS
      X(1)=Y1
      X(3)=Y2
      X(2)=1.-X1-Y1
      X(4)=1.-X2-Y2
      IC=3
      GO TO 100
  50  IC=0
 100  RETURN
      END
      SUBROUTINE UNRO(IC,X)
      DIMENSION X(6)
      COMMON I,J,K
      XX=X(1)
      GO TO (10,2,3)        ,IC
   2  X1S=X(1)
      X2S=X(3)
      X(7)=X(5)
      X(8)=X(6)
      X(5)=1.-X(5)-X1S
      X(6)=1.-X(6)-X1S
      X(9)=X1S
      X(10)=X1S
      X(1)=X(2)
      X(3)=X(4)
      X(2)=1.-X(2)-X1S
      X(4)=1.-X(4)-X2S
      IS=I
      I=J
      J=K
      K=IS
      GO TO 100
  10  CONTINUE
      X(7)=X(1)
      X(8)=X(1)
      X(9)=1.-X(5)-X(7)
      X(10)=1.-X(6)-X(8)
```

```
          GO TO 100
     3  Y1S=X(2)
        Y2S=X(4)
        Y5S=X(5)
        Y6S=X(6)
        X(5)=X(1)
        X(6)=X(1)
        X(7)=1.-Y5S-X(1)
        X(8)=1.-Y6S-X(1)
        X(9)=1.-X(5)-X(7)
        X(10)=1.-X(6)-X(8)
        X(2)=X(1)
        X(4)=X(3)
        X(1)=1.-Y1S-X(1)
        X(3)=1.-Y2S-X(3)
        IS=I
        I=K
        K=J
        J=IS
   100  RETURN
        END
        SUBROUTINE UNROX(IC,X)
        DIMENSION X(6)
        XX=X(1)
        GO TO (10,2,3)        ,IC
     2  X1S=X(1)
        X2S=X(3)
        X(1)=X(2)
        X(3)=X(4)
        X(2)=1.-X(2)-X1S
        X(4)=1.-X(4)-X2S
        GO TO 100
    10  CONTINUE
        GO TO 100
     3  Y1S=X(2)
        Y2S=X(4)
        X(2)=X(1)
        X(4)=X(3)
        X(1)=1.-Y1S-X(1)
        X(3)=1.-Y2S-X(3)
   100  RETURN
        END
        SUBROUTINE XAXB(A,B,SA,SB,N,X,IND)
        DIMENSION X(6)
        XA=0.
        XB=0.
        XAB=0.
        XBB=0.
        AMB=A-B
        ATB=A*B
        SAMSB=SA-SB
        IF(SAMSB)1000,400,1000
  1000  ATBD=AMB*SAMSB
        ABQM=AMB/SAMSB
```

```
      ARQP=(A+B)/SAMSB
      ARQMS=AHQM**2
      ROOT = SQRT(ABS(1.+2.*ARQP+ARQMS))
      X0 = 1. -ARQM
      IF (ATB) 101,6,6
C     A AND B OPPOSITE SIGNS
C     CASE I.
  101 IF(ROOT-.0005*ABS(X0)) 13,1,1
   13 XA = X0/2.
      XB = XA
      N= 1
      GO TO 300
    1 IF (ATBD) 2,12,12
C     DIFFERENT
    2 X0=(X0-ROOT)/2.
      GO TO 11
C     SAME
   12 X0=(X0+ROOT)/2.
   11 CONTINUE
 1100 XI = .5
      TA = SA-2.
      IF (TA) 1130,1130,1105
 1105 XI = (1.-SQRT(TA/SA))/2.
 1130 XAM = XI/10.
      CALL FABMIN(A,B,SA,XAM,XI,IND)
      IF (IND) 1110,1110,1120
 1110 XAM = 0.
      XA = AMIN1(1.E-8,XI/1000.)
 1120 IF (XI-.5) 11210,11200,11210
11200 XI = X0
11210 XBM = .5
      XIB = .5
      TB = SB-2.
      IF (TB) 1160,1160,1140
 1140 XIB = (1.-SQRT(TB/SB))/2.
      XBM = XIB/10.
      CALL FABMIN(0.,0.,SB,XBM,XIB,IND)
      XIB = 1.-XIB
      IF (IND) 1150,1150,1160
 1150 WRITE (6,1151)
 1151 FORMAT (1H0,30HERROR - CANT FIND BETA MINIMUM)
      GO TO 200
 1160 TA = 1.-XAM
      TB = 1.-XBM
      DIFF = A*TA+ALOG(TA/TB)-(SB*TB+ALOG(XBM/TB))*XBM
      IF (XAM) 1180,1180,1170
 1170 DIFF = DIFF+(B+SA*TA+ALOG(XAM/TA))*XAM
      GO TO 1185
 1180 XAM = XA
 1185 XA = .99999999*AMIN1(XAM,X0/2.)
      XB = .99999999*AMIN1(XBM,(XBM+X0)/2.)
      IF (DIFF) 1190,1195,1200
 1195 DIFF = -0.
 1190 XB = AMAX1(1.-XB,(X0+1.)/2.,(XIB+1.)/2.)
```

```
      XA = AMAX1(1.00000001*XAM,(XAM+X0)/2.)
1200 CONTINUE
   3 IF (A)5,5,4
   4 XX=XA
     XA=XB
     XB=XX
   5 CALL ZEROS(A,B,SA,SB,XA,XB,0.,AMIN1(X0,XI),AMAX1(X0,(.5-SIGN(.5,DI
    1FF))*XIB),1.-(.5+SIGN(.5,DIFF))*XIB,IND)
     N=2
     GO TO 300
C                CASE II
   6 C=ABS(A)+ABS(B)+SQRT(4.*ATB)
     XC=(ABS(A)+SQRT(ATB))/C
     IF(A) 7,7,8
   7 IF (ABS(ABS(SAMSB)-C)-.0005) 100,7000,7000
7000 IF(ABS(SAMSB)-C)200,100,8
   8 X0 = (X0-ROOT)/2.
     X0B = X0+ROOT
     IF (SB-2.) 2000,2000,200
2000 XI = X0
     XIB=X0B
     TA = SA-2.
     IF (TA) 2020,2020,2010
2010 XI = (1.-SQRT(TA/SA))/2.
     XIB=1.-XI
2020 XA = AMIN1(X0/2.,,XI/2.)
     XB=(X0+XC)/2.
     IF (A) 10,9,9
   9 XX=XB
     XB=XA
     XA=XX
  10 CALL ZEROS(A,B,SA,SB,XA,XB,0.,AMIN1(X0,XI),X0,X0B,IND)
     XAB = AMAX1(1.-XA,(X0B+1.)/2.,(1.+XIB)/2.)
     XBB = (XB+X0B)/2.
     CALL ZEROS(A,B,SA,SB,XAB,XBB,AMAX1(X0B,XIB)   ,1.,X0,X0B,IND)
     N=4
     IF (XB-XBB) 300,300,3000
3000 XAM = XI/10.
     CALL FARMIN(0.,,0.,,SA,XAM,XI,IND)
     XA = XAM
     XAB = 1.-XA
     XB = 1.E35
     XBB = XB
     N = 4
     GO TO 300
 100 TA = SA-2.
     IF (TA) 21000,21000,21100
21100 XI = (1.-SQRT(TA/SA))/2.
     GO TO 3000
21000 XA=XC
     XB = XC
     XAB = XC
     XBB = XC
     N=1
```

```
      GO TO 300
200   N=0
300   X(1)=XA
      X(2)=XB
      X(3)=XBB
      X(4)=XAB
      X(5)=X0
      X(6)=X0B
      RETURN
400   X0=ABS(A)/(ABS(A)+ABS(B))
      GO TO 11
      END
      SUBROUTINE ZEROS(A,B,SA,SB,X,Y,XLO,XHI,YLO,YHI,IND)
      COMMON /DPSI/ ITERS
      XMID = XHI+XLO
      XDIF = XHI-XLO
      YMID = YHI+YLO
      YDIF = YHI-YLO
      IND = 7
      DO 3 I=1,ITERS
      OLDX = X
      OLDY = Y
      T1=1.-X
      T2=1.-Y
      T3 = X-Y
      IF(T1.LT..0.OR.T2.LT.0..OR.X.LT.0..OR.Y.LT.0..OR.T3.EQ.0.)GO TO 3
      GA=A*ALOG( T1 )+SA*X**2-ALOG( T2 )-SB*Y**2
      GB=B*ALOG(X)+SA*( T1 )**2-ALOG(Y)-SB*( T2 )**2
      DX=X*T1*(Y*GB+T2*GA)/(1.-2.*SA*X*T1)/T3
9999  DY=Y*T2*(X*GB+T1*GA)/(1.-2.*SB*Y*T2)/T3
      X=X+DX
      Y=Y+DY
      IF (ABS(DX)-5.E-4*ABS(X)) 1,1,33
1     IF (ABS(DY)-5.E-4*ABS(Y)) 2,2,33
33    IF (X-XLO) 35,35,34
34    IF (X-XHI) 36,35,35
35    DX = DX/2.
      IF ((X-DX)-X) 355,365,355
355   X = X-DX
      GO TO 33
365   X = (2.*OLDX+XMID+SIGN(XDIF,DX))/4.
36    IF (Y-YLO) 38,38,37
37    IF (Y-YHI) 3,38,38
38    DY = DY/2.
      IF ((Y-DY)-Y) 39,300,39
39    Y = Y-DY
      GO TO 36
300   Y = (2.*OLDY+YMID+SIGN(YDIF,DY))/4.
3     CONTINUE
      IND = -7
2     RETURN
      END
      END
```

2. TERNARY MISCIBILITY GAP

The mainline program is called MIGAP within which option 1 calls BINOD to generate composition limits (x_1, y_1) and (x_2, y_2) for isolated miscibility gaps described by Eqs. (280)–(282). In addition, subroutine BIGAP (option 2) is first called by MIGAP to establish binary edge solutions [Eqs. (75), (76)] which are then used by BINOD to generate composition limits (x_1, y_1) and (x_2, y_2) of the miscibility gap described by Eqs. (280)–(282) for exited gaps which appear on the binary edges. This program is a special case of the previous program, TERNRY, requiring as input only a system identification label (CODE), the three binary interaction parameters E_{ij} (EIJ), E_{ik} (EIK), and E_{jk} (EJK) in calories per gram atom and the temperature in degrees Kelvin. In addition, the proper option (1 or 2) must be called. This choice is made on the basis of Meijering's discussion of ternary miscibility gaps (137) as follows.

First, the particular set of interaction parameters for the system under consideration are ordered such that $E_{jk} > E_{ij} > E_{ik}$. Binary critical temperatures $T_{cjk} > T_{cij} > T_{cik}$ are then computed on the basis of Eqs. (75)–(76). The next step is computation of the summit temperature of an isolated ternary gap which can occur in the ternary without being visible in the three constituent binary diagrams. The summit temperature, T_S, and composition (x_S, z_S), which must be positive to be considered, are given as follows (137):

$$T_S = -\frac{(E_{jk} + E_{ij} - E_{ik})^2 - 4E_{jk}E_{ij}}{8RE_{ik}}$$

$$x_S = \tfrac{1}{2} = x_J \quad \text{at the summit}$$

$$z_S = \frac{E_{ik} + E_{jk} - E_{ij}}{4E_{ik}} = z_I \quad \text{at the summit}$$

If T_S is greater than T_{jk}, then option 1 path is followed for $T_S > T$. In this calculation the interaction parameters are entered on the data cards in descending order with the largest value in the first position, the second largest in the next position and the smallest in the third position. In plotting the data the ternary triangle is set so that the apex is I, the right corner is J, and the left corner is K. Thus, the largest interaction parameter is associated with the JK side, the next largest interaction parameter is associated with the IJ, side and the smallest parameter with the IK side. In accordance with Fig. 141, $x = x_J$, $y = y_K$, and $z = z_I$. Thus, tie lines rotate from the edge with the " most positive " interaction parameter toward the edge with the next most-positive interaction parameter, as shown in Fig. 150.

At temperatures below T_{ik}, option 2 is employed. However, the maximum amount of information is obtained by employing all six possible arrangements of the interaction parameters by interchanging the positions of the interaction parameters on the data cards. In these cases, the data can be plotted by noting the edge solutions which are printed out. The appropriate values of the interaction parameter are assigned to each edge. The results can be readily displayed by noting that the tie lines rotate from the edge with the most positive interaction parameter toward the edge with the next " most-positive " interaction parameter. Examples of the ZrTaW exited gap shown in Fig. 150 and the isolated gap given in Fig. 20 of Meijering (137, Part II) are given below. The latter corresponds to $E_{jk} = E_{ij} = 0$, $E_{ik} = -19,870$ cal/g-atom at $T = 1000°K$.

MIGAP

```
// JOB
// FOR
*IOCS(CARD,TYPEWRITER,KEYBOARD,DISK)
*EXTENDED PRECISION
*LIST SOURCE PROGRAM
*NAME  MIGAP
      DIMENSION CODE(5)
   10 FORMAT(5A2,4F10.0,I30)
   15 FORMAT(/////'ISOLATED GAP - ',5A2,2X,'T=',F6.0,2X,'E1=',F7.0,
     12X,'E2=',F7.0,2X,'E3=',F7.0,//,'SUMMIT POINT - T=',F7.0,2X,'X1=',
     2F6.3,2X,'X2=',F6.3,2X,'X3=',F6.3)
   20 FORMAT(/ ,'CRITICAL TEMPERATURES - T1=',F7.0,2X,'T2=',F7.0,2X,
     1'T3=',F7.0)
   25 FORMAT(//,'NO SOLUTION - E3 = 0 OR POSITIVE')
   30 FORMAT(/,'INITIAL GUESS - ',6F10.3)
   35 FORMAT(//,4X,'TIE LINE COORDINATES',//,4X,'V1      V2      V3      U1
     1     U2      U3',/)
   40 FORMAT(/////'EXITED GAP - ',5A2,2X,'T=',F6.0,2X,'E1=',F7.0,2X,
     1'E2=',F7.0,2X,'E3=',F7.0)
   45 FORMAT(//,'EDGE 1 SOLUTION - ',2F8.3)
   50 FORMAT(//,'NO SOLUTION - WRONG STARTING EDGE')
   55 FORMAT(//,'EDGE 2 SOLUTION - ',2F8.3)
   60 FORMAT(//,'EDGE 3 SOLUTION - ',2F8.3)
   65 FORMAT(//,4X,'TIE LINE COORDINATES',//,4X,'U1      U2      U3      V1
     1     V2      V3',/)
   70 FORMAT(6F7.3)
   75 FORMAT(F7.3,2X,'NO SOLUTION')
   80 FORMAT(/,'      PROPOGATE IN OPPOSITE DIRECTION',/)
    1 READ(2,10) CODE,T,E1,E2,E3,IPATH
      KPATH=0
      R2=3.974
      T1=E1/R2
      T2=E2/R2
      T3=E3/R2
      EL=E1**2+E2**2+E3**2-2.*(E1*E2+E1*E3+E2*E3)
      TS=-EL/(4.*R2*E3)
      X1S=(E3+E1-E2)/(4.*E3)
      X2S=(E3+E2-E1)/(4.*E3)
      X3S=0.5
      STEP=.01
      S=(E1-E2+E3)*(R2*TS+E3-E1-E2)/((E1-E2-E3)*(E1-E2+E3)+2.*R2*TS*E3)
      SU10=X1S
      SU30=.99-X1S
      SV10=X1S*(1.+S)-.98*S
      SV30=.01
      GO TO (100,200),IPATH
  100 WRITE(1,15) CODE,T,E1,E2,E3,TS,X1S,X2S,X3S
```

```
      WRITE(1,20) T1,T2,T3
      IF(E3) 102,101,101
  101 WRITE(1,25)
      GO TO 1
  102 U1=SU10-STEP
      U30=SU30
      V10=SV10
      V30=SV30
      U2=1.-SU10-U30
      V2=1.-V10-V30



      WRITE(1,30)  SV10,V2,SV30,SU10,U2,SU30
      WRITE(1,35)
      KPATH=1
      GO TO 300
  200 WRITE(1,40) CODE,T,E1,E2,E3
      WRITE(1,20) T1,T2,T3
      CALL BIGAP(T,E1,U30,V30,KON)
      IF(KON) 201,201,202
  201 WRITE(1,45) U30,V30
      GO TO 203
  202 WRITE(1,50)
      GO TO 1
  203 CALL BIGAP(T,E2,Z1,Z2,KON)
      IF(KON) 204,204,205
  204 WRITE(1,55) Z1,Z2
  205 CALL BIGAP(T,E3,Z1,Z2,KON)
      IF(KON) 206,206,207
  206 WRITE(1,60) Z1,Z2
  207 U1=0.
      V10=STEP
      WRITE(1,65)
      WRITE(1,70) U1,V30,U30,U1,U30,V30
  300 CONTINUE
  301 U1=(U1+STEP)*1.0000001
      CALL BINOD(T,E1,E2,E3,U30,V10,V30,U1,U2,U3,V1,V2,V3,KON,NG)
      IF(KON) 302,302,304
  302 IF(NG) 303,303,304
  303 WRITE(1,70) U1,U2,U3,V1,V2,V3
      GO TO 301
  304 WRITE(1,75) U1
      IF(KPATH) 1,1,305
  305 WRITE(1,80)
      U1=SU10+STEP
      U30=SU30
      V10=SV10
      V30=SV30
      STEP=-STEP
      KPATH=0
      GO TO 300
      END
```

```
      SUBROUTINE BINOD(T,E1,E2,E3,U30,V10,V30,U1,U2,U3,V1,V2,V3,KON,NG)
      NG=0
      KON=0
      ITER=0
      RT=1.987*T
      A1=E1/RT
      A2=E2/RT
      A3=E3/RT
      SIG=A2-A1-A3
      A23=A2-A3
      A21=A2-A1
      U3=U30
      V1=V10
      V3=V30
    1 V2=1.-V1-V3
      U2=1.-U1-U3
      D1=U1-V1
      D3=U3-V3
      G=SIG*(U1*U3-V1*V3)+A1*(D3+V3**2-U3**2)+A3*(D1+V1**2-U1**2)
      F1=EALOG(U1/V1)+A23*D3-A3*D1-G
      F2=EALOG(U2/V2)+A1*D3+A3*D1-G
      F3=EALOG(U3/V3)+A21*D1-A1*D3-G
      GU3=SIG*U1+A1*(1.-2.*U3)
      GV1=-SIG*V3-A3*(1.-2.*V1)
      GV3=-SIG*V1-A1*(1.-2.*V3)
      F1U3=A23-GU3
      F1V1=-1./V1+A3-GV1
      F1V3=-A23-GV3
      F2U3=-1./U2+A1-GU3
      F2V1=1./V2-A3-GV1
      F2V3=1./V2-A1-GV3
      F3U3=1./U3-A1-GU3
      F3V1=-A21-GV1
      F3V3=-1./V3+A1-GV3
      DEL=F1U3*(F2V1*F3V3-F3V1*F2V3)-F2U3*(F1V1*F3V3-F3V1*F1V3)
     1    +F3U3*(F1V1*F2V3-F2V1*F1V3)
      DU3=-(F1*(F2V1*F3V3-F3V1*F2V3)-F2*(F1V1*F3V3-F3V1*F1V3)
     2    +F3*(F1V1*F2V3-F2V1*F1V3))/DEL
      DV1= (F1*(F2U3*F3V3-F3U3*F2V3)-F2*(F1U3*F3V3-F3U3*F1V3)
     3    +F3*(F1U3*F2V3-F2U3*F1V3))/DEL
      DV3= (F1*(F2V1*F3U3-F3V1*F2U3)-F2*(F1V1*F3U3-F3V1*F1U3)
     4    +F3*(F1V1*F2U3-F2V1*F1U3))/DEL
      IF(EABS(DU3)-1.E-04*EABS(U3)) 2,2,4
    2 IF(EABS(DV1)-1.E-04*EABS(V1)) 3,3,4
    3 IF(EABS(DV3)-1.E-04*EABS(V3)) 6,6,4
    4 V1=V1+DV1
      U3=U3+DU3
      V3=V3+DV3
      ITER=ITER+1
      IF(ITER-50) 1,1,5
    5 KON=1
      RETURN
    6 IF(U2) 13,13,7
```

```
   7 IF(U3) 13,13,8
   8 IF(V1) 13,13,9
   9 IF(V2) 13,13,10
  10 IF(V3) 13,13,11
  11 IF(U1-U2) 12,13,12
  12 U30=U3
     V10=V1
     V30=V3
     RETURN
  13 NG=1
     RETURN
     END



     SUBROUTINE BIGAP(T,E,SOL1,SOL2,KON)
     ITER=0
     KON=0
     IF(E-3.987*T) 9,9,100
 100 CONTINUE
     A=E/(1.987*T)
     X=EEXP(-A)
   1 XM=1.-X
     XC=(X*(1.-XM*(A+EALOG(X/XM))))/(1.-2.*A*X*XM)
     IF(EABS(XC-X)-1.E-06*EABS(XC)) 4,4,2
   2 ITER=ITER+1
     IF(ITER-50) 3,3,9
   3 X=XC
     GO TO 1
   4 SOL1=XC
     SOL2=1.-XC
     IF(SOL1) 9,9,5
   5 IF(SOL1-1.) 6,9,9
   6 IF(SOL2.) 9,9,7
   7 IF(SOL2-1.) 8,9,9
   8 IF(SOL1-0.5) 10,9,10
   9 KON=1
  10 RETURN
     END
```

EXITED GAP - ZRTAW3   T= 1873. E1= 14425. E2= 2026. E3= 8920.

CRITICAL TEMPERATURES - T1= 3629. T2=   509. T3= 2244.

EDGE 1 SOLUTION -   0.024   0.975

EDGE 3 SOLUTION -   0.171   0.828

   TIE LINE COORDINATES

U1	U2	U3	V1	V2	V3
0.000	0.975	0.024	0.000	0.024	0.975
0.010	0.965	0.024	0.059	0.027	0.912
0.020	0.955	0.024	0.121	0.032	0.846
0.030	0.946	0.023	0.185	0.036	0.778
0.040	0.936	0.023	0.250	0.042	0.706
0.050	0.927	0.022	0.316	0.049	0.633
0.060	0.918	0.021	0.382	0.057	0.559
0.070	0.909	0.020	0.446	0.065	0.487
0.080	0.900	0.019	0.507	0.075	0.417
0.090	0.891	0.018	0.563	0.085	0.350
0.100	0.883	0.016	0.613	0.096	0.289
0.110	0.874	0.015	0.657	0.107	0.234
0.120	0.866	0.013	0.696	0.118	0.184
0.130	0.859	0.010	0.730	0.129	0.139
0.140	0.851	0.008	0.759	0.140	0.100
0.150	0.843	0.006	0.784	0.150	0.064
0.160	0.836	0.003	0.806	0.160	0.032
0.170	0.829	0.000	0.825	0.170	0.004
0.180	NO SOLUTION				

ISOLATED GAP - Y Z X   T= 1000. E1=   0. E2=   0. E3=-19873.

SUMMIT POINT - T= 1250. X1= 0.250 X2= 0.250 X3= 0.500

CRITICAL TEMPERATURES - T1=   0. T2=   0. T3= -5000.

INITIAL GUESS -   0.615   0.375   0.010   0.250   0.010   0.740

TIE LINE COORDINATES

V1	V2	V3	U1	U2	U3
0.250	0.043	0.706	0.551	0.248	0.199
0.260	0.043	0.696	0.553	0.241	0.204
0.270	0.044	0.685	0.555	0.235	0.209
0.280	0.045	0.674	0.557	0.228	0.214
0.290	0.045	0.664	0.558	0.222	0.219
0.300	0.046	0.653	0.559	0.216	0.224
0.310	0.047	0.642	0.560	0.209	0.229
0.320	0.049	0.630	0.560	0.203	0.235
0.330	0.050	0.619	0.561	0.197	0.241
0.340	0.051	0.608	0.560	0.191	0.247
0.350	0.053	0.596	0.560	0.185	0.253
0.360	0.054	0.585	0.559	0.180	0.260
0.370	0.056	0.573	0.558	0.174	0.266
0.380	0.058	0.561	0.557	0.168	0.274
0.390	0.060	0.549	0.555	0.162	0.281
0.400	0.062	0.537	0.553	0.157	0.289
0.410	0.065	0.524	0.550	0.151	0.297
0.420	0.067	0.512	0.547	0.145	0.306
0.430	0.070	0.499	0.544	0.140	0.315
0.440	0.073	0.486	0.540	0.134	0.324
0.450	0.077	0.472	0.535	0.128	0.335
0.460	0.080	0.459	0.530	0.123	0.346
0.470	0.001	0.528	0.470	0.001	0.528
0.480	NO SOLUTION				

PROPOGATE IN OPPOSITE DIRECTION

V1	V2	V3	U1	U2	U3
0.250	0.043	0.706	0.551	0.248	0.199
0.240	0.042	0.717	0.549	0.255	0.195
0.230	0.042	0.727	0.546	0.262	0.191
0.220	0.041	0.738	0.543	0.269	0.186
0.210	0.041	0.748	0.540	0.277	0.182
0.200	0.041	0.758	0.536	0.284	0.179
0.190	0.041	0.768	0.532	0.292	0.175
0.180	0.042	0.777	0.527	0.300	0.171
0.170	0.042	0.787	0.522	0.309	0.168
0.160	0.043	0.796	0.517	0.318	0.164
0.150	0.044	0.805	0.511	0.327	0.161
0.140	0.045	0.814	0.504	0.337	0.158
0.130	0.046	0.823	0.496	0.347	0.155
0.120	0.048	0.831	0.488	0.358	0.152
0.110	0.051	0.838	0.479	0.370	0.150
0.100	0.054	0.845	0.468	0.383	0.148
0.090	0.059	0.850	0.455	0.397	0.146
0.080	0.065	0.854	0.441	0.413	0.145
0.070	0.074	0.855	0.422	0.432	0.144
0.060	0.089	0.850	0.399	0.454	0.146
0.050	0.115	0.834	0.363	0.484	0.151
0.040	NO SOLUTION				

3. TERNARY COMPOUND-SOLUTION PHASE EQUILIBRIA

This program solves by Newton–Raphson the pair of simultaneous equations (301) and (304), after first examining the stability of the compound phase over the allowed range of composition. The mainline program TERCP requires as input the following: system identification (CODE), x_* (XO), $T°K$ (T), $\Delta H_i^{\theta \rightarrow \Phi}$ (DHI), $\Delta H_j^{\theta \rightarrow \Phi}$ (DHJ), $\Delta H_k^{\theta \rightarrow \Phi}$ (DHK), $\Delta S_i^{\theta \rightarrow \Phi}$ (DSI), $\Delta S_j^{\theta \rightarrow \Phi}$ (DSJ), $\Delta S_k^{\theta \rightarrow \Phi}$ (DSK), E_{ij}^{Φ} (EIJ), E_{ik}^{Φ} (EIK), E_{jk}^{Φ} (EJK), ΔF_A^{Ψ} (DFA), and ΔF_B^{Ψ} (DFB).

All energy units are calories per gram atom, and temperatures are in degrees Kelvin. Output consists of a statement about the stability of the compound phase. When the compound phase is FULLY STABLE, the solutions (x_Φ, y_Φ) vs. y_Ψ are printed for two branches 1 and 2 in steps of y_Ψ from 0 to $1 - x_*$. When the compound phase is PARTIALLY STABLE, the solutions (x_Φ, y_Φ) vs. y_Ψ are printed in steps of y_Ψ from 0 to $y_{\Psi(\text{max})}$. Tie lines are generated by connecting points (x_Φ, y_Φ) on both branches with the associated point (x_*, y_Φ).

To illustrate the use of TERCP, output for the case of a Laves compound phase $Zr(W, Ta)_2$ $(x_* = \frac{1}{3})$ in equilibrium with the β (bcc) phase of W–Zr–Ta at 1873°K is included. This illustrates the case of a partially stable compound.

```
// JOB                    TERNARY COMPOUND-SOLUTION PHASE EQUILIBRIA
// FOR
*IOCS(CARD,TYPEWRITER,KEYBOARD,DISK)
*EXTENDED PRECISION
*ONE WORD INTEGERS
*LIST SOURCE PROGRAM
*NÁME  TERCP
      DIMENSION CODE(15)
    5 FORMAT(15A4)
    6 FORMAT(/,'SYSTEM    ',15A4,/)
   10 FORMAT(F5.4,4F6.0,3F6.2,3F8.0,/,2F8.0)
   20 FORMAT('XO=',F6.4,5X,'T=',F6.0,/,'DHI=',F6.0,' DHJ=',F6.0,' DHK=',
     1F6.0,' DSI=',F5.2,' DSJ=',F5.2,' DSK=',F5.2,/,'EIJ=',F8.0,' EIK=',
     2F8.0,' EJK=',F8.0,' DFA=',F7.0,' DFB=',F7.0)
   30 FORMAT(/,'X1=',F8.6,5X,'X2=',F8.6,5X,'  YL=0.',/)
   40 FORMAT(3(F6.4,2X))
   41 FORMAT(' XPHI     YPHI     YLAMBDA',/)
   42 FORMAT(' NO SOLUTION ON THE I-J EDGE',/)
   50 FORMAT('NON CONVERGENT AFTER 50 ITERATIONS')
   60 FORMAT(/,' SOLUTION ALONG BRANCH 1',/)
   70 FORMAT(/,' SOLUTION ALONG¢BRANCH 2',/)
   80 FORMAT(/,'COMPOUND PHASE FULLY STABLE',/)
   90 FORMAT(/,'COMPOUND PHASE PARTIALLY STABLE  YLAMBDA MAX=',F6.4,
     15X,'YC=',F6.4,5X,'YPHI MAX=',F6.4,/)
  100 READ(2,5) CODE
      WRITE(1,6) CODE
      READ(2,10)  XO,T,DHI,DHJ,DHK,DSI,DSJ,DSK,EIJ,EIK,EJK,DFA,DFB
      WRITE(1,20) XO,T,DHI,DHJ,DHK,DSI,DSJ,DSK,EIJ,EIK,EJK,DFA,DFB
      DFI=DHI-T*DSI
      DFJ=DHJ-T*DSJ
      DFK=DHK-T*DSK
      DELE=EIJ+EIK-EJK
      RT=1.987*T
      XMO=1.-XO
C TEST FOR POSSIBLE INTERSECTION OF FREE ENERGY CURVES, COMPUTE YC
      Q=XO*DFJ+XMO*DFI+RT*(XO*EALOG(XO)+XMO*EALOG(XMO))+EIJ*XO*XMO-DFA
      IF(Q) 101,101,102
  101 WRITE(1,42)
      GO TO 100
  102 S=DFK-DFI+EIK-XO*DELE+(DFA-DFB)/XMO
      ARG=(S/EIK)**2+4.*Q/EIK
      IF(ARG) 207,200,200
  200 ROOT=ESQRT(ARG)
C IF YC IS A REAL ROOT, THEN YLAMBDA MAX  IS CALCULATED, OTHERWISE XMO
      YCM=(S/EIK-ROOT)/2.
      YCP=(S/EIK+ROOT)/2.
      IF(YCM) 203,203,201
  201 IF(YCM-XMO) 202,203,203
  202 YC=YCM
      GO TO 208
  203 IF(YCP) 207,207,204
  204 IF(YCP-XMO) 205,207,207
  205 YC=YCP
      GO TO 208
  207 DYL=XMO/20.
      WRITE(1,80)
      GO TO 209
```

```
    208 YM=YC+.001
    104 UM=EEXP((Q+EIK*YM**2)/(XMO*RT))
        ZM=1.-XO-YM
        YLM=XMO-ZM*UM
        FUN=DFK-DFI-XO*DELE+(Q+DFA-DFB)/XMO+RT*EALOG(YM/YLM)+EIK*(YM**2/
       1XMO+1.-2.*YM)
        DYLM=UM*(1.-ZM*2.*EIK/(XMO*RT))
        DFUN=RT*(1./YM-DYLM/YLM)-2.*EIK*ZM/XMO
        CORR=-FUN/DFUN
        IF(EABS(CORR)-1.0E-06*EABS(YM)) 106,106,105
    105 YM=YM+CORR
        GO TO 104
    106 WRITE(1,90) YLM,YC,YM
        DYL=YLM/20.
    209 IPATH=0
C FIND EDGE SOLUTIONS X1 AND X2 AT Y=0
        X=1.0E-06
    300 XM=1.-X
        F=RT*(XMO*EALOG(XM)+XO*EALOG(X))+EIJ*(X**2+XO*(1.-2.*X))+XMO*DFI
       1+XO*DFJ-DFA
        DFDX=(X-XO)*(2.*EIJ-RT/(X*XM))
        CORR=-F/DFDX
        IF(EABS(CORR)-1.0E-06*EABS(X)) 302,302,301
    301 X=X+CORR
        GO TO 300
    302 IF(IPATH) 303,303,304
    303 X1=X
        IPATH=1
        X=1.0-1.0E-06
        GO TO 300
    304 X2=X
        WRITE(1,30) X1,X 2
C SET UP TO PROPAGATE SOLUTION ALONG BRANCH 1
        CONST=DFK+XO*(DFJ-DFI)+(XO*DFA-DFB)/XMO+RT*XMO*EALOG(XMO)
        IPATH=0
C INITIALIZE X AND Y FOR ITERATION LOOP
    700 Y=1.0E-05
        IF(IPATH) 401,401,402
    401 X=X1
        WRITE(1,60)
        WRITE(1,41)
        GO TO 403
    402 X=X2
        WRITE(1,70)
        WRITE(1,41)
    403 YL=0.
        DO 1000 I=1,20
C STEP YLAMBDA
        YL=YL+DYL
        KOUNT=1
        ZL=1.-XO-YL
        DFL=(ZL*DFA+YL*DFB)/XMO+RT*(YL*EALOG(YL)+ZL*EALOG(ZL)-XMO*EALOG(
       1XMO))
        TERM=ZL*DFI+XO*DFJ+YL*DFK-DFL
    404 Z=1.-X-Y
C CALCULATE EQUILIBRIA EQUATIONS AND DERIVATIVES
        H=DFK-DFI+(DFA-DFB)/XMO-DELE*X+EIK*(1.-2.*Y)+RT*EALOG(Y*ZL/(Z*YL))
        G=RT*(XO*EALOG(ZL*X/Z)+EALOG(Y/YL))+EIJ*(X**2+XO*(1.-2.*X))+EIK*
```

```
    1(1.-Y)**2+DELE*(Y*(X-XO)-X)+CONST
      DHDX=RT/Z-DELE
      DHDY=RT*(1.-X)/(Y*Z)-2.*EIK
      DGDX=RT*XO*(1.-Y)/(X*Z)+2.*EIJ*(X-XO)-DELE*(1.-Y)
      DGDY=RT*(Z+XO*Y)/(Y*Z)-2.*EIK*(1.-Y)+DELE*(X-XO)
      DENOM=DGDX*DHDY-DGDY*DHDX
      DELX=(H*DGDY-G*DHDY)/DENOM
      DELY=(G*DHDX-H*DGDX)/DENOM
C TEST FOR CONVERGENCE OF X AND Y
      IF(KOUNT-50) 500,500,504
  500 IF(EABS(DELX)-1.0E-06*EABS(X)) 501,501,502
  501 IF(EABS(DELY)-1.0E-06*EABS(Y)) 503,503,502
  502 X=X+DELX
      Y=Y+DELY
      KOUNT=KOUNT+1
      GO TO 404
  503 WRITE(1,40) X,Y,YL
      GO TO 1000
  504 WRITE(1,50)
 1000 CONTINUE
C SET UP TO PROPOGATE SOLUTION ALONG BRANCH 2
      IF(IPATH) 600,600,601
  600 IPATH=1
      GO TO 700
  601 GO TO 100
      END

FEATURES SUPPORTED
  IOCS
  ONE WORD INTEGERS
  EXTENDED PRECISION

CORE REQUIREMENTS FOR TERCP
COMMON       0    VARIABLES    240     PROGRAM   1542

END OF COMPILATION

// DUP

*STORE       WS   UA   TERCP

231D 0077
```

SYSTEM   W-ZR-TA   SOLUTION PHASE B, LAVES PHASE ZR(W,TA)2

XO=0.3333   T= 1873.
DHI=-2000. DHJ= 1030. DHK=-1500. DSI= 0.00 DSJ= 0.90 DSK= 0.80
EIJ=   14425. EIK=   2026. EJK=   8920. DFA= -1975. DFB=   1399.

COMPOUND PARTIALLY STABLE   YLAMBDA MAX=0.0743   YC=0.1822   YPHI MAX=0.3429

X1=0.017403   X2=0.979317   YL=0.

SOLUTION ALONG BRANCH 1

XPHI	YPHI	YLAMBDA
0.0186	0.0169	0.0037
0.0199	0.0341	0.0074
0.0213	0.0515	0.0111
0.0229	0.0692	0.0148
0.0246	0.0871	0.0185
0.0265	0.1052	0.0223
0.0287	0.1235	0.0260
0.0310	0.1420	0.0297
0.0336	0.1606	0.0334
0.0366	0.1793	0.0371
0.0399	0.1981	0.0409
0.0436	0.2170	0.0446
0.0479	0.2359	0.0483
0.0528	0.2547	0.0520
0.0586	0.2735	0.0557
0.0655	0.2921	0.0595
0.0738	0.3106	0.0632
0.0845	0.3287	0.0669
0.0992	0.3464	0.0706
0.1230	0.3634	0.0743

SOLUTION ALONG BRANCH 2

XPHI	YPHI	YLAMBDA
0.9765	0.0025	0.0037
0.9737	0.0050	0.0074
0.9707	0.0077	0.0111
0.9677	0.0105	0.0148
0.9646	0.0133	0.0185
0.9613	0.0162	0.0223
0.9579	0.0193	0.0260
0.9544	0.0224	0.0297
0.9508	0.0257	0.0334
0.9469	0.0291	0.0371
0.9429	0.0327	0.0409
0.9387	0.0364	0.0446
0.9343	0.0404	0.0483
0.9296	0.0445	0.0520
0.9247	0.0488	0.0557
0.9194	0.0534	0.0595
0.9138	0.0583	0.0632
0.9077	0.0636	0.0669
0.9011	0.0692	0.0706
0.8938	0.0754	0.0743

Page Index for Computed Properties of Binary and Ternary Systems

System	Regular solution interaction parameters	Phase diagram	Vapor pressure–composition curves	Free energy–composition curves	Congruency conditions
Al–Ti	188	192	—	—	—
Al–V	188	194	—	—	—
Co–Ti	188	194	—	—	—
Cr–Ti	188	193	—	—	—
Cr–V	188	—	—	—	—
Cu–Mn	207	206	—	—	—
Cu–Ti	188	195	—	—	—
Hf–Ir	89	102, 103, 148	169	—	—
Hf–Mo	89	96–99, 142–145	168	143	—
Hf–Nb	89	34, 94–98	—	—	—
Hf–Os	89	100, 101, 147	—	—	—
Hf–Pd	89	104, 105, 151	—	—	181
Hf–Pt	89	104, 105, 151	—	—	181
Hf–Re	89	123, 152, 155	175	154	—
Hf–Re–Mo	247	248–255	—	—	—
Hf–Rh	89	102, 103, 148	—	—	181
Hf–Ru	89	100, 101, 147	—	—	—
Hf–Ta	89	34, 94–98	168	35, 36	—
Hf–Ti	188	191	—	—	—
Hf–W	89	96–99, 142	—	—	—
Ir–Hf	89	102, 103, 148	169	—	—
Ir–Mo	90	37, 58, 116, 117	—	—	181

APPENDIX 3 (continued)

System	Regular solution interaction parameters	Phase diagram	Vapor pressure–composition curves	Free energy–composition curves	Congruency conditions
Ir–Nb	89	110, 111, 158	—	—	181
Ir–Os	90	63, 130, 131	—	—	—
Ir–Pd	90	134, 135	—	—	—
Ir–Pt	90	134, 135	178	—	—
Ir–Re	90	61, 127	176	—	—
Ir–Ru	90	63, 130, 131	—	—	—
Ir–Ta	89	110, 111, 158, 159	172	—	181
Ir–W	90	37–58, 116, 117	174	—	—
Ir–Zr	89	102, 103, 148, 150	—	149	—
Fe–Ru	42–45	26–28	—	—	—
Mn–Cu	207	206	—	—	—
Mo–Hf	89	96–99, 142, 145	168	143	—
Mo–Hf–Re	247	248–255	—	—	—
Mo–Ir	90	37, 58, 116, 117	—	—	181
Mo–Nb	89	106, 107	170	—	—
Mo–Os	90	57, 114, 115	—	—	—
Mo–Pd	90	59, 118, 119	—	—	—
Mo–Pt	90	59, 118, 119	174	—	—
Mo–Re	90	56, 125	—	—	—
Mo–Rh	90	37, 58, 116, 117	—	—	—
Mo–Ru	90	51, 57, 114, 115	173	—	—
Mo–Ta	89	106, 107	—	—	—
Mo–W	238	—	—	—	—
Mo–W–Os	238	239–240	—	—	—
Mo–Zr	89	96–99, 142	—	—	—
Nb–Hf	89	34, 94–98	—	—	—
Nb–Ir	89	110, 111, 158	—	—	181
Nb–Mo	89	106, 107	170	—	—
Nb–Os	89	55, 108, 109, 156	—	—	181
Nb–Pd	89	112, 113, 160	—	—	—
Nb–Pt	89	112, 113, 160	172	—	181
Nb–Re	89	56, 124, 152	—	—	—
Nb–Rh	89	110, 111, 158	—	—	—
Nb–Ru	89	55, 108, 109, 156	—	—	181
Nb–Ti	188	192	—	—	—
Nb–W	89	106, 107	—	—	—
Nb–Zr	89	34, 94–98	—	—	—
Ni–Ti	188	195	—	—	—
Os–Hf	89	100, 101, 147	—	—	—
Os–Ir	90	63, 130, 131	—	—	—
Os–Mo	90	57, 114, 115	—	—	—
Os–Mo–W	238	239, 240	—	—	—
Os–Nb	89	55, 108, 109, 156	—	—	181

APPENDIX 3 (continued)

System	Regular solution interaction parameters	Phase diagram	Vapor pressure–composition curves	Free energy–composition curves	Congruency conditions
Os–Pd	90	62, 132, 133	—	—	—
Os–Pt	90	62, 132, 133	—	—	—
Os–Re	90	126	—	—	—
Os–Rh	90	130, 131	—	—	—
Os–Ta	89	55, 108, 109, 156	—	—	181
Os–Ti	188	191	—	—	—
Os–W	90	57, 114, 115	173	—	—
Os–Zr	89	100, 101, 147	—	—	—
Pd–Hf	89	104, 105, 151	—	—	181
Pd–Ir	90	134, 135	—	—	—
Pd–Mo	90	59, 118, 119	—	—	—
Pd–Nb	89	112, 113, 160	—	—	—
Pd–Os	90	62, 132, 133	—	—	—
Pd–Re	90	`61, 128	177	—	—
Pd–Rh	90	134, 135	—	—	—
Pd–Ru	90	62, 132, 133	—	—	—
Pd–Ta	89	112, 113, 160	—	—	—
Pd–Ti	188	195	—	—	—
Pd–W	90	59, 118–121	—	120	—
Pd–Zr	89	104, 105, 151	—	—	181
Pt–Hf	89	104, 105, 151	—	—	181
Pt–Ir	90	134, 135	178	—	—
Pt–Mo	90	59, 118, 119	174	—	—
Pt–Nb	89	112, 113, 160	172	—	181
Pt–Os	90	62, 132, 133	—	—	—
Pt–Re	90	61, 128	—	—	—
Pt–Rh	90	134, 135	—	—	—
Pt–Ru	90	62, 132, 133	178	—	—
Pt–Ta	89	112, 113, 160	—	—	—
Pt–Ti	188	195	—	—	—
Pt–W	90	59, 118, 119	—	—	—
Pt–Zr	89	104, 105, 151	170	—	181
Re–Hf	89	123, 152, 155	175	154	—
Re–Hf–Mo	247	248–255	—	—	—
Re–Ir	90	61, 127	176	—	—
Re–Mo	90	56, 125	—	—	—
Re–Nb	89	56, 124, 152	—	—	—
Re–Os	90	126	—	—	—
Re–Pd	90	61, 128	177	—	—
Re–Pt	90	61, 128	—	—	—
Re–Rh	90	61, 127	—	—	—
Re–Ru	90	126	177	—	—
Re–Ta	89	56, 124, 152	175	—	181

APPENDIX 3 (*continued*)

System	Regular solution interaction parameters	Phase diagram	Vapor pressure–composition curves	Free energy–composition curves	Congruency conditions
Re–Ti	188	193	—	—	—
Re–W	90	56, 125	176, 179	—	—
Re–W–Ta	243	244–247	—	—	—
Re–Zr	89	123, 152	—	—	—
Rh–Hf	89	102, 103, 148	—	—	181
Rh–Nb	89	110, 111, 158	—	—	—
Rh–Mo	90	37, 58, 116, 117	—	—	—
Rh–Os	90	130, 131	—	—	—
Rh–Pd	90	134–135	—	—	—
Rh–Pt	90	134, 135	—	—	—
Rh–Re	90	61, 127	—	—	—
Rh–Ru	90	130, 131	—	—	—
Rh–Ta	89	110, 111, 158	—	—	—
Rh–Ti	188	194	—	—	—
Rh–W	90	37, 58, 116, 117	—	38	—
Rh–Zr	89	102, 103, 148	—	—	181
Ru–Fe	42–45	26–28	—	—	—
Ru–Hf	89	100, 101, 147	—	—	—
Ru–Ir	90	63, 130, 131	—	—	—
Ru–Nb	89	55, 108, 109, 156	—	—	181
Ru–Mo	90	51, 57, 114, 115	173	—	—
Ru–Pd	90	62, 132, 133	—	—	—
Ru–Pt	90	62, 132, 133	178	—	—
Ru–Re	90	126	177	—	—
Ru–Rh	90	130, 131	—	—	—
Ru–Ta	89	55, 108, 109, 156	171	—	—
Ru–Ti	188	191	—	—	—
Ru–W	90	57, 114, 115	—	—	—
Ru–Zr	89	100, 101, 147	169	—	—
Ta–Hf	89	34, 94–98	168	35, 36	—
Ta–Ir	89	110, 111, 158, 159	172	—	181
Ta–Mo	89	106, 107	—	—	—
Ta–Os	89	55, 108, 109, 156	—	—	181
Ta–Pd	89	112, 113, 160	—	—	—
Ta–Pt	89	112, 113, 160	—	—	—
Ta–Re	89	56, 124, 152	175	—	181
Ta–Re–W	243	244–247	—	—	—
Ta–Rh	89	110, 111, 158	—	—	—
Ta–Ru	89	55, 108, 109, 156	171	—	—
Ta–Ti	188	192	—	—	—
Ta–W	89	106, 107	—	—	—
Ta–W–Zr	241	242, 243	—	—	—
Ta–Zr	89	34, 94–98	—	—	—

APPENDIX 3 (continued)

System	Regular solution interaction parameters	Phase diagram	Vapor pressure–composition curves	Free energy–composition curves	Congruency conditions
Ti–Al	188	192	—	—	—
Ti–Co	188	194	—	—	—
Ti–Cr	188	193	—	—	—
Ti–Cu	188	195	—	—	—
Ti–Hf	188	191	—	—	—
Ti–Ir	188	194	—	—	—
Ti–Mo	188	193	—	—	—
Ti–Nb	188	192	—	—	—
Ti–Ni	188	195	—	—	—
Ti–Os	188	191	—	—	—
Ti–Pd	188	195	—	—	—
Ti–Pt	188	195	—	—	—
Ti–Re	188	193	—	—	—
Ti–Rh	188	194	—	—	—
Ti–Ru	188	191	—	—	—
Ti–Ta	188	192	—	—	—
Ti–V	188	192	—	—	—
Ti–W	188	193, 196	—	—	—
Ti–Zr	188	191	—	—	—
W–Hf	89	96–99, 142	—	—	—
W–Ir	90	37–58, 116, 117	174	—	—
W–Mo	238	—	—	—	—
W–Mo–Os	238	239, 240	—	—	—
W–Nb	89	106, 107	—	—	—
W–Os	90	57, 114, 115	173	—	—
W–Pd	90	59, 118–121	—	120	—
W–Pt	90	59, 118, 119	—	—	—
W–Re	90	56, 125	176, 179	—	—
W–Re–Ta	243	244–247	—	—	—
W–Rh	90	37, 58, 116, 117	—	38	—
W–Ru	90	57, 114, 115	—	—	—
W–Ta	89	106, 107	—	—	—
W–Ta–Zr	241	242, 243	—	—	—
W–Ti	188	193, 196	—	—	—
W–Zr	89	96–99, 142	—	—	—
V–Al	188	194	—	—	—
V–Cr	188	—	—	—	—
V–Ti	188	192	—	—	—
Zr–Ir	89	102, 103, 148, 150	—	149	—
Zr–Mo	89	96–99, 142	—	—	—
Zr–Nb	89	34, 94–98	—	—	—
Zr–Os	89	100, 101, 147	—	—	—
Zr–Pd	89	104, 105, 151	—	—	181

APPENDIX 3 (*continued*)

System	Regular solution interaction parameters	Phase diagram	Vapor pressure–composition curves	Free energy–composition curves	Congruency conditions
Zr–Pt	89	104, 105, 151	170	—	181
Zr–Re	89	123, 152	—	—	—
Zr–Rh	89	102, 103, 148	—	—	181
Zr–Ru	89	100, 101, 147	169	—	—
Zr–Ta	89	34, 94–98	—	—	—
Zr–Ta–W	241	242, 243	—	—	—
Zr–Ti	188	191	—	—	—
Zr–W	89	96–99, 142	—	—	—

APPENDIX 4

Page Index for Computed Properties of Binary and Ternary Compounds

Compound	Heat of formation	Crystal structure	Compound	Heat of formation	Crystal structure
AlTi	189	189	HfRh$_2$	163	140, 163
Al$_3$Ti	189	189	HfRu$_2$	163	140, 163
Al$_3$V	189	189	HfW$_2$	163	140, 163
Al$_8$V$_5$	189	189	Ir$_3$Hf	163	150, 163
CNb	75	75	Ir$_2$Hf	163	140, 163
CZr	75	75	IrMo$_3$	164	164
CHf	75	75	IrNb$_2$	164	164
CoTi	189	189	Ir$_3$Nb	164	75, 151, 164
Co$_2$Ti	189	189	IrTa$_2$	164	164
Co$_3$Ti	189	189	Ir$_3$Ta	164	151, 164
Cr$_2$Ti	189	189	Ir$_3$Ti	189	189
CuTi	189	189	IrW$_3$	164	164
HfC	75	75	Ir$_2$Zr	163	140, 163
HfIr$_2$	163	140, 163	Ir$_3$Zr	163	150, 163
HfIr$_3$	163	150, 163	Mo$_2$Hf	163	140, 163
HfMo$_2$	163	140, 163	Mo$_3$Ir	164	164
HfOs$_2$	163	140, 163	Mo$_5$Os$_3$	164	164
HfPd$_2$	163	140, 163	MoRe$_3$	164	164
HfPd$_3$	163	150, 163	Mo$_2$Re$_3$	164	164
HfPt$_2$	163	140, 163	(Mo, Re)$_2$Hf	247	247
HfPt$_3$	163	150, 163	Mo$_5$Ru$_3$	164	164
HfRe$_2$	163	140, 163	Mo$_2$Zr	163	140, 163
Hf(Re, Mo)$_2$	247	247	NbC	75	75

APPENDIX 4 (continued)

Compound	Heat of formation	Crystal structure	Compound	Heat of formation	Crystal structure
NbIr$_3$	164	75, 151, 164	Re$_{24}$Ti$_5$	189	189
Nb$_2$Ir	164	164	ReW	164	164
NbPd$_3$	164	153, 164	Re$_3$W	164	164
NbPt$_3$	164	75, 153, 164	Re$_2$Zr	163	140, 163
NbRe$_3$	163	153, 163	Rh$_2$Hf	163	140, 163
NbRe$_4$	75	75	Rh$_3$Hf	163	150, 163
NbRh$_3$	164	151, 164	RhNb$_2$	164	164
Nb$_2$Rh	164	164	Rh$_3$Nb	164	151, 164
NbRu$_3$	163	153, 163	RhTa$_2$	164	164
NiTi	189	189	Rh$_3$Ta	164	151, 164
Ni$_3$Ti	189	189	Rh$_3$Ti	189	189
Os$_2$Hf	163	140, 163	Rh$_2$Zr	163	140, 163
Os$_3$Mo$_5$	164	164	Rh$_3$Zr	163	150, 163
OsTa	163	153, 163	Ru$_3$Mo$_5$	164	164
OsTa$_2$	163	153, 163	Ru$_3$Nb	163	153, 163
OsTi	189	189	RuTa	163	153, 163
Os$_3$W$_5$	163	163	RuTi	189	189
Os$_2$Zr	163	140, 163	Ru$_3$W$_5$	164	164
Pd$_2$Hf	163	140, 163	RuZr	—	146
Pd$_3$Hf	163	150, 163	Ru$_2$Zr	163	140, 163
Pd$_3$Nb	164	153, 164	TaIr$_3$	164	151, 164
Pd$_3$Ta	164	153, 164	Ta$_2$Ir	164	164
Pd$_3$Ti	189	189	TaOs	163	153, 163
PdZr	163	163	Ta$_2$Os	163	153, 163
PdZr$_2$	163	140, 163	TaPd$_3$	164	153, 164
Pd$_2$Zr	163	148, 150, 163	TaPt$_3$	164	153, 164
Pd$_3$Zr	163	148, 150, 163	TaPt$_3(\alpha, \beta)$	164	153, 164
Pt$_2$Hf	163	140, 163	TaRe$_3$	163	153, 163
Pt$_3$Hf	163	150, 163	TaRh$_3$	164	151, 164
Pt$_3$Nb	163	75, 153, 164	Ta$_2$Rh	164	164
Pt$_3$Ta	164	153, 164	TaRu	163	153, 163
Pt$_3$Ta(α, β)	164	153, 164	(Ta, W)Re$_3$	243	243
Pt$_3$Ti	189	189	(Ta, W)$_2$Zr	241	241
PtZr	163	163	Ta$_2$Zr	241	241
PtZr$_2$	163	163	TiAl	189	189
Pt$_2$Zr	163	140, 163	TiAl$_3$	189	189
Pt$_3$Zr	163	148, 151, 163	Ti$_2$Al	189	189
Re$_2$Hf	163	140, 163	TiCo	189	189
Re$_3$Mo	164	164	TiCo$_2$	189	189
Re$_3$Mo$_2$	164	164	TiCr$_2$	189	189
(Re, Mo)$_2$Hf	247	247	TiCu	189	189
Re$_3$Nb	163	153, 163	TiIr$_3$	189	189
Re$_4$Nb	75	75	TiNi	189	189
Re$_3$Ta	163	153, 163	TiNi$_3$	189	189
Re$_3$(Ta, W)	243	243	TiOs	189	189

APPENDIX 4 (continued)

Compound	Heat of formation	Crystal structure	Compound	Heat of formation	Crystal structure
$TiPd_3$	189	189	$ZrMo_2$	163	140, 163
$TiPt_3$	189	189	$ZrOs_2$	163	140, 163
Ti_5Re_{24}	189	189	$ZrPd$	163	163
$TiRh_3$	189	189	$ZrPd_2$	163	148, 150, 163
$TiRu$	189	189	$ZrPd_3$	163	148, 150, 163
W_2Hf	163	140, 163	Zr_2Pd	163	163
W_3Ir	164	164	$ZrPt$	163	163
W_5Os_3	164	164	Zr_2Pt	163	163
WRe	164	164	$ZrPt_2$	163	140, 163
WRe_3	164	164	$ZrPt_3$	163	148, 151, 163
W_5Ru_3	164	164	Zr_2Pt	163	163
$(W, Ta)Re_3$	243	243	$ZrRe_2$	163	140, 163
$(W, Ta)Zr_2$	241	241	$ZrRh_2$	163	140, 163
W_2Zr	163	140, 163	$ZrRh_3$	163	150, 163
VAl_3	189	189	$ZrRu$	—	146
V_5Al_8	189	189	$ZrRu_2$	163	140, 163
ZrC	75	75	$ZrTa_2$	241	241
$ZrIr_2$	163	140, 163	ZrW_2	163	140, 163
$ZrIr_3$	163	150, 163	$Zr_2(W, Ta)$	241	241

Author Index

Numbers in parentheses are reference numbers and indicate that an author's work is referred to although his name is not cited in the text. Numbers in italics show the page on which the complete reference is listed.

321

Subject Index

Symbols used for elements are alphabetized following the entry for the element.

A

Activity, 40, 41, 196, 205–208
  gradients, 217
  of iron in Fe–Ru alloys, 42
Allotropism, 5–34, 213, *see also*
  Polymorphism
Aluminum, 183, 186
  heat of vaporization, 186
  lattice stability parameters, 183, 185
  pseudopotential calculations for, 31,
    185, 186
  volume per gram atom, 186
Al–Cu, 212
Al–Mg, 183
Al–Ti, 183, 190, 197
AlTi, 190, 197
Al$_3$Ti, 190, 197
Al–V, 183, 196, 197
Al$_3$V, 189
Al$_8$V$_5$, 189
Al–Zn, 44, 45, 209
Atmospheric pressure, 6, 7, 9, 215
Atomic size, 53
Avogadro's number, 78

B

Band structure, 29
Barium
  lattice stability parameters, 15
  volume difference per gram atom, 15
Base phase, 41, 139–143, 189

Beryllium, 183, 215
  lattice stability parameters, 15, 185
  volume difference per gram atom, 15
Binary system
  computed phase diagrams for, 92–160,
    191–195
  thermodynamics of, 69–91
Bismuth, 11, 12
  lattice stability parameters, 11
  volume differences per gram atom, 11
Body-centered cubic (bcc), 2, 5, 12–14, 18,
    33–40, 63–139, 213
Brillouin zone, 4, 212

C

Cadmium, lattice stability parameters, 47, 48
Cd–Ag, 212
Cd–Zn, 1
Calcium, 215
  lattice stability parameters, 15
Calories
  per gram atom, 20
  per mole, 6, 20
Carbides
  CHf, 75, 161
  CNb, 75, 161
  CZr, 75
Carbon, 15, 75
CsCl structure, 146, 188
Chemical potential, 36
Chromium, 60, 64–67, 213
  heat of vaporization, 186

U

V

W

X

Y

Z